Biology of the Cyclostomes

Biology of the Cyclostomes

M. W. Hardisty D. Sc.

Emeritus Professor of Zoology
School of Biological Sciences
University of Bath

LONDON
Chapman and Hall

First published 1979
by Chapman and Hall Ltd.
11 New Fetter Lane
London EC4P 4EE

© *1979 M. W. Hardisty*

Printed in Great Britain at the
University Press, Cambridge

ISBN 0 412 14120 5

British library cataloguing in publication data

Hardisty, Martin Weatherhead
  Biology of the cyclostomes.
  1. Cyclostomata (Polyzoa)
  I. Title
  597′.2   QL398.C9   79–40803

  ISBN 0–412–14120–5

Contents

Preface

The proliferation of scientific texts and their rapidly escalating costs demands of an author some justification for the production of yet another specialised volume; particularly one that treats of a relatively obscure group of animals — the Cyclostomes — whose significance is little appreciated outside the circle of professional biologists. Yet, within the zoological literature this group of vertebrates has always commanded a degree of attention, quite disproportionate to the comparatively small numbers of species involved or their economic importance. This special interest stems in the main from their unique phylogenetic status. As jawless vertebrates the hagfish and the lamprey are regarded as the sole survivors of a once flourishing group of Palaeozoic vertebrates — the Agnathans — amongst which are numbered the first vertebrates to appear in the fossil record. Because of this relationship to the fossil agnathans it was inevitable that past discussion of the phylogenetic significance of the cyclostomes should have been dominated by comparative anatomists and palaeontologists, although in recent years their unique evolutionary position has increasingly attracted the interest of comparative physiologists and students of molecular evolution.

Within the last fifteen years both the hagfish and the lamprey have been the subject of separate publications describing in detail many aspects of their morphology, physiology and life cycles (Brodal, A. and Fänge, R., *The Biology of* Myxine, 1963; Hardisty, M. W. and Potter, I. C., *The Biology of Lampreys*, 1971–72.), but neither of these set out to present a synoptic view of the cyclostomes as a whole or to examine in depth the interrelationships of the lampreys and hagfishes and their relevance to the early evolution of the vertebrates. In approaching these problems in the wider perspectives of more recent comparative and mainly non-morphological information, it is hoped that the present volume may achieve a more balanced and broadly based assessment of the cyclostome group, which in addition to its interest for zoologists may also prove helpful to workers in other fields of biology whose studies may have phylogenetic and evolutionary implications.

With a single modestly-sized volume, the attempt to range over the entire spectrum of the biological sciences from palaeontology to molecular evolution carries with it the dangers of superficiality in the treatment of specialised topics. It is for this reason that I have dispensed with the details of descriptive morphology. Information of this kind is already available, not only in the volumes already referred to but also in a number of comprehensive treatises such as the *Handbuch der Zoologie*, (Pietschmann, 1934) the *Traité de Zoologie* (Grassé, 1958) or in the liberally illustrated anatomical descriptions of Marinelli and Strenger (1956). For similar reasons, the sheer immensity of the historical literature of the cyclostomes has made it quite impracticable fully to document all the material referred to in the text, although every effort has been made to cite more recent work and more especially research that has been published since the appearance of Brodal and Fänge (1963) or Hardisty and Potter (1971–72).

Because of the gaps that exist in our knowledge, a degree of imbalance in the treatment of particular aspects of cyclostome biology has been unavoidable. Certain areas of hagfish physiology remain as yet almost totally unexplored. For example, whereas extensive investigations have been carried out on the neuromuscular system of the lamprey, hardly any comparable electrophysiological studies have been made on the hagfish. Similarly, while we now have quite detailed information on the life cycles, ecology and reproductive habits of at least a few species of lamprey, we still remain in almost total ignorance of the life of hagfishes in their natural and inaccessible environment. On the other hand, in the field of comparative endocrinology the two cyclostome groups have received more equal attention and what might be regarded as the undue weight given to these topics is only a reflection of the intensity of the research effort devoted to them in recent years.

In examining the nature of the relationship between the two cyclostome groups and the gnathostomes in the light of our present biochemical and physiological information, I am aware that by adopting the methodology of cladism I may have tended to give insufficient weight to the fossil agnathans. Although it may be admitted that these methods of analysis may perhaps be less objective than has sometimes been claimed, the techniques of phylogenetic classification (which can only be applied with their full rigour to living forms), nevertheless have at least the merit of being able to embrace the quantitative data of the molecular evolutionist. In any event, the palaeontological evidence has already been exhaustively discussed by many other authors much more competent than myself to assess the significance of these aspects of cyclostome relationships.

The increasing fragmentation of biology has been reflected in the growth of multi-author publications in which a number of specialists make independent contributions to a particular topic. This procedure has some very obvious advantages in the authority and discrimination that experts are able to bring to bear in their chosen fields, but volumes of this kind may tend to suffer from the absence of overall objective and the lack of integration between the individual contributions. In attempting the daunting task of examining information gathered over such wide and varied areas of the biological sciences, I can only hope that my own shortcomings, be they exhibited as misinterpretation or as errors of fact, may to some extent be compensated for by greater consistency of aims and approach.

In the preparation of the manuscript certain chapters have been read by Lord Richard Percy and Dr R. Morris, to whom I am grateful for their comments. I should also like to thank Professor W. H. Beamish and those other authors who have allowed me to see manuscript copies of their work before publication. I am especially grateful to my former colleague Professor I. C. Potter for his help with the section on lamprey karyotypes. Finally, I should like to thank the School of Biological Sciences of the University of Bath who have continued to support me by providing essential services and also the Librarian and his staff, who at all times have been ready to help in the provision of the necessary literature.

Acknowledgements

My thanks are due to the authors, editors and publishers of the following books or journals for permission to make use of text-figures or Tables.

Acta physiologica Scandinavica; Acta Regiae Societas Scientarum et Litteratum, Gothenbergensis; Acta Zoologica (Stockholm); *Biological Reviews of the Cambridge Philosophical Society* (Cambridge University Press); *Cell and Tissue Research* (Springer-Verlag); *Chordate Morphology*, Jollie 1972 Reprint edition (Robert E. Krieger Publishing Co. Inc.); *Comparative Biochemistry and Physiology* (Pergamon Press); *General and Comparative Endocrinology* (Academic Press); *Journal of Anatomy* (Cambridge University Press); *Journal of Experimental Biology* (Company of Biologists Ltd.); *Journal of the Fisheries Research Board of Canada; Journal of Zoology*, Zoological Society of London (Academic Press); *Progress in Brain Research* (Elsevier-North-Holland Biomedical Press); *Scientific American* (Scientific American Inc.); *The Biology of Lampreys* (Eds. Hardisty and Potter) (Academic Press); *The Biology of Myxine* (Eds. Brodal and Fänge) (Universitetsforlaget, Oslo); *Textbook of Zoology*, Parker and Haswell (Macmillan and Co. Ltd.); *Transactions of the American Fisheries Society; Transactions of the New York Academy of Sciences*; United States Fish and Wildlife Service; *Zoologica Scripta*, Royal Swedish Academy of Sciences (Almqvist and Wiksell International); *Australian Journal of Science* (Australian and New Zealand Association for the Advancement of Science).

Among the authors whose illustrations have been used as a basis for text-figures I am especially indebted to the following: Prof. F. W. H. Beamish, Prof. A. Gorbman, Dr. P. R. Flood, Dr. J. A. James, Prof. P. Janvier, Dr. D. B. McMillan, Dr. R. Morris, Dr. Y. Östberg, Prof. A. Peters, Prof. I. C. Potter, Prof. J. Pikkarainen, Dr. R. Strahan, Prof. R. Nieuwenhuys and Dr. B. R. Smith.

1
Introduction

The Cyclostomes, comprising the lampreys and the hagfishes or myxinoids, are a group of eel-like aquatic vertebrates, not widely known outside zoological circles, for whom they have a special interest as the sole survivors of an extinct, but once flourishing, group of jawless or agnathan Palaeozoic 'fishes' known as the Ostracoderms.

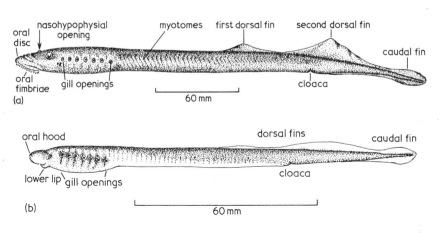

Fig. 1.1 The European river lamprey, *Lampetra fluviatilis* (a) and an ammocoete (b).

The entirely marine hagfish, once classified as an 'intestinal worm' is a blind scavenger, living on the sea bed in comparatively deep water and attracting the attention of commercial fishermen only because they have a habit of devouring their bait. Their colloquial name 'hag' is an expression of the repulsion that they tend to arouse; an aversion that owes much to their 'soft, naked body, blotched by the sub-cutaneous blood sinuses, the lack of eyes and a characteristic clinging odour' (Strahan and Honma, 1960). Since they are

1

said to be rubbery in texture, tasting strongly of fish oils (Jensen, 1966) it is hardly surprising that they have little or no economic value and, except in one locality in Japan, have not been exploited as human food.

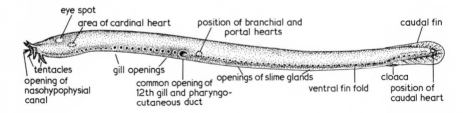

eye spot
area of cardinal heart   position of branchial and   caudal fin
                                     portal hearts

tentacles   gill openings   openings of slime glands   cloaca
opening of   common opening of   ventral fin fold   position of
nasohypophysial   12th gill and pharyngo-   caudal heart
canal   cutaneous duct

Fig. 1.2 A polybranchiate hagfish, *Eptatretus stouti*. Redrawn from Jensen, 1966.

Lampreys, although primarily freshwater animals, may spend part of their adult life in the sea, before finally returning to the rivers to spawn and die. Here the egg develops into a blind, larval form – the ammocoete (Fig. 1.1b) which burrows into the silt banks of the river, where it remains for several years, growing to a maximum length of about eight inches and feeding on microscopic organisms, filtered from the water. During this part of their life cycle, they are seen only rarely, but may be dredged up in large numbers from the mud in which they are hidden, or caught by the use of electrical fishing gear, which causes them to emerge from their burrows. The transformation of the fully grown ammocoete into an adult lamprey (Fig. 1.1a) occurs during late summer or early autumn and, after metamorphosis, the further development of the young adult or macrophthalmia (so called because of its relatively large eyes) may follow either one of two divergent patterns. In many species – the so called brook lampreys – the adult never feeds and cannot therefore, be any larger than the ammocoete at the end of its larval life. These dwarf lampreys have an adult life span of only 6–9 months, after which they spawn and die. Other lampreys, once their metamorphosis has been completed, become parasitic on fish, either in freshwater or in the sea, attaching themselves to their prey by their toothed oral disc or sucker (Fig. 1.3) and feeding on blood or other tissues.

Where river systems remain relatively free from pollution, lampreys and their larvae may occur in very large numbers but, because of their unobtrusive habits, are likely to be detected only by an experienced observer. They are of no interest to the angler (except as bait) and, because of their jawless mouth and parasitic feeding habits, cannot be taken by rod and line. Adult lampreys are usually only seen when they are ascending the rivers on their spawning migration. Even then, they are mainly nocturnal and daytime activity is generally restricted to a very brief period in spring or early summer while

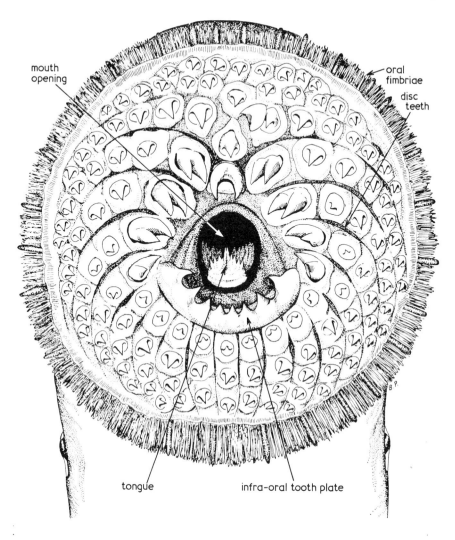

mouth
opening

oral
fimbriae

disc
teeth

tongue

infra-oral tooth plate

Fig. 1.3 The oral disc and teeth of a lamprey, *Petromyzon marinus* From Hubbs and
Potter, 1971.

spawning is in progress. However, although rarely observed and even then
often confused with eels, lampreys have acquired a widespread and traditional
reputation as a delicacy and smoked lamprey is particularly valued in the
Baltic countries of Eastern Europe, where commercial lamprey fisheries are
still of considerable economic importance. This esteem has some scientific
backing, since at the start of their migration (when their flesh is most

palatable), the high fat content of the muscle tissue rivals that of the eel, while their extraordinarily high levels of vitamin A accounts for their use in Japan as a remedy for night blindness (Higashi *et al.*, 1958; Shidoji and Muto, 1977).

In spite of far-reaching differences in their internal organization, lampreys and hagfishes are superficially similar and sinuous creatures, swimming gracefully by eel-like undulatory waves. The largest species of both groups reach lengths approaching one metre and body weights of about one kilogram. In keeping with their more active habits, the median fins are better developed in the lamprey, which usually has two distinct dorsal fins (Fig. 1.1a). Moreover, the propulsive tail region behind the cloacal opening is relatively much longer in the lamprey and although both have symmetrical caudal fins, that of the hagfish is more rounded and spatulate in shape (Fig. 1.2). Behind the branchial region the hagfish may have a small ventral fin fold, but this lacks internal cartilaginous supports (fin rays) and rarely projects more than one or two millimetres from the surface. Whereas the hagfish body tends to taper towards the anterior end, the branchial and head region of the lamprey is deeper than the trunk and bears a constant number of seven circular gill ports, the first a short distance behind its prominent paired eyes. In this respect, hagfishes are highly variable. The Pacific hagfish, *Eptatretus stouti* in Fig. 1.2 has 12 separate gill openings, placed much further back than in a lamprey, but as many as 15 of these openings may be found in some species. On the other hand in the Atlantic hagfish, *Myxine glutinosa*, there are only five pairs of gills, whose external ducts open at a common pore (Fig. 2.5). Peculiar to the hagfish is a pharyngo-cutaneous duct, usually opening in common with the last gill on the left side and connecting internally with the pharynx. Characteristic of the myxinoids are the large slime glands opening by rows of ducts along the ventro-lateral margins of the body. The specific name *glutinosa* given to the Atlantic hagfish enshrines the ability of these animals to throw off large volumes of slime when they are irritated. For example, when stimulated by handling or by dilute formaldehyde, *Myxine* is capable of producing about half a litre of slime, consisting of a tangled fibrous network, holding imbibed water in its meshes and converting it into a gelatinous mass (Strahan, 1959).

The tapering head of the hagfish carries three pairs of sensory tentacles and between the base of the upper pair is the so-called 'nostril' or nasohypophysial opening (Fig. 1.4), leading internally to a duct, passing backwards below the brain to join the pharynx (Fig. 1.5). Through this system, the respiratory water current passes backwards to the gills, propelled by the movements of a complex scroll-like velum (Fig. 5.4) housed in a chamber just in front of the junction with the pharynx. The opening of the mouth is somewhat ventral and overhung by a forwardly projecting rostrum. The eye of *Myxine* is extremely small and degenerate and covered by a layer of muscle so that it cannot be seen from the surface, but in *E. stouti*, where it is rather more developed, its

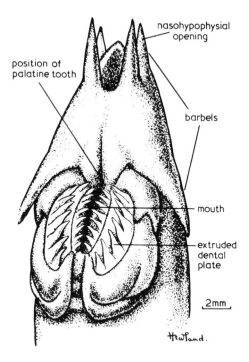

Fig. 1.4 Ventral surface of the head of *Myxine glutinosa* showing the dental plate in its extruded position. From Dawson, 1963.

position is marked by a small area of skin, relatively free of pigmentation (Fig. 1.2).

The mouth of the lamprey is surrounded by the expanded sucker-like oral disc, covered on its internal surface by small disc teeth (Fig. 1.3). Above and behind the oral disc is the nasohypophysial opening and behind this is a small unpigmented area of skin marking the site of the pineal organs. Internally, the head region shows marked differences from the hagfish (Fig. 1.5), of which the most important is the separation of the food and respiratory passages. The gill pouches open internally into a separate blind-ended respiratory tube or *Wassergang*, whose entrance into the mouth cavity is guarded by a very small valve-like velum. The food passage or oesophagus lies above the water tube and opens separately into the mouth. In the lamprey, the nasohypophysial tract conveys water to the olfactory organ, proceeding backwards to end blindly in a dilated sac above the anterior gill pouches.

Although they lack the paired fins of other vertebrates and their skeleton is entirely cartilaginous, the organization of the cyclostome body conforms to the basic vertebrate or, more appropriately, craniate pattern, in which the

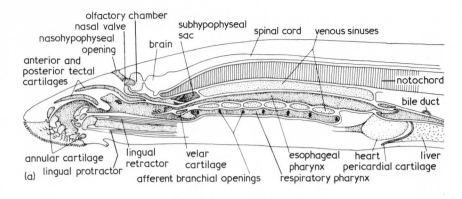

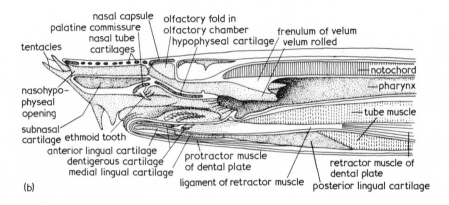

Fig. 1.5 Sagittal section through the head and branchial region of a lamprey (a) and the head of a hagfish (b). From Jollie, 1962.

main axial skeleton consists of a notochord, stretching from below the brain to the tip of the tail, but which in these animals is unconstricted by the development of jointed vertebrae. Immediately below the notochord lies the main distributive artery – the dorsal aorta (Fig. 2.3), flanked on either side by the main venous channels – the cardinal veins. Above it is the flattened spinal cord passing forwards to the brain, here housed in an incomplete cranium, with only a membranous roof. Immediately behind the branchial region is the heart. In the lamprey, this is contained within a closed pericardium, whereas in the hagfish, it is open to the general body cavity behind. Through the latter runs a straight and almost uniform intestine, opening together with the kidney ducts at the cloaca.

In its external appearance, internal organization and mode of life, so great is the contrast with an adult lamprey, that for a long time the true larval status of the ammocoete was not appreciated and although Baldner in the 18th century had recognized these 'blind lampreys' as belonging to the same species as the adults, his observations were generally ignored, until their true relationship was finally established by Müller (1856). Lacking the suctorial disc of the adult, the mouth is overhung by the flexible oral hood (Fig. 1.6) and on the dorsal surface of the head is a nasohypophysial depression leading, at this stage, to a solid cord of cells and not to a hollow duct. Behind this depression is a translucent spot marking the position of the pineal, but the paired eyes are buried below the skin surface and are not visible from the exterior. The larval pharynx is an undivided tube into which open the seven gill pouches and at whose entrance are the muscular velar paddles, which aid in maintaining the respiratory current. On the floor of the pharynx is a ciliated groove and below it lies a complex tubular endostyle opening by a median duct (Fig. 11.2).

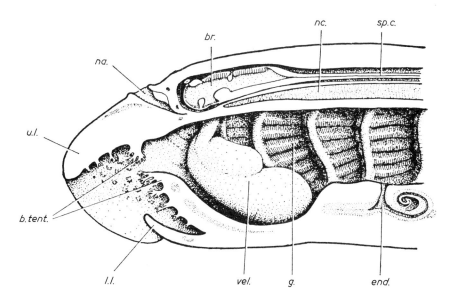

Fig. 1.6 Diagrammatic half section of the head of an ammocoete of *Geotria*. From Strahan (in Parker and Haswell, Vol. 2. 1966).
br. brain; b. tent. buccal tentacles; end. endostyle; g. gill; l.l. lower lip; na. nasohypophysial opening; nc. notochord; sp. c. spinal cord; u.l. upper lip of oral hood; vel. velum.

Restricted as they were to mainly morphological criteria and impressed by the many apparently primitive characters that were shared by the lampreys and hagfishes, the earlier comparative anatomists had generally favoured a system of classification that grouped them together in a distinct vertebrate

Class, the Cyclostomata or Marsipobranchii; the latter referring to their pouch-like gill sacs (Fig. 6.1). Within this class the two groups were given the status of orders or subclasses as the Hyperoartii (lampreys) and Hyperotreti (hagfishes) or as the Petromyzontia and Myxinoidea. As may be seen from the morphological comparisons in Table 1.1, many of the features that distinguish the Cyclostomes from the higher vertebrates (Gnathostomes) relate to the absence in the hagfish or lamprey of structures that have been developed by other vertebrates such as paired fins, hinged jaws, hard tissues, jointed vertebral column, stomach, horizontal semicircular canal, sympathetic nerve

Table 1.1 Morphological comparisons of cyclostomes and gnathostomes.

	Cyclostomes	Gnathostomes
Skeleton and fins	No paired fins	Pectoral and pelvic fins
	No hard tissues	Bone and dentine
	Unconstricted notochord	Development of arcualia
	Ribs absent	Present
	Incomplete cranial roof	Complete
	No occipital region of cranium – IX nerve emerges behind skull	IX nerve intra-cranial
	Branchial skeleton attached to cranium	Independent
	Visceral arches external to gills	Internal
	Biting or rasping tongue mechanism	Hinged jaws
Muscles	No transverse septum in myotomes	Present
	No internal muscles of accommodation in eye	Present
	4th somite present	Disappears
Gills	Pouch-like gill sacs	
	5–15 gills vagally innervated	5 functional gills
Gut	Velum present	Absent
	Single 'nostril' opening into naso-hypophysial tract	Paired nostrils opening into mouth
	No differentiated stomach	Present
	No discrete exocrine pancreas	Present
Vascular system	Extensive sinus system but no true lymphatics	Lymphatic system
	No capillary fenestrations	Present
	No plasma cells	Present
	Single Cuvierian duct	Paired ducts
Nervous system and sense organs	Flattened spinal cord	Cylindrical (except in *Latimeria*)
	Nerve fibres non-medullated	Medullated and non-medullated fibres
	No sympathetic chain	Present
	Post-trematic innervation of gills	Pre- and post-trematic innervation
	No horizontal semicircular canal	Present
Excretory and reproductive systems	Persistence of pronephros (only in larval lamprey)	Embryonic or larval stages only
	No genital ducts	Present

chain or genital ducts. Although in many cases we may be confident that the absence of this or that feature represents a primitive condition that must have existed in the common vertebrate ancestor, in other cases it is possible that structures originally present in the vertebrate line may have been lost secondarily by the cyclostomes. Examples of this are the absence of jointed vertebrae and the failure of the cyclostomes to develop bony structures. Here, although we cannot exclude the possibility that these are primitive conditions (especially in the hagfish), there are some grounds for a belief that the absence of these structures may be the result of secondary reduction from the extensive exoskeleton that existed in the fossil agnathans (Chapters 3 and 7). Similarly, it has even been argued on the basis of inferences from the fossil cephalaspids that lampreys are descended from ancestors with mobile paired fins (Janvier, 1978). On the positive side are structures peculiar to the cyclostomes, but which in some instances are known to have been present in the related fossil agnathans (Chapter 3). These include the relatively large numbers of gill pouches, a nasohypophysial tract with a single external opening and possibly a velum. Above all, the cyclostomes differ radically from the gnathostomes in the arrangement and relationships of their branchial skeleton, which is fused into a continuous structure, placed externally to the gills (Figs. 7.2 and 7.3). Indeed so fundamental is this difference that at one time it was proposed as a basis for dividing the vertebrates into two distinct phyla – the Endobranchiata and the Ectobranchiata – according to whether the gills were endodermal or ectodermal in origin and placed inside or outside the visceral arches.

Yet, in spite of these common 'cyclostome characteristics' and the superficial resemblance between the lampreys and hagfishes, it is increasingly apparent that a considerable gulf separates the two groups; a separation that becomes more pronounced as we learn more of their physiology and biochemistry. This has led to the general abandonment of the term 'cyclostome' as a taxonomic unit and to the ranking of the hagfishes and lampreys as distinct Classes or subclasses within higher taxa that also include the various groups of fossil agnathans (Chapter 3). Some idea of the extent of this divergence may be gauged from the morphological comparisons in Table 1.2, summarizing many of the points that arise in subsequent chapters. To this may be added an almost equally impressive list of biochemical or physiological criteria (Table 14.1) which are examined in the appropriate sections of this volume. At the same time, in our eagerness to document the phyletic distance that appears to separate these animals, we must not overlook the fact that just as it is possible to discern an 'agnathan grade' of vertebrate organization, so too the cyclostomes undoubtedly share many distinctive biochemical and physiological characteristics marking them off from the higher vertebrates. (Table 14.2).

A continually recurring problem is to decide how far the apparently simpler organization of the myxinoid is to be regarded as a primitive condition or alternatively as a direct or indirect result of its habits and mode of life. In the

Table 1.2 Morphological comparisons of lampreys and hagfishes.

Lampreys	Hagfishes
External features	
One, or more usually two dorsal fins with fin rays	At most a small skin fold without fin rays
Oral funnel	Absent
No sensory tentacles	Three pairs of sensory tentacles
Skeleton and muscles	
Neurocranium with incomplete cartilaginous roof	Roof of neurocranium entirely membranous
Rudimentary 'vertebral' elements	Absent
Radial fin muscles	Absent
Parietal muscles meet ventrally	Parietal muscles do not reach ventral surface
Parietal muscle 'boxes' surrounded on all except medial surface by slow fibres	Slow fibres on ventral and lateral surface only
Dual innervation of central fibres	Innervation at one end of fibre only
Nervous system and sense organs	
Dorsal and ventral spinal roots not united	United
Müller and Mauthner neurons present	Absent
Cerebellum present	Absent
Choroid plexuses present	Absent
Well-developed brain ventricles	Brain ventricles reduced
Heart innervated by vagus	Aneural heart
First gill innervated by IX	Vagal innervation of all gills
Posterior branchial nerve forms a hypoglossal	Absent
Functional paired eyes	Eyes degenerate
Extrinsic eye muscles present	Absent
Pineal complex present	Absent
Labyrinth with two semicircular canals	Simple toroidal labyrinth
Ciliated chambers in labyrinth	Absent
Lateral line system present	Absent
Skin photoreceptors innervated by lateralis nerve	Innervation by spinal nerves
Neurulation with formation of solid neural keel	Neurulation involves formation of neural canal
Accessory olfactory organ	Absent
Retinal receptors with synaptic ribbons	Spherical synaptic bodies[1]
Normal ciliary pattern $(9+2)$	Olfactory and kinocilia of labyrinth lack central filaments $(9+0)$[2]
Distinct fat column above spinal cord	Absent
Cardiovascular system	
Single dorsal aorta	Paired and median aortae
Single left Cuvierian duct	Single right Cuvierian duct
No accessory hearts	Cardinal, portal and caudal hearts
Afferent and efferent arteries supply hemibranchs of adjacent gill pouches	Each pouch supplied by single afferent
Blood volume not more than 10%	Blood volume greater than 10%

Table 1.2 *(contd.)*

Lampreys	Hagfishes
Gut and respiratory system	
Gill sacs of adult open into separate water tube	Pharynx not divided
Seven gill pouches	Number varies from 5–14
1st gill close to head	Branchial zone posterior
Nasohypophysial opening dorsal	Terminal
Nasohypophysial tract ends blindly	Opens into pharynx
No pharyngo-cutaneous duct	Present
'Protopancreas' in ammocoete	Zymogen cells not concentrated in diverticulum
No bile duct or gall bladder in adult	Present
Ciliated intestine with typhlosole	Not ciliated and without typhlosole
Excretory organs	
Pronephric tubules atrophy at metamorphosis	Persistent pronephros
Differentiated kidney tubule with collecting ducts	Rudimentary tubule only
Reproduction	
Sex dimorphism at maturity	No sex dimorphism
Small isolecithal eggs	Large yolky eggs with horny shell
Holoblastic cleavage	Discoidal cleavage
Prolonged larval stage	No larval stage
Determinate life span – die after single spawning	Eggs produced throughout life.

(1) Holmberg and Ohman (1976); (2) Lowenstein (1973); Theisen (1976)

extent of the degeneration in its sensory equipment the hagfish has retreated so far down the road of evolutionary regression that it has been described as little more than a 'vertebrate worm' (Ross, 1963) and there can be little doubt that this degeneration has had wider repercussions on other functional systems. But, although we can have little difficulty in relating the degeneration of the eyes and lateral line system or the reduction in skin pigmentation to their deep water habitat and burrowing habits, there remain many other features, whose apparent simplicity is more difficult to evaluate. Among these are the rudimentary kidney tubule (Section 9.6.2), the simple ring-like labyrinth (Section 4.4), the absence of a cerebellum (Section 8.2.2), the undifferentiated state of the adenohypophysis (Section 10.2.1) or even the absence of extrinsic eye muscles and their motor nerves (Section 3.3). Are we to regard these as primitive features or is their apparent simplicity or total absence a result of evolutionary regression? Where we lack guidance from related fossil forms, the answers to questions such as these can be no more than a balance of probabilities, based on such evidence as we can find from the mode of life of the hagfish and a more detailed knowledge of its physiology and development.

The idea that the agnathans were ancestral to the higher jawed vertebrates is one that is still reflected in the biological literature. This concept of an agnathan-gnathostome evolutionary sequence arose quite naturally from the fact that the jawless vertebrates appear to precede the gnathostomes in the geological record. However, quite apart from the conceptual problems of deriving the hinged jaws and internal visceral skeleton of the gnathostome from a continuous external branchial skeleton of the cyclostome type, many would now regard this as a misleading oversimplification, based on a fundamental misconception of the manner in which new species or higher taxa arise. For adherents of the cladistic view, these are formed by the splitting up of an ancestral species or group resulting in the appearance of two or more sister groups and the disappearance of the ancestral form. Accordingly, throughout this volume, the agnathans and gnathostomes are regarded as sister groups of equal antiquity, derived from an unknown and extinct common ancestor. At the same time this in no way diminishes the evolutionary significance of the cyclostomes. Even though they and their fossil relatives may no longer be thought of as being in the direct line of descent of the higher vertebrates, a close study of their biology may still enable us to make some reasonable inferences on conditions that may have existed in the remote common ancestor of both jawless and jawed vertebrates.

2

Distribution, variety and life cycles

2.1 Global distribution

The known species of cyclostomes are partitioned between the two Hemispheres and separated by a tropical and subtropical zone, where for the most part they are absent. There can be no doubt that the main limiting factor in this antitropical distribution is temperature and their northward or southward range is said to be restricted by the annual 20° isotherm. Only two species of lamprey appear to depart from this pattern. These are the two Mexican freshwater forms, *Tetrapleurodon spadiceus* and *T. geminis*, but these occur at high altitudes where water temperatures are unlikely to exceed the levels that lampreys are able to tolerate. At least during their long larval life, a lethal temperature limit of 29–31°C has been established for four different species of ammocoete (Potter and Beamish, 1975). The extreme northward or southward penetration of lampreys in the Northern or Southern Hemispheres cannot be defined with precision since little is known of their oceanic feeding ranges, but in the Northern Hemisphere lampreys certainly occur off the coasts of Northern Scandinavia, Siberia and Alaska, while in the Southern Hemisphere, *Geotria australis* is known to feed in Antarctic waters (Section 12.4.3).

Being mainly confined to relatively deep water, our knowledge of the distribution of the hagfish is less complete and rests, in many cases, on the recovery of only isolated specimens. At least two species, *Eptatretus springeri* and *Myxine circifrons* have been recorded from tropical or subtropical regions, but these are deep water forms; the latter species having been recovered from a depth of over 1300 m in the Gulf of Panama. In view of the inability of the hagfish to osmoregulate in water of varying concentrations (Chapter 9), salinity could obviously be a more important factor in their distribution and may explain their absence from the less saline waters of the Baltic.

In the distribution of known species between the two Hemispheres, there is

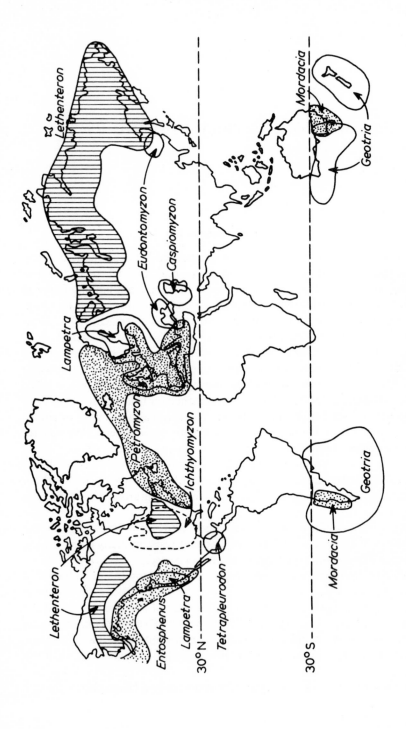

Fig. 2.1 Global distribution of the main genera and sub-genera of lampreys. Based on Hubbs and Potter, 1971.

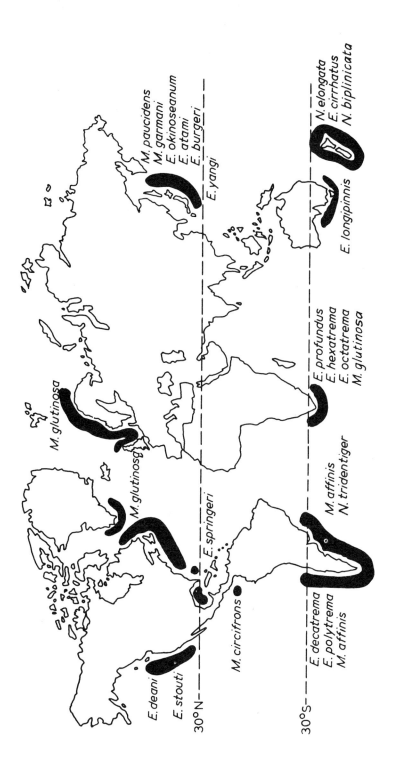

Fig. 2.2 Global distribution of hagfishes. Based on known species listed by Adam and Strahan (1963) and Strahan (1975).

a marked contrast between the lampreys and the hagfishes. Thus, of the 38 species of lampreys listed in Table 2.1, only four belong to the Southern Hemisphere, whereas the numbers of myxinoid species are almost equally divided between the two Hemispheres (Table 2.2). A notable feature of lamprey distribution is their absence from southern Africa, whereas two species of hagfish have been recorded from the seas off the Cape. Since the availability of suitable freshwater breeding areas must ultimately limit the distribution of the lamprey, it is possible that their absence from South Africa as well as their limited speciation in the Southern continents may be due to the relatively small number of river systems in the temperate areas of the southern continents where conditions are suitable for spawning and larval life.

Table 2.1 Systematic list of lamprey species, their biological characteristics and geographical distribution.

FAMILY	PETROMYZONIDAE	Biological type	Area of distribution
Genus	*Ichthyomyzon*		
Species	*I. unicuspis*	Fluviatile, parasitic	From Hudson Bay drainage to Mississippi Basin
	I. fossor	Non-parasitic	Similar
	I. castaneus	Fluviatile, parasitic	Western Manitoba, Iowa, Wisconsin to Alabama
	I. gagei	Non-parasitic	Similar
	I. bdellium	Fluviatile, parasitic	Ohio drainage system
	I. greeleyi	Non-parasitic	Similar
Genus	*Petromyzon*		
Species	*P. marinus*	Anadromous parasitic	In W. Europe from Barents Sea to W. Mediterranean. N. America Labrador – Florida
		Also landlocked form	Great Lakes System
Genus	*Caspiomyzon*		
Species	*C. wagneri*	Anadromous, parasitic	Caspian basin
Genus	*Eudontomyzon*		
Species	*E. danfordi*	Fluviatile, parasitic	Danube basin and some tributaries of Baltic and Black Seas
	E. vladykovi	Non-parasitic	Similar
	*E. lanceolata	Non-parasitic	Black Sea drainage of N. Turkey
	E. mariae	Non-parasitic	Similar to *danfordi*
	E. morii	Non-parasitic	Korea, Manchuria
	*E. gracilis	Non-parasitic	Slovakia
Genus	*Tetrapleurodon*		
Species	*T. spadiceus*	Fluviatile, parasitic	Mexico
	T. gemini	Non-parasitic	Mexico
Genus	*Lampetra*		
Subgenus	*Entosphenus*		

Table 2.1 *(contd.)*

FAMILY	PETROMYZONIDAE	Biological type	Area of distribution
Species	Lampetra (Entos-phenus) tridentata	Anadromous, parasitic Also dwarf landlock-ed forms	Alaska to California and Japan British Columbia, California
	*L. (Entosphenus) minima (extinct)	Dwarf parasitic land-locked form	Oregon
	*L. (Entosphenus) folletti	Non-parasitic	California
	L. (Entosphenus) lethophaga	Non-parasitic	N. E. California and Oregon
	L. (Entosphenus) hubbsi	Non-parasitic	South Central California
Subgenus	Lethenteron		
Species	L. (Lethenteron) japonica	Anadromous, parasitic	White Sea to N. Pacific. Alaska
Subspecies {	L. j. japonicus, L. j. septentrionalis	Anadromous, parasitic	
	L. j. kessleri	Probably non-parasitic	Siberia
	L. (Lethenteron) lamottenii	Non-parasitic	N. America Gt. Lakes, St. Laurence basin, Alaska, Mississippi drainage and Atlantic Coastal regions
	L. (Lethenteron) reissneri	Non-parasitic	Japan and Amur basin
	L. (Lethenteron) misukurii	Non-parasitic	
	L. (Lethenteron) zanandreai	Non-parasitic	N. Italy. Po basin
	*L. (Lethenteron) meridionale	Non-parasitic	Eastern Gulf drainage. N. America
Subgenus	Lampetra		
Species	L. fluviatilis	Anadromous, parasitic. Landlocked forms also	Western Europe and Baltic W. Mediterranean
	L. planeri	Non-parasitic	Similar
	L. aepyptera	Non-parasitic	Gulf drainage N. America. Maryland, Virginia and N. Carolina
	L. ayresii	Anadromous, parasitic	British Columbia to California
	L. richardsoni	Non-parasitic	British Columbia to Oregon
	*L. pacifica	Non-parasitic	California and Oregon

Table 2.1 *(contd.)*

FAMILY	PETROMYZONIDAE	Biological type	Area of distribution
FAMILY Genus Species	GEOTRIIDAE *Geotria* G. *australis*	Anadromous, parasitic	Western and South Australia, Victoria, W. Coast of Tasmania, New Zealand, Chile, Argentine
FAMILY Genus Species	MORDACIIDAE *Mordacia* G. *mordax* G. *praecox* M. *lapicida*	Anadromous, parasitic Non-parasitic Anadromous, parasitic	S. E. Australia, Tasmania New South Wales W. Coast of S. America

*These are new species that have been described since the publication of Hubbs and Potter (1971). Within the subgenus *Entosphenus* two additional non-parasitic forms have been recognised – *E. hubbsi* (Vladykov and Kott, 1976b) and *E. folletti* in addition to the dwarf and reputedly extinct parasitic species, *E. minima* (Bond and Kan, 1973). To the subgenus *Lethenteron* has been added a further non-parasitic species *L. meridionale* (Vladykov *et al.*, 1975) and the new non-parasitic species *L. pacifica* is presumably a derivative of *L. ayresii*. The new non-parasitic lamprey described by Kux and Steiner (1971/72) from the Black Sea drainage in Northern Turkey was designated *Lampetra lanceolata*, but since it occurs in area where *L. fluviatilis* is absent it should probably be allotted to the genus *Eudontomyzon* which is strongly represented in this region of the Black Sea basin. A further non-parasitic member of this genus has been described from Slovakia and designated *E. gracilis* (Kux, 1965).

2.2 The life cycles of lampreys

All lampreys breed in freshwater and pass the major part of their life cycle in a freshwater larval stage, before their final transformation to the adult state. With respect to their life after metamorphosis, the various lamprey species fall into three categories; non-parasitic or brook lampreys, freshwater parasitic species and anadromous parasitic forms. The non-parasitic lampreys constitute a little over a half of all the known species (Table 2.1) and are to be found throughout the entire range of the group. Although, during metamorphosis, they develop the necessary mechanisms for parasitic feeding, these are never employed; the gonads proceed to mature almost immediately after transformation, the intestine atrophies and the newly formed adult foregut usually remains a solid rod without an internal lumen (Fig. 2.3). In the absence of feeding and growth after metamorphosis, these lampreys remain dwarf forms; slightly smaller than the fully grown ammocoete and within 6–9 months after metamorphosis, they reach sexual maturity, spawn and die. Of the parasitic lampreys, five species are entirely freshwater forms, confined to

Table 2.2 Systematic list of hagfish species* and their geographical distribution. From Adams and Strahan, 1963, and Strahan, 1975.

Class Myxini, Order Myxiniformes (Myxinoidea)

FAMILY MYXINIDAE			FAMILY EPTATRETIDAE		
Genera	Species	Distribution	Genera	Species	Distribution
Myxine	*M. glutinosa*	Eastern and Western N. Atlantic. North wards to Greenland and southwards to Carolina and Mediterranean. Also reported from Cape of Good Hope†	*Eptatretus*	*E. polytrema*	Chile
				E. deani	Alaska – British Columbia
				E. stouti	California
				E. decatrema	Chile
				E. okinoseanum	Japan
				E. octatrema	S. Africa
				E. cirrhatus	S. Pacific
	M. garmani	E. coast of Japan		*E. hexatrema*	S. Africa
	M. paucidens	E. coast of Japan		*E. burgeri*	E. coast Japan
	M. affinis	S. Chile to S. Argentine		*E. profundum*	Cape of Good Hope
	M. circifrons	Gulf of Panama		*E. atami*	Japan
				E. yangi	Formosa
Notomyxine	*N. tridentiger*	Argentine to Patagonia		*E. springeri*	Gulf of Mexico
Neomyxine	*N. biplinicata*	New Zealand		*E. longipinnis*	Australia
Nemamyxine	*N. elongata*	New Zealand			

*Two further species, not yet described, have been mentioned by Jespersen (1975) as having been taken off the coast of Southern California; a white-bordered hagfish (*Eptatretus* sp.) in shallow water of 110 m and a deep water bathybial hagfish (*Myxine* sp.) at depths of 1554–2012 m.

†*M. glutinosa* has now been recorded in Western Antarctic waters (Permitin, 1977).

the river systems in which they breed and pass their larval life. This group includes the three parasitic members of the genus *Ichthyomyzon* of the Mississippi Basin, the river lamprey of the Danube, *Eudontomyzon danfordi* and the Mexican river lamprey *T. spadiceus* (Table 2.1). These fluviatile species tend to be smaller than the typical anadromous lampreys and their adult life is more restricted. After metamorphosis, the anadromous lampreys move downstream towards the sea or estuary, where they feed voraciously and grow rapidly, eventually returning to the rivers on their upstream spawning migration after a period of some 1–3 years. Like all other lampreys, death follows shortly after the completion of the spawning act.

Apart from the fluviatile species which have probably always been freshwater forms, there are others in which the freshwater feeding habit has clearly been a comparatively recent innovation. Of these, the best-known example is that of the dwarf, landlocked form of the sea lamprey, *Petromyzon marinus*, which occurs throughout the Great Lakes system of North America

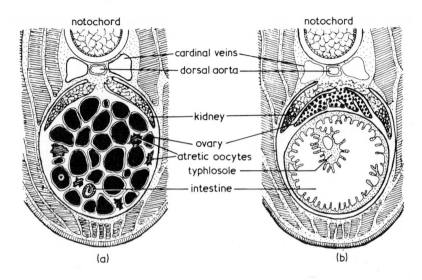

Fig. 2.3 Comparisons of the condition of the ovaries and intestine as seen in transverse
sections of a non-parasitic (*L. planeri*) lamprey (a) and a parasitic form (*L.
fluviatilis*) (b) at the completion of their metamorphosis (macrophthalmia
stage).

In (a) note the small numbers of large yolky oocytes, many in various stages of
degeneration and the shrunken intestine. In (b) much larger numbers of oocytes
are present, but these remain very much smaller in size, while the intestine is large
with a folded epithelium and well-developed typhlosole.

as well as in the smaller lakes of New York State (Section 4.3.1). This race was
almost certainly derived from the larger anadromous sea lamprey (probably
during Glacial or immediate post-Glacial times) which still enters the
St. Lawrence and other rivers on the eastern seaboard of North America on its
spawning migration. Other landlocked freshwater forms, derived from
anadromous lampreys are known to exist within the species *Lampetra
fluviatilis*, *L. japonica* and *L. (Entosphenus) tridentata*.

2.2.1 Paired species

The widespread occurrence of non-parasitic lampreys presents a number of
interesting evolutionary problems. In their classical morphological analysis of
the genus *Ichthyomyzon*, Hubbs and Trautman (1937) had shown that each of
the three non-parasitic species, *I. fossor, I. greelyi* and *I. gagei* possessed
characteristics that could be matched against the respective parasitic species,
I. unicuspis, I. bdellium and *I. castaneus*. Since the area of distribution of each
of the brook lampreys lay within that of the related parasitic species, it was a
reasonable inference that the dwarf non-parasitic forms had evolved from the

corresponding parasitic ancestral species (Zanandrea, 1954; Hardisty and Potter, 1971c). This concept has now been extended to most of the known non-parasitic lampreys and examples of these 'paired species' are now recognized in all lamprey genera with the exception of *Geotria*, *Petromyzon* and *Caspiomyzon*. In a few exceptional cases, a non-parasitic species may occur in an area where there is now no parasitic form to which it can be related. For example, the brook lamprey of the Po Basin in North Italy, *L. zanandreai* is restricted to an area where the European river lamprey, *L. fluviatilis* is no longer found. Similarly, the non-parasitic *L. (Okkelbergia) aepyptera* does not appear to have any parasitic counterpart within the southern States of the USA or the eastern drainage of the Gulf of Mexico, where it now occurs. A plausible explanation is that these isolated species are relict forms, remaining in areas from which their corresponding parasitic ancestors have since disappeared.

In a number of genera, there appears to have been a tendency for several distinct brook lamprey species to evolve from a common parasitic progenitor in different areas of its range and probably at different period. For example, the large anadromous Pacific lamprey, *L. (Entosphenus) tridentatus*, which spawns in the rivers of the Pacific coast of North America from Alaska to California, appears to have given rise to landlocked forms as well as to several brook lamprey species. These include *L. (Entosphenus) minima* now said to be extinct and reputed to have been a scavenging, parasitic or even cannibalistic form, but whose dwarf body size would be more consistent with a non-parasitic life cycle. Within the drainage of the same river system from which *minima* was reported, two further non-parasitic species have been described, *L. (Entosphenus) lethophaga* and *L. (Entosphenus) folletti*, while yet another species *L. (Entosphenus) hubbsi* has been recorded in south central California. A further example of this multiple speciation may be drawn from the European genus, *Eudontomyzon*, within which the parasitic *E. danfordi* is thought to have given rise to *E. vladykovi*, *E. mariae* and possibly to two recently described species, *E. lanceolata* and *E. gracilis*.

In general, the extent of the geographical dispersion of brook lampreys tends to be related to the area covered by their presumed parasitic ancestor and where the latter is an anadromous form, this will obviously tend to be more extensive than in cases where the primary species is confined to fresh water. Thus, the non-parasitic species in the genera, *Ichthyomyzon* or *Eudontomyzon* tend to have a more localized distribution than the brook lamprey, *L. planeri*, whose parasitic and anadromous progenitor *L. fluviatilis*, ranges widely throughout Europe, from the Baltic in the east to the Western Atlantic seaboard and the Mediterranean. The remarkable parallelism in the evolution of non-parasitic forms is also illustrated by the presence in the genus *Lampetra* of two related river lampreys, *L. fluviatilis* in Europe and *L. ayresii* in North America, each of which has given rise to corresponding brook lampreys – *L.*

planeri and L. richardsoni. Amongst the lampreys of the Southern hemisphere, it is only in the genus *Mordacia* that a non-parasitic form has so far been recognized and this species *M. praecox* has been found in only two river systems in New South Wales (Potter, 1970).

Looked at in a wider perspective, the persistent tendency in most lamprey genera for the evolution of non-parasitic derivatives may be seen as only one of the many viable alternative survival strategies that have been adopted by these highly adaptable animals. At one extreme of the spectrum are the largest anadromous species committed to an extended, wide ranging oceanic migration, and to maximum body size and fecundity. At the opposite pole come the brook lampreys. These have abandoned a migratory feeding stage, reduced the duration of adult life, and accepted a body size no larger than that of the ammocoete with vastly reduced fecundity. Between these two extremes, come the majority of parasitic lampreys, some fresh water, some anadromous, which could be regarded as having adopted an intermediate style of life, probably involving an adult feeding stage of shorter duration with a correspondingly smaller body size and fecundity. Significantly, it is mainly from this less 'committed' group of species that the majority of non-parasitic species appear to have originated, while on the other hand, the largest lampreys, *Geotria australis* and the anadromous *P. marinus* have failed to produce brook lamprey derivatives.

The widespread occurrence of the non-parasitic type of life cycle and the fact that these forms must have arisen repeatedly in the history of lampreys, suggests that its genetic basis must be common to most lamprey populations. In searching for a physiological explanation for this phenomenon, it was natural that attention should have been focused on the apparently precocious sexual maturation of the brook lamprey and on the possibility that accelerated gonadal development might be controlled by the pituitary. This view of the brook lamprey as a paedomorphic form was reinforced by the discovery in a single stream in Northern Italy of neotenous female ammocoetes of the non-parasitic species, *L. zanandreai*, in which apparently mature eggs were already present in the body cavity of animals showing no indications of metamorphosis (Zanandrea, 1956, 1958). Comparisons of the life cycles of the paired species, *L. planeri* and *L. fluviatilis* (Fig. 2.4) suggest a different interpretation of the origin of non-parasitic forms (Hardisty and Potter, 1971a). Here, the crucial difference between the two species is not that sexual maturity occurs earlier in the brook lamprey (in fact, in relation to the life cycle it takes place at the same point in time), but rather that metamorphosis is delayed by about two years relative to its position in the life cycle of the parasitic species (Fig. 2.4). On this interpretation, the extension of the larval period in *L. planeri* has been balanced by an equal curtailment of adult life, leaving the total life span unchanged with an average duration of seven years. Since the larval stage is relatively protected and mortality low, these modifications to

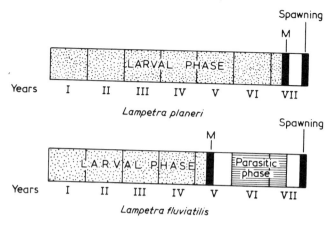

Fig. 2.4 Diagrammatic presentation of the life cycles of the paired species of lamprey, *L. planeri* (non-parasitic) and *L. fluviatilis* (parasitic). Unshaded areas represent periods when the animals are not feeding.
M. Metamorphosis.

the life cycle would have obvious selective advantages under conditions which made the feeding migration difficult or impossible, or where the supply of suitable host fishes was insufficient to permit the full realisation of the growth potential and fecundity of the adult lamprey.

2.3 Systematics, phylogeny and species distribution

2.3.1 Lampreys

A recent review of lamprey taxonomy (Hubbs and Potter, 1971) has placed the greatest emphasis at the generic level on the character of the dentition. The resulting scheme differs in only minor respects from that adopted by Vladykov and Kott (1976a). Whereas the latter authors recognize *Lampetra*, *Lethenteron* and *Entosphenus* as distinct genera, these were listed by Hubbs and Potter as subgenera of *Lampetra*; a view that is consistent with the identical amino acid composition of the haemoglobins of *L. (Lethenteron) japonica* and *L. fluviatilis* (Section 13.1.1). In addition, Hubbs and Potter accorded generic status to *Okkelbergia*, but others have preferred to allocate this isolated North American brook lamprey to the genus *Lampetra*.

In their general analysis of phyletic trends, Hubbs and Potter regarded the dentition of *Ichthyomyzon* as the most primitive type and the same pattern was also traced, with some modifications in *Petromyzon* and *Caspiomyzon*. The present day members of the genus *Ichthyomyzon* still retain the freshwater habit that probably characterized the ancestral lamprey stock (Chapter 9) and display further primitive features in the confluence of the two dorsal and caudal fins. The remaining Northern hemisphere lampreys have been grouped into two main lineages – one leading to *Eudontomyzon*, *Tetrapleurodon* and *Okkelbergia* (all freshwater forms), and the other to the *Lampetra* complex, consisting of *Entosphenus*, *Lampetra* and *Lethenteron*. The position of the southern genera, *Geotria* and *Mordacia* poses a number of questions. Although differing in many respects from the Holarctic lampreys, *Geotria* still retains a basic common pattern in the dentition of the oral disc. On the other hand *Mordacia* shows a number of quite fundamental divergencies from all other lampreys, of which the disparity in karyotype is perhaps the most significant (Section 13.9), and it is impossible to avoid the conclusion that their separation from the main lamprey stock is of great antiquity and certainly much older than the point of divergence of *Geotria*. In addition, the patterns of distribution of the two genera are difficult to explain. *Geotria* occurs on both coasts of South America (its northern limit being approximately 33°S), in New Zealand, in Western Australia and the West Coast of Tasmania. *Mordacia* on the other hand is found only on the West Coast of South America and the south-eastern areas of Australia (Fig. 2.1). Thus, its distribution is considerably more restricted than that of *Geotria*, perhaps reflecting a less extensive range of oceanic migration.

2.3.2 Hagfishes

Perhaps because of the constancy of their environment and their somewhat unadventurous habits, speciation in myxinoids appears to have been more restricted than in the more adaptable lampreys, although we should make allowance for the possibility that because of their inaccessibility, many hagfish species may remain to be identified. The work of the myxinoid taxonomist has been complicated by the great individual variability that exists in those morphological characters he commonly employs; a variability that has been related to the range of variability in their karyotypes (Chapter 13). However, in at least one respect, hagfishes seem to fall into two, if not three, natural groups based on the arrangement of their efferent gill ducts. Thus, in the *Myxinidae*, there is a single external opening for all these ducts, whereas in the *Eptatretidae*, they open by separate pores to the exterior (Fig. 2.5).

Among the polybranchiate hagfishes with separate gill openings are three species in which these pores are crowded together towards the posterior end of the branchial zone; a condition that has been regarded as intermediate between that of the *Myxinidae* and the *Eptatretidae*. However, although these species had previously been assigned to a distinct genus (Adam and Strahan, 1963) *Paramyxine*, more recently Strahan (1975) has preferred to regard the latter as a junior synonym of *Eptatretus*; a point of view that has been adopted in the list of species in Table 2.2.

Our knowledge of the distribution of hagfishes may be biased by the fact

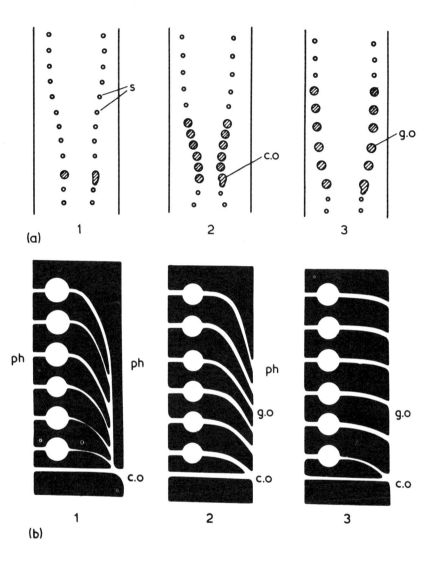

Fig. 2.5 Diagrammatic representation of the branchial regions of hagfishes.
(a) Ventral surfaces showing the slime gland pores (s) and branchial openings
(g.o). Common opening of the last gill and the pharyngo-cutaneous duct (c.o).
(b) The arrangement of the efferent gill ducts, with the pharynx (ph) to the left.
1   Condition in *Myxine*.
2   Posterior crowding of the gill pores in *Eptatretus* (*Paramyxine*) *atami*.
3   Separate spaced gill pores as in *Eptatretus burgeri*.

that these animals are most likely to be recovered in areas where commercial fisheries are more active. This may account for the impression that certain species, such as *Myxine glutinosa* or *E. stouti* are more numerous (and also for their dominance in the scientific literature). The same factors may also be at least partly responsible for the large number of species recorded from Japanese waters and Japan is the only country where the hagfish (*E. atami*) has been fished commerically and has been exploited as human food. It has been estimated that at least 20 hagfish species remain to be described (C. L. Hubbs, personal communication), many of them from the eastern Pacific, particularly between California and Baja California, Mexico. Even for the known species it is clear that there is a great disparity in the diversity of the hagfish fauna of the North or South Atlantic when compared to the Pacific, perhaps reflecting the more recent geological age of the Atlantic oceans.

In their body size, the hagfishes also appear to be more uniform than lampreys. For the recognized species in Table 2.2, the maximum lengths are those recorded for *E. okinoseanum* (800 mm) and *M. glutinosa* (790 mm), but an undescribed species, widely distributed in the Pacific is said to reach lengths of at least one metre (C. L. Hubbs, personal communication). The smallest hagfish appears to be *E. yangi* from Formosa, with an average length of 230 mm and a maximum of 250 mm.

3
Perspectives and relationships

3.1 Cyclostomes and the fossil agnathans

Appearing for the first time in Upper Cambrian deposits (Repetski, 1978), the fossil agnathans underwent extensive evolutionary radiation, flourishing in the Silurian then decreasing in numbers, finally to disappear from the fossil record at the close of the Devonian. Since the first jawed vertebrates, the acanthodians were not to be found earlier than the upper Silurian, it is hardly surprising that the agnathans and gnathostomes should have been regarded as a phyletic sequence, representing the evolution of hinged jaws from the more primitive jawless condition of some agnathan vertebrate ancestor. Many now reject this point of view (not least because of the difficulties of deriving the visceral arches of the gnathostomes from the fundamentally different structures of the agnathans) preferring instead to regard the agnathans and gnathostomes as sister groups of equal antiquity, but sharing a hypothetical and more remote common ancestor.

In the 19th century it already had been appreciated that the fossil agnathans were jawless creatures, like the living cyclostomes and were divisible into two broad groupings, the Osteostraci and the Heterostraci, on the basis of the structure of their hard tissues. Thanks very largely to Stensiö's (1927, 1958), brilliant reconstructions of the head of the cephalaspids (Osteostraci) it became apparent that this group of fossils shared many of the structural features of the lamprey and this is now recognized by grouping them in the Cephalaspidomorphi, together with the Anaspida. A second grouping, the Pteraspidomorphi, including the Heterostraci (Pteraspida) with the imperfectly known Thelodonti, is now generally adopted, although there has been less agreement on the morphological criteria on which this should be based. Controversy has centred around the nature of the olfactory organs of the agnathans and their relation to the development of the snout region. For example, a majority of palaeozoologists have followed the example set by Kiaer (1924) who first suggested that the Heterostraci had paired nasal sacs and nostrils (diplorhinal) like the gnathostomes, whereas in the cephalaspids and anaspids, as well as in living cyclostomes only a single 'nasal' opening is present (monorhinal). Basing his views on the embryology of the hagfish and lamprey, Stensiö (1968), on the other hand, has argued that all the agnathans are primarily diplorhinal and that it is essential to distinguish between the true 'nostril' of the cyclostomes (the primary openings of the nasal sacs) and the external opening of the nasohypophysial tract (Fig. 3.10). Thus, both in the lamprey and hagfish embryos, the nasal sac develops as a bilobed structure, whose two halves open separately into the nasohypophysial tract (prenasal sinus); the latter representing a secondarily developed extra-cephalic space to whose external opening the term nostril should not be applied (Section 3.3).

In the embryo lamprey, the dorsal position of the nasohypophysial complex is achieved by the development and growth of a post-hypophysial fold, initially separating the combined olfactory and hypophysial anlage from the stomodaeal depression (Fig. 3.1). As a result of the hypertrophy of this fold, the prenasal part of the snout is shifted backwards and upwards, bringing the nasohypophysial opening into the dorsal position that it also occupies in the fossil cephalaspidomorphs. A similar development of the post-hypophysial fold does not occur in the myxinoid embryo, with the result that the prenasal region is not displaced posteriorly, remaining in position as the dorsal part of the snout, while the nasohypophysial canal is situated between the prenasal and posthypophysial components. This divergence in ontogenetic growth patterns can readily be accepted as a consequence of differences in the functional adaptations of the feeding and respiratory mechanisms of hagfishes and lampreys, but its direct relevance to conditions in the pteraspidomorphs or their relationships with the myxinoids is more debatable. On the one hand, Stensiö's controversial and myxinoid-like reconstructions of the heterostracan head (which he believed would have developed in a similar way to that of the hagfish embryo) were used as a justification for his view that the myxinoids descended from pteraspidomorph ancestors and should be classified accordingly. However, in this group of fossils, the nasohypophysial tract or its opening cannot be identified and on the contrary, a majority of authors have inferred from impressions on the dorsal shield that the heterostracans would have possessed paired nasal sacs, presumably opening into the mouth (Heintz, 1963).

3.1.1 The cephalaspidomorphs
The cephalaspids (Osteostraci) are known from the upper Silurian to the upper Devonian and are characterized by their possession of a bony head shield, the rest of the trunk being covered by articulated scales (Fig. 3.2). These are three-layered, with an outer layer of dentine-like material, a middle vascular layer and a basal region containing bone cells. A pore canal system is present which may be part of a lateral line system.

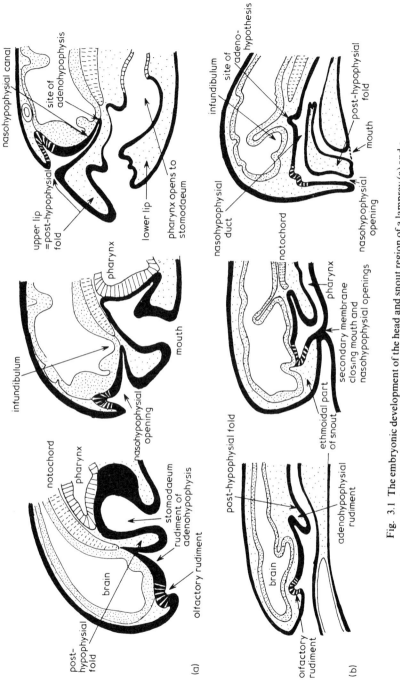

Fig. 3.1 The embryonic development of the head and snout region of a lamprey (a) and a hagfish (b). Based on Stensiö, 1968.

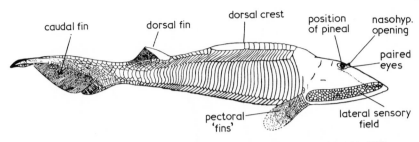

Fig. 3.2 A reconstruction of a cephalaspid, *Hemicyclapsis*. After Stensiö, 1968.

The large paired eyes are placed dorsally and between them is a pineal foramen. In front of this is the nasohypophysial opening and a projecting rostrum, representing as in the lamprey the preoral region of the snout. The wide ventral mouth opens into a large buccal cavity, separated from the branchial chambers by a ridge, possibly carrying a velum and representing the mandibular arch. Behind the mouth there are usually 9–10 pairs of gill chambers (the first corresponding to the hyoidean pouch), with separate ventral openings (Fig. 3.3). As in the lamprey, the otic capsule contained only the two vertical canals. In its general pattern the brain is similar to that of the lamprey, although it differs in the presence of a bilobed cerebellum.

Judging from the dorso-ventrally flattened head and the ventrally placed gill openings, it is reasonable to suppose that the cephalaspids were bottom-living forms, perhaps feeding by sucking into the mouth organic debris or small organisms from the bottom silt (Denison, 1961). Speculations on their feeding mechanisms have varied from a unidirectional current produced by a velum or by ciliary tracts, to a pumping action like that of the adult lamprey, involving alternate elevation and depression of the floor of the mouth and branchial regions. However, in animals that may have measured as much as 30 cm in width, velar contractions, even if combined with ciliary mechanisms might have been inadequate. Watson (1954) has suggested that the flexible branchial floor could have been raised by muscles passing from it to the interbranchial ridges during the expiratory phase; the inspiratory phase as in present-day lampreys being produced by elastic recoil. At the same time, it has been argued that the relatively weak suction produced by elastic forces during inspiration would probably have been inadequate to sustain a predatory mode of life, much less a parasitic method of feeding (Lessertisseur and Robineau, 1970). However, the presence in the cephalaspid *Boreaspis* of small depressions on the supra-oral region has led to a suggestion that they may have possessed a bilobed tongue, similar to the apical region of the lamprey piston (Janvier, 1977).

The anaspids from the Silurian and Devonian are fish-like in shape, with a fusiform, laterally compressed body rarely exceeding 15 cm in length (Fig. 3.4). The dermal skeleton consisted of rows of scales of a bone-like material,

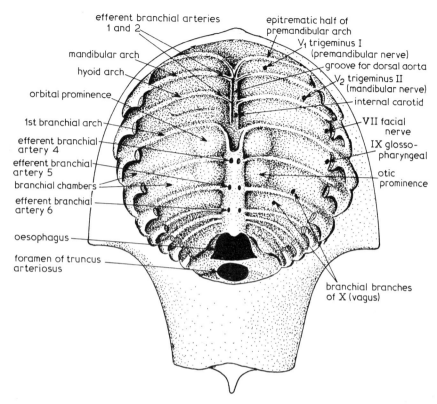

Fig. 3.3 Ventral view of the cephalic shield of a cephalaspid, *Kieraspis* with some of the structures identified by Stensiö. Redrawn from Stensiö, 1968.

The numbering of arteries, nerves and arches is based on Stensiö's interpretations of these structures.

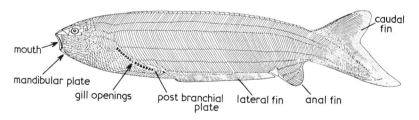

Fig. 3.4 Reconstruction of an anaspid, *Pharygolepis*. From Ritchie, 1964.

but lacking enclosed bone cells and some forms show a reduction in dermal bone. The head region may either be naked or covered with plates. The terminal mouth has been interpreted as a vertical, oval slit-like aperture. The paired eyes are laterally placed and between them is a pineal foramen. The number of gill openings may vary from 6 to 15 in different species and these are often arranged in a ventrally slanting row. In marked contrast to the cephalaspids, the branchial region is positioned posteriorly and the first gill pouch lies some distance behind the eye. In this respect therefore, the anaspid is closer to the lamprey condition than the cephalaspid. In place of separate pectoral and pelvic fins, the anaspids possessed a continuous lateral fin fold, which unlike the less mobile 'pectoral fins' of the cephalaspids, is thought to have been a flexible structure, provided with muscles attached to internal skeletal rods or radials. Views on their probable feeding habits have varied from ciliary mechanisms (Heintz, 1958) to suctorial detritus feeding or even planktonic feeding (Whiting, 1972). The latter author has related the restricted size range of anaspids to their retention of a microphagous feeding mechanism similar to that of the ammocoete, suggesting that the enlarged chin may have contained a velar pumping mechanism. Other authors have suggested that this enlarged prebranchial region might have housed a rasping tongue (Parrington, 1958), foreshadowing the predaceous feeding mechanism of the lamprey (Stensiö, 1958).

Because of its possible relevance to the phylogeny of the petromyzonids, the anaspid *Jamoytius* has attracted considerable attention (Fig. 3.5). This form,

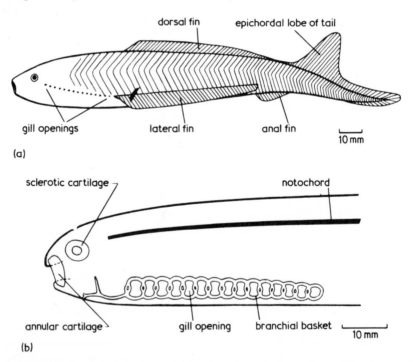

Fig. 3.5 Reconstructions of *Jamoytius*. From Miles, 1971, after Ritchie.
(a) General morphology. (b) Head and branchial region.

from Scottish Silurian deposits was 18–20 cm long, with a tubular trunk and bluntly rounded head. The mouth was surrounded by an annular cartilage like that of the lamprey, but this lay closer to the surface of the snout, precluding the presence of an oral funnel of the petromyzonid type. The body surface was covered by thin flexible scales, possibly of a horny epidermal nature, corresponding in their arrangement to the underlying myotomes. Continuous lateral fin folds appeared to extend backwards from the branchial region and behind these are paired anal folds. A continuous dorsal fin was present and the caudal fin was of the hypocercal type. The large paired eyes were latero-dorsal in position and surrounded by sclerotic rings. Behind the annular cartilage, a nasohypophysial opening has been identified. A significant feature is the presence of a branchial basket of the petromyzonid type, with longitudinal epi- and hypo-trematic bars above and below the gill openings, connected by vertical elements representing the branchial bars. The total number of these arches cannot be determined precisely, but is considered to be not less than 15. Unlike the adult lamprey, there appears to be no connection between the branchial skeleton and the notochord, neither are the two halves of the branchial basket united in the mid-ventral line. Ritchie (1968) has suggested that ciliary feeding would be unlikely in an active animal of this size and the shape of the mouth would seem to preclude planktonic feeding. A more likely possibility is that *Jamoytius* employed some form of suctorial feeding, browsing on epiphytic or epizootic organisms, perhaps with the aid of a rasping tongue mechanism.

3.1.2 The pteraspidomorphs
The heterostracans are first represented by isolated exoskeletal fragments from the Upper Cambrian of North America (Repetski, 1978). They are therefore the oldest vertebrate group in the fossil record. The head and anterior trunk were enclosed by bony plates (Figs. 3.6 and 3.12) composed of a characteristic material called aspidin. Internal bone is absent and the lack of an ossified endocranium is responsible for our ignorance of the internal structure of the head, although some inferences have been drawn from impressions left on the inner surfaces of the dorsal and ventral shields (Fig. 3.7).

The mouth, usually anterior and ventral, lies below the edge of the dorsal shield, typically drawn out into a long, pointed rostrum. The gill pouches lie within the cephalic shield and open, as in *Myxine*, by a common duct. No nasohypophysial complex was present and impressions thought to be those of paired nasal sacs suggest that the nostrils may have opened into the mouth. The eyes were latero-dorsal in position and a depression between them on the inner surface of the shield has been identified as the pineal, although the covering of bone would presumably have precluded a light-sensitive function. In the ear only two semicircular canals can be identified, but the possibility

Fig. 3.6 A reconstruction of pteraspids and their mode of feeding. From Janvier, 1974.

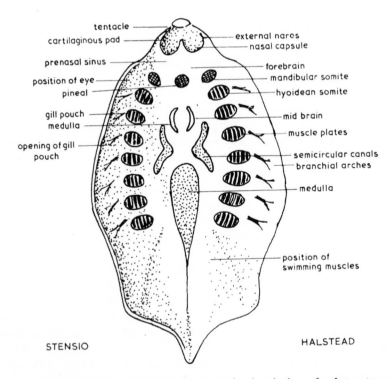

Fig. 3.7 Interpretations of the impressions on the dorsal plate of a heterostracan according to Stensiö (left) and Halstead (right). Redrawn from Halstead, 1973.

that a third horizontal member may have been present cannot be entirely excluded (Halstead, 1973b). The well-defined lateral line system is said to be of a generalized and primitive pattern, from which both the gnathostome and cyclostome condition could be derived.

Stensiö's reconstructions of the heterostracan brain differ radically from that proposed by Whiting and Tarlo (1965) and Halstead (1973a, b). These authors believe that the brain bore a marked similarity to that of the lamprey, showing the same linear arrangement of its component regions, a rudimentary cerebellum (represented by a transverse commissure), a mid-brain choroid plexus and an extended medulla. These primitive features would be such that the heterostracan brain could be the starting point from which both gnathostome or cephalaspidomorph brains could have been derived.

Equally fundamental are differences in the interpretations of the impressions on the inner surfaces of the heterostracan head shield (Fig. 3.7). Paired oval impressions on either side of the midline were identified by Stensiö, Kiaer, Watson and others as gill pouches but, more recently, Tarlo and Whiting (1965) have compared these to head myotomes, of which the first two are identified with mandibular and hyoid segments. If this interpretation is accepted, it may be inferred that the movements of the eyeball could only have been achieved by extrinsic eye muscles derived from only the first (premandibular segment) and not, as in existing vertebrates, from each of the first three pro-otic segments (Section 3.2). Below these structures are markings, identified by Stensiö as gill pouches, but which these authors have re-interpreted as gill arches showing the typical gnathostome pattern of anterior and posterior hemibranchs, rather than the continuous branchial basket of the petromyzonids. The possibility that an endostyle may have existed is suggested by an elongated groove sometimes seen on the ventral shield. Thus, these recent views of heterostracan morphology suggest a more primitive type of organization than had previously been postulated in Stensiö's reconstructions. Rather than pointing towards myxinoid affinities, they would indicate that the heterostracans combined some petromyzonid characteristics, with others from which a gnathostome type might be derived. Halstead (1971) has even argued that perforations in the dorsal shield of the amphiaspids might represent spiracles and that these heterostracans may therefore have taken the first step in the transformation of the mandibular arch into jaws of the gnathostome type. This would presumably have begun with the loss of gill function by the hyoid pouch. Should any, or all, of these newer interpretations of heterostracan anatomy be correct, we should have to regard this group of agnathans as close to the ancestral line which gave rise to the gnathostomes.

3.1.3 *Mayomyzon* – a fossil lamprey

The discovery of the first fossil lampreys in upper Carboniferous deposits confirmed the great antiquity of this vertebrate group and at the same time showed how little their general organization has changed over a period of about 280 million years (Bardack and Zangerl, 1971).

In its general body form, *Mayomyzon* is strikingly similar to present-day lampreys, with an eel-like scaleless body devoid of dermal armour or paired fins, but with a continuous dorsal fin like that of the ammocoete, separated only by a notch from the caudal fin. A striking feature of these fossils is their small size, varying in total length from only 30–60 cm.

In the head region it has been possible to make out some of the cranial cartilages, which show strong resemblances to those of adult lampreys (Fig. 3.8). Of the internal organs, only the liver and intestine can be identified, but the position of the heart is suggested by a concavity on the anterior face of the liver. There are large paired eyes and a single, dorsally placed nasohypophysial opening. Seven pairs of gill sacs are indicated whose position is of especial interest, although there is no reason to assume, as suggested by Moy-Thomas and Miles (1971) that the first in the series represents the spiracular or hyoid pouch. The first two gill sacs lie immediately below the otic capsule, in contrast to the position in living lampreys, where the first pouch is situated some distance behind the auditory region. As Damas (1944) and Strahan (1958) have shown, the branchial region of the lamprey is shifted posteriorly during ontogenesis. Thus, in a 50 somite embryo, figured by Damas, the first gill sac lies immediately below the auditory capsule and is similar in position to the first gill sac of *Mayomyzon*. Since, on other grounds, there is every reason to believe that the fossils represent post-metamorphic stages (if indeed these

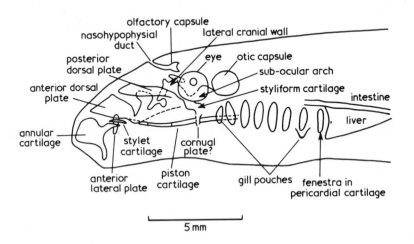

Fig. 3.8 A reconstruction of the head skeleton and associated structures of the fossil lamprey, *Mayomyzon*. Redrawn from Bardack and Zangerl, 1971.

This may be compared with the cranial skeleton of a modern lamprey, Fig. 7.2.

animals possessed a larval stage), this anterior position of the gills must be regarded as a primitive feature, resembling embryonic or early larval stages of present-day lampreys.

In other respects also, *Mayomyzon* may be regarded as being less specialized than modern lampreys. Thus, compared with the adult lamprey, the frontal region of the *Mayomyzon* head appears to lie much closer to the annular cartilage or the anterior dorsal plate, suggesting that the extension of the preoral region had not progressed to the same extent as in present-day lampreys. In place of the expanded oral funnel of the lamprey, the snout is tubular in shape. The auditory capsules are not united to the cranium as in the lamprey or hagfish and the nasal capsule is placed some distance in front of the skull. These conditions recall the kind of relationships that exist today only in the embryonic or early larval stages of the lamprey.

The piston mechanism and medio-ventral cartilages are present in *Mayomyzon*, although these are shorter than in modern lampreys. This is correlated with the more anterior position of the branchial region. No trace of horny disc teeth can be seen and this, combined with the shape of the mouth and the small body size, makes it most unlikely that these animals had developed the predaceous habits of present-day lampreys. Perhaps, like hagfishes they were scavengers or carrion feeders; a habit apparently still retained by some at least of our existing species of lampreys (Section 4.3.1).

3.2 Cranial organization and metamerism

High among the preoccupations of 19th century zoology was the analysis of the segmentation of the vertebrate head and the relationships between the jaws and visceral arches. Although the results of these studies and the hypotheses on which they were based have now passed out of the field of controversy into the textbooks of comparative anatomy, they are still an essential basis for discussions on the relationships between fossil agnathans, cyclostomes and gnathostomes and for attempts to establish morphological homologies.

The classical picture of the vertebrate head is that of a segmented structure, with myotomes and corresponding dorsal and ventral nerve roots; the latter belonging to the same series as the segmental spinal nerves and represented here by certain cranial nerves (Fig. 3.9). The visceral clefts show a similar serial arrangement, but in this case the openings lie between adjacent segments. Between the visceral clefts, the pharyngeal wall is supported by a series of jointed skeletal elements – the visceral arches – and, while the fossil record has supplied no direct evidence, it was clear from its structure, ontogeny and relationships, that the gnathostome jaw had originally belonged to anterior members of this visceral arch system, retaining in its innervation by the mandibular and maxillary branches of the trigeminal nerve (V), a similar

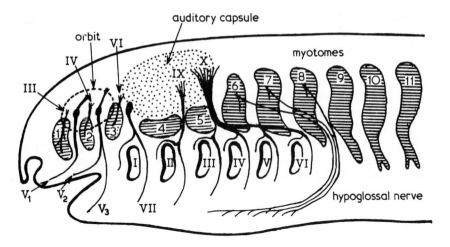

Fig. 3.9 The classical scheme of vertebrate head segmentation as based on the dogfish embryo. After Goodrich.

pattern to that of the nerves supplying the branchial arches. Here, each of the branchial nerves divides into a pretrematic and post-trematic branch, running through the visceral arches in front of and behind each gill chamber. Further confirmation came from the embryologists, who showed that like the visceral arches, the jaw cartilages (palato-quadrate and mandibular) were derived from ectomesenchyme (Platt, 1894); a skeletogenous tissue originating in the neural crest.

Although, from time to time, additional segments have been postulated for the ancestral vertebrate head, existing vertebrates show only three, more or less, complete pro-otic segments – the premandibular, mandibular and hyoid. From the myotomes of these somites are derived the extrinsic eye muscles, innervated by ventral nerve roots, represented here by the oculomotor (III), trochlear (IV) and abducens (VI). The dorsal roots corresponding to these three segments would be the ophthalmicus profundus (V^1), the maxillary (V^2) and mandibular branches (V^3) of the Vth cranial nerve – and the facial nerve (VII). The sclerotomes of the three pro-otic segments (corresponding to the arcual segmental elements of the vertebral axis) also contribute to the development of the base of the skull. Thus the antotic pila, acrochordals and anterior parachordals have been attributed to the sclerotomes of the mandibular and hyoid segments (Bjerring, 1968). According to this author, the myotome of the second, mandibular somite also gives rise to the polar cartilage, lying between the anterior end of the parachordals behind and the trabeculae in front (Fig. 7.1). The trabeculae have usually been regarded as the remnant of a premandibular arch. In this case, like the other visceral arches, they should be of neural crest or ectomesenchymal origin, but a number of authors believe that they have a somitic or mesodermal origin.

In a hypothetical protovertebrate stage, some earlier anatomists had postulated the existence of two additional pro-otic segments in front of the premandibular arch, with corresponding visceral arch elements and a recent review has suggested an additional terminal segment in front of the premandibular (Bjerring, 1977). Even if we disregard the more extreme of these earlier views it is hardly surprising that the general picture of a phylogenetic reduction or transformation in the more anterior visceral arches should have led to a search for traces of more primitive conditions in the agnathans and that interpretations of their cranial structure should have tended to conform to this idealized ancestral pattern. In fact, when Stensiö's classical reconstructions of cephalaspid material appeared, these showed almost precisely the kind of head organization and visceral arch system that we should have been expected to find in an ancient and ancestral vertebrate stock. Thus, complete premandibular and mandibular arches were identified with functional mandibular and hyoidean (spiracular) branchial chambers (Fig. 3.3). On this analysis, therefore, the cephalaspid would be truly primitive, possessing two functional gill sacs anterior to those of any existing vertebrates including the lamprey, where a hyoid pouch is present only in the embryo, but disappears in the adult.

Stensiö's identification of these anterior arches (Fig. 3.3) was based on the general topography of the epibranchial region and especially on the relations between the epibranchial visceral arch ridges and the corresponding blood vessels and cranial nerves. A re-examination of these structures in the light of conditions in embryonic, larval and adult cyclostomes has led a number of authors to quite different conclusions. Thus, Allis (1931) considered that the profundus V^1 and V^2 nerves of Stensiö were in fact the two branches of the trigeminal (maxillary and mandibular) of gnathostomes. After an exhaustive study of the cranial nerves of the cyclostomes, Lindström (1949) also accepted the absence of a profundus (V^1) nerve in the cephalaspid material, attributing this to its superficial course over the surface of the head, where it would not be preserved in fossil material. The same author also doubted whether a differentiation of maxillary and mandibular branches of the trigeminal nerve was practicable in the cyclostomes, preferring to regard Stensiö's first nerve V^1 as the trigeminal proper and his V^2 as the facial VII. More recently, Whiting (1972, 1977) has compared the course of the cranial nerves in larval lampreys with the situation in cephalaspids and has found this to be consistent with Lindström's analysis.

These revised interpretations of cranial structure in the cephalaspids have the merit of bringing the branchial and cranial anatomy of these animals more into line with that of the surviving agnathans. Thus, Damas (1954) believes that the most anterior of Stensiö's 'branchial' pouches — the mandibular — was really an ectodermal structure comparable with the posterior region of the buccal cavity in the ammocoete. This is separated from the endodermal pharynx by the velum; a muscular structure attached to the mandibular arch. The existence of a velum in cephalaspids had also been proposed by both Stensiö and Wangjio (1952), but attached in this case to the premandibular rather than to the mandibular arch – a position where it is difficult to see how it could have functioned in a similar way to the velum of ammocoetes or hagfishes.

In the head of embryo lampreys, as in other craniates, three pro-otic segments can be distinguished – premandibular, mandibular and hyoid. In the usual way, the myotomes of these segments give rise to the extrinsic eye muscles, innervated by cranial nerves, III, IV and VI, although Bjerring (1977) now regards the lamprey oculomotor as representing the ventral roots of the first two pre-otic segments; its dual nature being shown by its two separate nerve bundles. The mandibular myotome may also give rise to a polar cartilage, represented by at least part of the so-called trabecula (Bjerring, 1968). From the sclerotomes of the mandibular and hyoid somites are derived the cartilages around the anterior end of the notochord which form the basis of the cranial floor – the parachordals, antotic pila and acrochordal cartilages (Chapter 7). It is questionable whether the first or premandibular segment is complete in the sense of giving rise to the normal range of differentiated elements and Damas (1944) has doubted whether it has any skeletal significance. Jollie (1962) also points out that the mandibular artery of the embryo lamprey is later related to the velum, marking the anterior end of the pharynx and that there would therefore be no room for a more anterior arch. The possibility of visceral elements representing a premandibular arch in the lamprey has aroused considerable controversy and widely divergent views have been expressed, depending on whether the presence of a true trabecula is accepted and whether this structure is regarded as a visceral (ectomesenchymal) or somitic (mesodermal) derivative of the first or second somite (Chapter 7).

As a result of his studies of the embryological development of the arterial system in lampreys, Claydon (1938) was unable to find any indications that a premandibular artery, pouch or arch had ever existed in ancestral forms and Damas (1954) has also rejected the existence of any true arch elements in front of the hyoid arch. Allis (1924), on the other hand, recognized the trabecula as a premandibular element and a number of other authors also believed that traces of the same arch can be detected in the lamprey cranial skeleton (Stensiö, 1927; Holmgren, 1946; Johnels, 1949).

Some of the difficulties involved in the attempt to homologize the cranial structures of cyclostomes and gnathostomes are illustrated in Stensiö's analysis (1968) of the ontogenesis of the cyclostome snout. As described earlier (Section 3.1), this involves the hypertrophy of the anterior part of the embryonic post hypophysial fold, which is said to belong to the epitrematic part of the premandibular segment. From this, Stensiö infers that the sub-nasal and pre-nasal cartilages (Chapter 7) of the petromyzonids and myxinoids were derived from the premandibular arches. In the absence of

information on the ectomesenchymal or mesodermal origin of these structures, these assumptions seem to prejudge the issue. In addition, as Damas
(1944) emphasized, most of the cartilages in the splanchnocranium of the
adult lamprey are not derived directly from mucocartilaginous precursors in
the ammocoete and their true lineage and relationships cannot therefore be
determined. What is more, while certain regions of the larval head can be
defined broadly as premandibular or mandibular areas, it is impossible to
identify the corresponding arch elements.

In myxinoids, the primary segmentation of the head is distorted by the
degeneration of the eyes and the disappearance of their extrinsic muscles and
corresponding cranial nerves. In addition, the branchial region has been
shifted posteriorly away from the head. In lampreys, the first and embryonic
visceral pouch (hyoid) is innervated by the facial nerve (VII), the first of the
seven persistent gill sacs by the glossopharyngeal (IX) and the remaining six
by branches of the vagus. In myxinoids, on the other hand, all the gill sacs
appear to have a vagal innervation and the existence of a separate
glossopharyngeal nerve has been disputed. In *Eptatretus*, where the number of
gill sacs varies between 10 and 14, a hyoid pouch is present in embryonic
stages, but this and the two following gill sacs disappear in later development.
In *Myxine*, where only six functional gill sacs occur, the condition of the
embryonic arterial system (Holmgren, 1946) suggests that the number may
originally have been similar to that of the *Eptatretidae* and that the more
anterior gill sacs have been lost. This backward migration of the branchial
region of the myxinoids is a distinctive and adaptive specialization which
marks them off from the petromyzonids.

3.3 Fossil agnathans and the phylogeny of cyclostomes

However difficult it may be to assess the precise lineage of the petromyzonids
or the point at which they may have separated from a cephalaspidomorph
stock, the palaeontological and anatomical evidence leaves little room for
doubt of their affinity with this group of agnathans. The anaspid with its
reduced exoskeleton, lamprey-like branchial region and more active habits
has often been considered a more likely candidate for close petromyzonid
affinities than the heavily armoured and benthic cephalaspids, although our
knowledge of their internal anatomy does not match the detailed information
on the cranial morphology of the cephalaspids. Since Stensiö's reconstruction
of the cephalaspid head, more recent studies, using techniques of serial
sectioning, have disclosed further remarkable parallels in the detailed
anatomy of the nasohypophysial complex and eyes of cephalaspids and
lampreys (Janvier, 1971, 1974, 1975, 1976).

As in the ammocoete, the primary nasohypophysial opening of the
cephalaspid was situated at the base of a depression on the surface of the head.
In the cephalaspid, this primary opening was keyhole-shaped, probably with

an anterior hypophysial and a posterior nasal opening, separated in some cases by the fusion of its lateral borders (Fig. 3.10). In the adult lamprey, the external or secondary nasohypophysial pore leads to a nasohypophysial duct, passing downwards and backwards towards the olfactory organ. The duct and its external opening are regarded by Janvier as secondary or specialized features, which were not developed in the more primitive cephalaspid, and are not present in the larval lampreys. Their development may be associated with the forward extension in the adult lamprey of the parietal musculature, which might have followed the reduction and final loss of a cephalic shield of the cephalaspid type. The true primary nasohypophysial opening of the adult lamprey is thus the so-called nasohypophysial valvule, separating the nasohypophysial atrium from the external duct. This is a membranous, not a muscular structure, conical in shape and bearing at its apex the primary nasohypophysial opening. This, like its counterpart in the cephalaspid, has a similar keyhole shape with an anterior hypophysial and posterior nasal aperture. Its function is assumed to be sensory, detecting foreign particles entering the tract which can then be expelled by sudden contractions of the dilated nasohypophysial sac. The backward extension of the nasohypophysial tract in the adult lamprey reaches the dorsal roof of the anterior gill sacs. Ventilation of the olfactory tract is therefore achieved by rhythmical compressions of the nasohypophysial sac through the alternating contractions and relaxations of the gill pouches. In cephalaspids there is no extension of the nasohypophysial sac and, in any case, a similar mechanism to that of the lamprey would have been precluded by the complete ossification of the endocranium. In all probability the ventilatory current would have been produced as in present-day ammocoetes, by the ciliated epithelium of the nasal tract. Except for the apparent absence of an accessory olfactory organ, other parts of the nasohypophysial complex show similar close parallels with the condition in lampreys, particularly the shape of the olfactory capsule and its relation to the telencephalon. How are we to account for the possession by all the cephalaspidomorphs of this dorsally placed nasohypophysial complex? Watson (1954) believed this position to be due to the development of the dorsal lip as part of the larval feeding mechanism; a view that is inconsistent with the inferences drawn by Strahan (1958) from his analysis of morphogenetic trends in the agnathan head. A more likely explanation is that the dorsal nasohypophysial opening was an adaption to benthic feeding habits, which would have made it advantageous to shift the opening into the olfactory organ away from the mouth, thus preventing its obstruction by silt particles. On the other hand, the terminal site of the nasohypophysial opening of the myxinoids (and according to Stensiö and Janvier of the pteraspids also) could have been associated with more selective and active feeding habits.

In the details of the extrinsic eye muscles and the shape of the eyeball, the cephalaspids show further close parallels with the petromyzonids (Janvier, 1975). Thus the disposition and insertion of these muscles appears to have

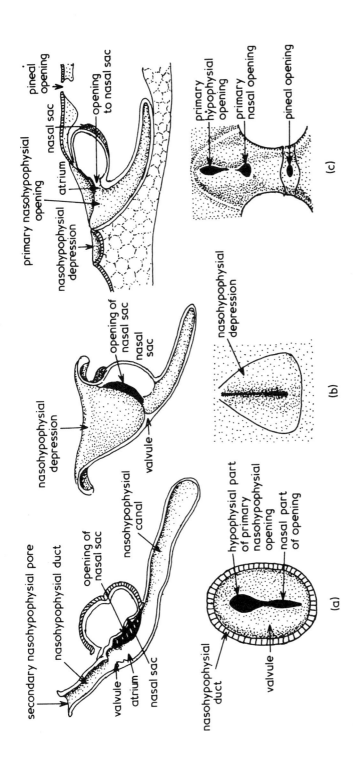

Fig. 3.10 Comparisons of the nasohypophysial tracts of an adult (a) and larval lamprey (b) and a cephalaspid (c). Redrawn from Janvier, 1974.

been identical in both groups. However, the lamprey's peculiar method of accommodation by means of a corneal muscle, derived partly from dorsal somatic muscle and partly from visceral muscle, does not appear to have been present in the cephalaspids. In any case, the myotomes of the cephalaspid did not extend dorsally over the surface of the head as in lampreys and if a corneal muscle were present, it could not have developed in the same way that it does in the lamprey. Janvier (1978) considers that the corresponding muscles of cephalaspids would have been used to operate mobile paired fins, innervated like the prebranchial muscles of the lamprey by the first two spinal nerves, forming a 'spino-occipital plexus'. The same author also believes that a corneal muscle could have been present in the anaspids and might be represented by an assemblage of scales posterior to the orbit, perhaps corresponding to underlying muscles continuous with the dorsal myotomes.

Attention has also been drawn to interesting parallels between the membranous labyrinth of the lamprey and the canal system connecting the cavity of the cephalaspid bony labyrinth with the so called 'electric fields' of the lateral and dorsomedial areas of the cephalic shield (Jarvik, 1965). These areas, now considered to represent a stato-acoustic sensory system, take the form of depressions roofed over by small bony plates. From each of the lateral fields a series of five wide canals, believed to have carried endolymph, lead towards the labyrinth; opening into its ventro-lateral area where there are distinct dilations (Fig. 3.11). From both sides of the median dorsal field, a wide canal joins the dorsal and medial edge of the vestibular cavity. At the point where this canal enters the vestibular compartment, it is joined by a posterior-ly directed canal containing the ductus endolymphaticus and a blood vessel.

The membranous labyrinth of lampreys also shows five similar sac-like evaginations. Of these, the two anterior members are regarded as evaginations of the anterior ampulla and the posterior pair as extensions of the posterior ampulla. The median sac has been referred to as an evagination of the lagena. In addition, two further blind-ended dorsal appendages are present; the more ventral of these has been regarded as the ductus endolymphaticus, while the dorsal extension carrying a sensory ending is considered to be unique to the lamprey labyrinth. The lamprey labyrinth therefore contains five ventral and one dorsal extension, which are not found in any other existing vertebrate and which correspond quite closely to the position and arrangement of the canals of the cephalaspid sensory fields and their vestibular cavities. In view of these parallels, it is difficult to resist Jarvik's argument that these structures in present-day cyclostomes are relics of the more extensive sensory system of its cephalaspidomorph ancestors.

Equally remarkable are correspondences in certain features of internal anatomy. Among these are the larger size of the right habenular ganglion (suggesting the existence of a reduced parapineal), reduction of the left Cuvierian duct, the large, irregular veins and probably high blood volumes, the asymmetrical position of the oesophagus on the left side of the body cavity

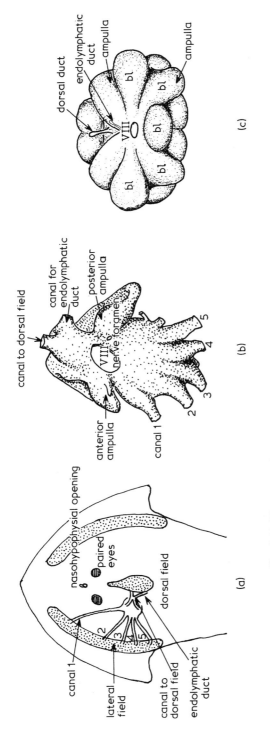

Fig. 3.11 The sensory fields (a) of a cephalaspid and its connections with the labyrinth (b). This may be compared with a medial view of the lamprey labyrinth (c), with its five dilations (bl). **Redrawn from Jarvik, 1965.**

and the displacement of the heart to the right (Watson, 1954).

Although in certain respects, such as the degree of concentration and backwards migration of the branchial region, the anaspids may be regarded as more specialized than the cephalaspids, they nevertheless exhibit trends that may have been involved in the phylogeny of the petromyzonids. The loss of more anterior gills, their small external openings and the shape of the gill pouches, together with the enlargement of the stomadeum, were almost certainly connected with a change in feeding habits from the silt or detritus feeding habits of the cephalaspids perhaps, as suggested by Whiting (1972), to surface planktonic feeding.

Although, in general, the anaspids may be too specialized to have served as the ancestral stock for the lampreys, this is less applicable to *Jamoytius*, where the backward movement of the branchial region had not proceeded as far as in other representatives of the group. Here the first gill pouch (believed to represent the hyoid and perhaps innervated by the VII nerve) lay just behind the orbit. The reduced exoskeleton, and the presence of a lamprey-like branchial basket are other significant features. The fact that the mouth was supported by an annular cartilage is strongly suggestive of a suctorial method of feeding, possibly assisted by a rasping tongue mechanism. Given the loss of the scaly exoskeleton, further development of the oral funnel and prenasal region and a reduction of the more posterior gill sacs, it would not seem a far cry from a form like *Jamoytius* to the fossil lamprey, *Mayomyzon*. However, a quite different interpretation of this fossil has been put forward by Wickstead (1969). Following earlier suggestions made by White (1946) this author considers that certain similarities between the branchial basket and lateral fins of *Jamoytius* and developmental stages of *Branchiostoma*, make the fossil form a suitable candidate for an ancestral acraniate perhaps representing only a larval or metamorphosing stage.

The relationships and phylogenetic position of the myxinoids is a much more intractable problem than that of the lampreys. In the lampreys, the evidence from *Mayomyzon* gives us confidence that we are dealing with relict forms, and the palaeontological evidence points unmistakeably to their cephalaspidomorph affinities. For the hagfish we have no such secure guide lines. What is more, discussion of this problem is bedevilled by the controversial features of heterostracan head structure. Although opinion has been divided over Stensiö's reconstructions of the pteraspid head, (Fig. 3.12), further detailed arguments in support of heterostracan-myxinoid affinities have been produced by Janvier (1974, 1975) in comparisons of the eye, prenasal sinus, and feeding mechanisms. From the shape of the orbit of the pteraspid, he has inferred that the eyeball was elongated and conical in shape like that of the hagfish, *Eptatretus burgeri*, a species in which the degeneration of the eye has not proceeded as far as in *Myxine*. Since this shape would preclude the rotation of the eyeball, Janvier has inferred that, as in present-

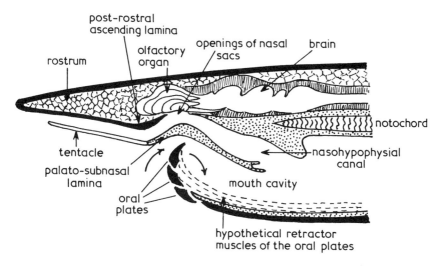

Fig. 3.12 Sagittal section through the head of a heterostracan, based on Stensiö's reconstructions. Redrawn from Janvier, 1974.

day hagfishes, the extrinsic eye muscles would have been absent or reduced. This would conform to Tarlo and Whiting's (1965) interpretations of heterostracan head anatomy, which suggest that the somites of segments 2 and 3 were still complete and had not yet given rise to ocular muscles. In addition, Janvier has detected a median ridge on that part of the sub-rostral region in front of the olfactory capsule, which like Stensiö, he regards as the roof of the prenasal sinus (Fig. 3.12). A similar structure is also said to be present in the myxinoid embryo. This is seen as evidence for the inherently paired nature of the prenasal sinus, reflecting that of the olfactory organ itself. Other myxinoid characteristics identified by Janvier in the heterostracans are grooves indicating the supposed position of sensory tentacles and further markings on the lateral parts of the dorsal shield, which he interprets as the sites of mucous glands.

Without implying strict homologies between the structures concerned, Janvier has drawn a parallel between the probable feeding mechanisms of pteraspids and those of myxinoids (Fig. 3.6). The hagfish tears off pieces of food by alternate eversion and retraction of its horny teeth, involving the separation and bringing together of the tooth apices (Section 5.1.1). The mouth of the pteraspid is limited posteriorly by rows of plates bearing a toothed lamina. These could have operated in a similar way to those of the myxinoid, expanding like a fan when everted and coming into contact again during retraction.

Later heterostracans – the amphiaspids from the lower and middle

Devonian — are thought to have changed to a suctorial method of feeding. Lateral openings on either side of the eyes and considered by many to be spiracles or prespiracles, are regarded by Janvier as the openings of paired nasal sinuses. However, as Janvier points out, there are difficulties in postulating the presence in a mud-feeding animal of unprotected nasal sacs opening into the mouth, where a majority of authors believe the nasal openings of heterostracans to have been situated. Both in lampreys and hagfishes, the prenasal sinus has sensory innervation from branches of the profundus (V^1) and is thus able to protect the nasal epithelium from the entrance of foreign particles. In lampreys, this is achieved by reflex contractions of the branchial region; in the hagfish by movements of the velum. Such functions serve to remind us of the advantages of dorsally placed nasal openings in bottom-feeding forms.

Acceptance of the views of Stensiö or Janvier would thus imply that many of the specialized characteristics of myxinoids had already been developed by the Silurian or Devonian heterostracans. Among such common features might be listed the similarities in feeding mechanisms, the structure of the ethmoidal region and prenasal sinus, the tendency to concentrate the gill openings, the shape of the eye, the absence of extrinsic eye muscles, the presence of tentacles on the sub-nasal cartilages and the tracts of mucous glands. Such an impressive assemblage of morphological characters would indeed be strong arguments in favour of a direct derivation of myxinoids from the Palaeozoic heterostracans, and such differences as exist might well be attributable to the mode of life and deep sea environment of the hagfish. For example, the much greater anterior extension of the prenasal sinus in the myxinoid would be of obvious advantage in an animal with a burrowing habit.

The divergent views of heterostracan morphology lead to almost diametrically opposed interpretations of the phylogenetic significance of these fossils, the first vertebrate group to appear in the geological record. Acceptance of the fundamentally myxinoid type of cranial organization would at least have the merit of bringing the heterostracans more into line with the other agnathan groups, making it easier to visualise the derivation of the pteraspidomorphs and cephalaspidomorphs from a remote common ancestor, as well as explaining the possession by both myxinoids and petromyzonids of certain features common to the whole agnathan group. The view adopted by Stensiö's critics would have the effect of setting the heterostracans apart from the rest of the agnathans, with little or no relevance to the evolutionary history of the myxinoids. What is more, the interpretations given by Halstead and Whiting of heterostracan head structures, including the external features of the brain, would give to this group a more central position in the gnathostome lineage and closer to the main stream of vertebrate evolution, leaving unresolved the problem of myxinoid origins.

Jarvik's (1968) interpretations of early vertebrate phylogeny (reproduced in

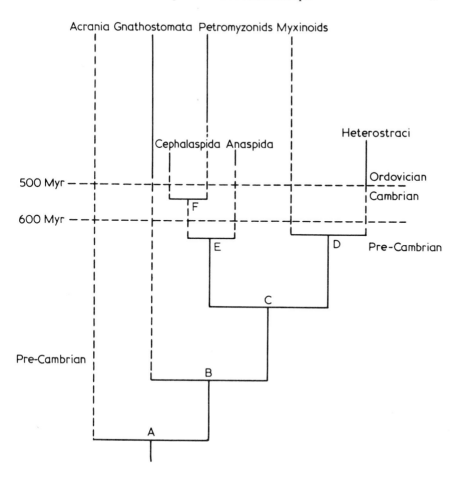

Fig. 3.13 A cladogram based on Jarvik's (1968) views on the phylogeny of the cyclostomes and their relationships to the fossil Agnatha.
(a) Common chordate ancestor. (b) Common craniate ancestor. (c) Common agnathan ancestor. (d) Common ancestor of myxinoids and heterostracans. (e) Common cephalaspidomorph ancestor. (f) Common cephalaspid- petromyzonid ancestor.

a modified form as a cladogram in Fig. 3.13) probably represent the closest approximation to the interrelationships of the cyclostomes and the fossil agnathans that is possible with our present knowledge. Here, the cephalaspidomorphs are represented as a sister group to the myxinoids and heterostracans, sharing with them a common ancestor at C, while the lampreys and cephalaspids constitute the sister groups of the anaspids. Thus, like Strahan (1958), Jarvik has preferred to relate lampreys more closely to the cephalaspids than to the anaspids. At the same time, although this view has much to recommend it in the light of the striking parallels in the nasohypophysial complex, eyes, labyrinth and central nervous systems of the two groups, we must at least allow the possibility that our conclusions could be biased by the much more detailed knowledge that exists on the internal anatomy of the cephalaspids and our comparative ignorance of the internal structure of the anaspids.

Although Jarvik has not accepted completely Stensiö's view that the myxinoids were directly derived from the heterostracans, he nevertheless considers them to be more closely related to the pteraspidomorphs than to any other group of agnathans. Accordingly, they are shown in Fig. 3.13 as sharing with them a common ancestor at D. However, these interpretations of the relationships of the two cyclostome groups to the gnathostomes will need some modification in view of the evidence that the lampreys are more closely allied to the higher vertebrates than the hagfishes and that the separation of the myxinoids from the early vertebrate stock may have preceded the origin of the petromyzonids (Chapters 13 and 14).

4
Ecology and behaviour

4.1 Habitats

4.1.1 Myxinoids

Characteristically, hagfishes are adapted for life at considerable depths at high salinities and low light intensities, but within the group there are distinct species differences, particularly in the depths to which they penetrate. *Myxine glutinosa* normally occurs between 50–100 m, but has been recorded down to 1100 m off the Norwegian coast and in Japanese waters *M. garmani* is found between 400–800 m. In some inner fiords, *M. glutinosa* may occur in much shallower water at about 30 m, but in these areas there may be marked stratification with the freshwater overlying water of high salinity (Tambs-Lyche, 1969). *Myxine* is apparently less tolerant of reduced salinity than some species and this may explain its absence from the Baltic. From its distribution, it has been inferred that the upper temperature limit for this species is between 10–13°C. When hauled to the surface in a trap from deep water, the animals are stressed by the rapid change in salinity and exude large quantities of mucus.

Some at least of the members of the genus *Eptatretus* are able to tolerate conditions in much shallower water, with higher temperatures and lower salinity. For example, the populations of *E. burgeri* studied by Fernholm (1974) in Japanese waters were living at depths of only 10 m and in other areas the same species has been found at 6–8 m. A similar shallow water habit is also characteristic of the New Zealand hagfish, *E. cirrhatus* which occurs at depths from 4–100 m. Shallow water species such as *E. burgeri* must be able to tolerate considerably higher temperatures than those that limit the distribution of *Myxine*. Thus, in the area where these animals were studied by Fernholm, the average winter temperature was 13°C rising to 24°C in July. As far as is known this species is unique in showing a seasonal migration to deeper water, where spawning is believed to occur. This takes place in late summer and autumn when the shallow areas reach their maximum temperature (25°C) and the animals return again in late autumn when temperatures are once again beginning to fall. Off the Californian coasts, *E. stouti* has been reported at depths varying from 45–1000 m and a temperature of 10°C is said to be close to its upper limit (McInerney and Evans, 1970). However, it is generally regarded as a relatively shallow water species and has been seen swimming near the surface where salinities may be as low as 24‰. This would be consistent with the tolerance that this species has shown to low salinity in laboratory experiments (Section 9.3.1). *E. deani*, on the other hand is a deep water species and is reported to be a dominant element in the fish populations of the San Diego Trough 12 km off the California coast at depths of 1200 m (Jensen, 1966; Smith and Hessler, 1974). Evidence cited by Strahan and Honma (1960) suggests that other members of the *Eptatretidae* may also be more tolerant to reduced salinities than *Myxine*, since on the Japanese coasts, *E. (Paramyxine) atami* is said to be found around river mouths. Similarly the New Zealand hagfish, *Nemamyxine* has been taken within an estuary.

Myxine displays the most extreme adaptations to deep water life in the almost complete absence of skin pigmentation and in the extent of its optic degeneration (Fernholm and Holmberg, 1974). On the other hand, in the shallower water species of the genus *Eptatretus*, the skin is less transparent (except over the eyes), a vitreous body is present in the eye and retinal structures are more highly differentiated. However, the extent of the degeneration in the eye of *Myxine* does not necessarily imply that its deep sea habits are of very great geological antiquity and the existence of similar atrophy in cave-dwelling vertebrates indicates that regression may, in an evolutionary context, take place over relatively short periods.

Myxine inhabits areas with a soft bottom, often on the continental shelf adjacent to estuaries where extensive silt deposits are most likely to occur. This preference for a soft substrate has been illustrated by Cole (1913) who described the movements of a population of hagfishes on the N.E. coast of England several miles inshore, towards an area that had previously had a hard bottom, but which had been covered with mud dropped from the hoppers of dredgers. Given these conditions, local populations of *Myxine* were said to be as numerous as earthworms in a fertile soil. Earlier reports had indicated that some *Eptatretus* species favoured hard or rocky bottoms and this led Strahan (1963a, b) to suggest that these forms were not so firmly committed to the burrowing habit as the members of the genus *Myxine*, where the fusion of the efferent branchial ducts could be regarded as an adaptation to this mode of life. However, observations in the field have shown that both *E. stouti* and *E. burgeri* inhabit areas of soft mud and the burrowing habits of the latter species have been confirmed by skin divers.

4.1.2 Lampreys
The marine phase of the lamprey is almost certainly a secondary introduction

into the life cycle of a primarily freshwater group and even today represents no more than a third to a quarter of the total life span. The ultimate success of a lamprey population must therefore depend to a large extent, on the quality of its freshwater environment. This must provide appropriate conditions for the development of the eggs and the survival of the larval stage, together with a quite different set of environmental conditions that are necessary for the completion of spawning. Evolved as a means of exploiting the organic rich deposits of the river systems, the ammocoete phase requires relatively stable silt beds into which the larvae can burrow, together with water of sufficient productivity to supply adequate amounts of algal food. Within a given river system, the distribution of the larval population results from the interaction of the characteristic behaviour of ammocoete and adult; the mainly passive downstream drift of the larva and the rheotactic upstream migration of the spawning adult. Thus, throughout the larval period the ammocoete population will tend to move downstream towards the middle and lower reaches of the river system, but this movement is counteracted each year by the ascent of the spawning adults towards the higher reaches (Hardisty and Potter, 1971a).

A typical ammocoete habitat is an area where the current is slack and so placed that it is protected from major fluctuations in water levels or stream velocities. Such conditions are most likely to be found in eddies or backwaters or at bends in the river bed, where there are likely to be sufficient accumulations of soft silt and sand to provide a suitable substrate for the burrowing ammocoete. In such places, often partially shaded by trees, diatoms may form an incrustation or 'polster' on the interface between the silt and the water and this may be an important factor in the stability of these ammocoete 'beds'. Current velocities over these sites have been put at an average of 0.4–0.5 m s^{-1} and limiting velocities are said to be about 0.6–0.8 m s^{-1}; rates of flow that are of the same order as the maximum swimming speeds of which a large ammocoete is capable.

For spawning lampreys, the important requirements are first and foremost a gravel bed, moderately fast currents and cool, well-oxygenated water. In general, current velocities on the spawning grounds will tend to be greater than the maximum swimming speeds of which lampreys are capable. As a result, once the animals move out of the nest (which protects them from the current) they are only able to maintain their position by anchoring to the substrate with their suctorial discs.

4.2 The burrowing habit

In their anatomical and physiological adaptations to a burrowing habit there are remarkable parallels between larval lampreys and hagfishes. These need not necessarily have any phylogenetic significance, although it is possible that both hagfishes and lampreys have been derived from benthic ancestors with mud-feeding habits. Similar common adaptations to burrowing are to be found in a wide variety of vertebrates, where they have arisen independently by convergent evolution.

4.2.1 Larval lampreys

When about to burrow, the ammocoete points its head region downwards towards the substrate and once it has made contact, the normal swimming movements are replaced by violent whiplash contractions of the tail, propelling the head and branchial region below the surface. At this point, the vigorous contractions of the tail subside and this region is now laid more or less horizontally over the surface. Final submergence is achieved by contractions of the more anterior trunk regions using the expanded oral hood as an anchor below the surface, literally pulling the tail region into the mud. Once buried, the animal assumes an arched posture forming a U-shaped burrow which can be detected by a conical depression on the surface of the mud, with a central hole marking the position of the mouth (Fig. 4.1). These openings are said to lie at an angle of 60–70° to the mud surface in the direction opposite to the current flow. The depth of the burrow varies with the size of the ammocoete and when stimulated mechanically or by an increase in light intensity, they retreat to greater depths, in the case of large ammocoetes, up to 18 cm below the surface (Sterba, 1962). The walls of the burrow are cemented

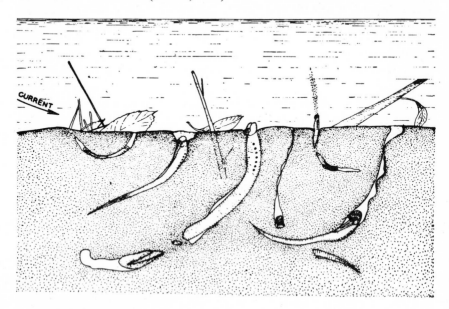

Fig. 4.1 Burrowed ammocoetes of the sea lamprey, *Petromyzon marinus*. From Applegate, 1950.

by mucous secretions, particularly from the dorsal columns of the endostyle (Section 11.1.2), facilitating the passage of the exhalant current from the gills.

Although they may change their position during the daytime, on each occasion forming a new mucus tube, ammocoetes normally leave their burrows only at night (Enequist, 1937). During the daytime, any disturbance in the water only has the effect of driving them deeper into the burrow or at least causes the withdrawal of the snout, which normally protrudes from the entrance hole. Under simulated field conditions, ammocoetes show a diurnal rhythm of activity, leaving their burrows for periods during the night to swim freely in the surrounding water. Under laboratory conditions, this activity has been found to be most intense in the second hour of a 12 hour dark-light cycle, with a second smaller activity peak during the eighth hour of the dark period. This pattern is quite similar to the diurnal activity rhythm of the adult lamprey (Section 4.2.3). Except in the genus *Mordacia*, ammocoetes also show a diurnal rhythm in the pigmentation of the skin. This darkens during the day through the expansion of the melanophores and pales at night when the pigment cells contract. If the ammocoete is kept in continuous darkness, this rhythm will persist for a time, eventually becoming irregular and finally disappearing to leave the animal permanently dark. Permanent darkening can also be produced by subjecting the animals to constant illumination or by removal of the pineal (Section 8.2.4). Although these pigmentary responses might at first sight appear too trivial to have much adaptive value in an animal that spends nearly all the daytime hours burrowed beneath the mud surface, pigmentary protection from light is fundamental to a wide variety of animals. The importance of this protection can be seen in *Myxine*, where the skin is almost devoid of pigment and quite transparent (Fernholm and Holmberg, 1974). When subjected to continuous illumination on a white background, *Myxine* dies within a matter of days, showing every sign of being severely stressed (Holmberg 1968, 1971).

In their natural habitat, ammocoetes are likely to be affected by seasonal changes in the oxygen content of the water. When first exposed to low oxygen tensions under laboratory conditions, the ammocoete first tends to protrude its head or even the whole of the branchial region above the surface of the substrate, thus tending to inhale water which would have a higher oxygen and lower CO_2 content than that within the burrow. At even more extreme levels of oxygen depletion approaching lethal concentrations, the larvae may emerge from the burrow completely, displaying restless movements punctuated by bursts of swimming which, in their natural habitats, might be expected to move them towards more favourable conditions (Potter *et al.*, 1970).

Physiological adaptations to burrowing include the responses of ammocoetes to light and to contact stimuli. In a partly illuminated aquarium, ammocoetes respond to the light stimulus by random swimming, which ceases when they arrive in the shaded area. Although the spinal cord itself is sensitive to light, the tail region is especially so and contains photoreceptors innervated by the lateral line nerves. Light responses from this area are obviously significant in ensuring the complete withdrawal of the ammocoete below the substrate. In the aquarium, free-swimming ammocoetes usually settle down with their lateral surfaces in contact with the bottom. Equally, they will lie quietly in an artificial substrate of glass beads or inside glass tubes. This response to contact stimuli (thigmokinesis), combined with a reduction in light intensity below the threshold level that induces swimming, are no doubt important factors in reducing the daytime movement of burrowed ammocoetes.

4.2.2 Hagfishes

Like the ammocoete, the hagfish remains for the most part, buried in the mud during the daytime, emerging at night, either to rest on the surface or to engage in short bursts of swimming. However, it is not known whether this pattern of behaviour has an innate rhythmic basis, like the diurnal rhythms of lampreys or is simply due to the animal's photokinetic responses. Even if such rhythms are present they could not be regulated in the same manner that is thought to operate in the lamprey, where the pineal is believed to be involved in the control of metabolic rhythms (Section 8.2.3).

In Japanese waters, the commercial fishery for *E. atami* is normally carried out at night (Strahan and Honma, 1960) although, where fishing is operated at depths below 200 m (under conditions of very low light intensity), there are apparently no differences in the size of day or night catches. Similar conditions have also been reported from Norway, where daytime catches of *Myxine* in shallow water (30 m) are said to be poor, whereas large numbers are caught in daytime in deep water (120 m) (Foss, 1968). These examples suggest that environmental changes in light intensity may be more important than innate circadian rhythms in determining the levels of activity of hagfishes. In spite of their degenerate eyes, hagfishes, like ammocoetes, will congregate in the shaded part of a partially illuminated aquarium. Again like the larval lamprey, they possess skin photoreceptors, concentrated mainly in the head region and around the cloaca, although these differ from those of the ammocoete in being innervated by spinal nerves rather than by a lateral line nerve. In *Myxine*, these dermal receptors show their greatest spectral sensitivity to light of 500–520 nm, corresponding to the wavelengths of the light that would penetrate into their deep water habitat (Steven, 1955). Their responses to light are characterized by very long reaction times. These vary from 10 seconds at the maximum intensities employed to between 2 and 5 minutes at threshold intensities, similar to those that the animal would experience in its natural environment (Newth and Ross, 1955).

In its method of burrowing the hagfish does not differ greatly from the ammocoete. With powerful swimming movements it dives head first into the substrate at an angle of 45–90°. According to Fernholm (1974), swimming movements continue until the animal is completely buried, but Strahan (1963) and Foss (1968) have described a gliding movement, rather like that of an ammocoete, which draws the hinder part of the body under the mud, after the head and anterior regions have submerged. Eventually, the head may reappear some distance (50–100 cm) from the point where the animal had first entered. Shaped like a shallow U, the entrance to the burrow may show a conical depression, marking the position of the head and nasohypophysial opening, while a small elevation some distance away, indicates the point where the exhalant current emerges. The head and tentacles normally project from the burrow and during the night many animals have been seen with the anterior third of their body exposed. Although Foss has described a copious secretion of slime immediately after the animal has submerged, the walls of the burrow are said to be unsupported by mucus and collapse once the hagfish has left.

In the Hardanger Fjord there are shallow areas where *Myxine* abounds at depths of only 30 m, making observation by diving practicable. Here the sea bed shows large numbers of hillocks, like miniature volcanoes, with a crater-like hole at the top. In an earlier series of observations, hagfishes were seen to enter these mounds through the top or sides (Foss, 1963), but later investigations failed to provide direct evidence that they were inhabited by hagfishes, although their structure was suggestive of burrowing activity (Foss, 1968). Observations by divers on *E. burgeri* have failed to reveal similar structures (Fernholm, 1974) and their relevance to the normal life of the hagfish remains in doubt.

Laboratory observations had suggested that *E. stouti* was a species that preferred to rest on a hard bottom in a coiled position and because of this behaviour, Strahan believed that the members of this genus were less adapted to burrowing than *Myxine* and that the coiled position was a way of ensuring maximum surface contact on a hard and rocky sea bed. This has now been discounted by bathyscaph observations on the muddy terraces and deep water canyons off the Pacific Coasts of North America that are the habitats of *E. stouti* and *E. deani* (Jensen, 1965). These have shown that both species burrow into the substrate in the same way as *Myxine* and similar observations have been made on *E. burgeri* in its natural environment (Fernholm 1974). Like the ammocoete, *Myxine* normally rests on its side on the hard surface of an aquarium and the same position is often adopted by adult lampreys at times when they are no longer attached by their suctorial oral disc. In the case of *Myxine*, this behaviour has been linked with the peculiar orientation of its labyrinth, so placed that with the animal lying on its side, the two toroidal rings and their maculae would lie one above the other in or near the horizontal plane (Lowenstein and Thornhill, 1970). Another possibility, considered by the same authors, is the presence in *Myxine* of a specialized skin region over the site of the degenerate eye and innervated by the V nerve, which responds to pulsed mechanical stimuli of a similar frequency range to those that excite the vibration receptors of the lamprey labyrinth. However, in view of the fact that this behaviour is shared by the ammocoete, it seems more likely that it reflects the importance of skin contact stimulation in reducing the activity of a burrowed animal.

4.2.3 Adult lampreys

Although characteristic of the ammocoete, burrowing responses do not entirely disappear after metamorphosis and the macrophthalmia of parasitic lampreys may remain for considerable periods buried in the bottom of the stream, emerging mainly during the night (Fig. 4.2). It is probably during these phases of nocturnal activity that they are carried passively downstream, especially during periods of flood. In non-parasitic species, the burrowing

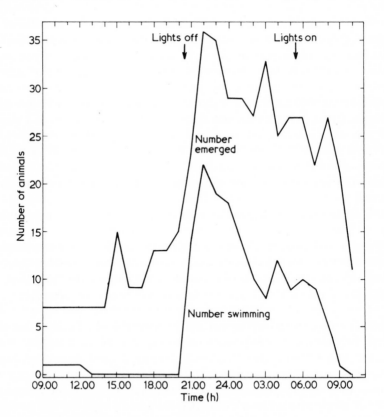

Fig. 4.2 Nocturnal activity of recently metamorphosed (macrophthalmia) river lampreys *L. fluviatilis* in an aquarium. From Potter and Huggins, 1973.

habit and light-avoiding responses are retained for longer periods, only to be lost when spring water temperatures approach the levels at which spawning behaviour is initiated. At these times the adults begin to show a preference for lighted areas and the spawning grounds are often exposed to direct sunlight.

Throughout the earlier stages of the spawning migration, anadromous lampreys avoid the light during the daytime, hiding under rocks or river banks and resuming their upstream movement only during the hours of darkness. These responses are exaggerated in the southern lamprey, *Mordacia mordax*, which is said to burrow deeply into the river bed during its long upstream migration and whose unique dorso-laterally placed eyes are thought to be a specialized adaptation to this mode of life (Potter *et al.*, 1968).

In spite of the striking contrast in the habits of ammocoetes and adult lampreys, it is interesting that both should show rather similar innate patterns of activity. That the migrant lamprey is mainly active at night has always been known to lamprey fishermen and there has been a widespread belief that their upstream movements tend to be inhibited during periods of bright moonlight. This light-avoiding response has been exploited in commercial fisheries of Eastern Europe, where rivers may be illuminated by electric lamps, leaving only a dark corridor through which the lampreys swim to enter the traps (Abakumov, 1956). Under laboratory conditions, both the degree of activity and its diurnal pattern has been shown to vary throughout the period covered by the upstream migration in *L. fluviatilis* (Claridge *et al.*, 1973). The greatest activity occurred in November and December soon after the animals would normally have entered freshwater (Fig. 4.3). Under a light-dark cycle of 11/13 hours, peak activity was observed at 19.00–23.00 hours, just after the lights had been switched off and again at 01.00–03.00 hours. Activity then declined, remaining at a low level until just before the next dark period. In January and February there was less activity, but its diurnal distribution was similar. In March, activity again increased but its phasing changed. There were still two peaks of activity, but these now occurred two or three hours later, at 22.00–01.00 hours and 03.00–05.00 hours. Finally, in April, the sexually mature animals showed almost as much activity in the light as in the dark and this reached its peak at about the beginning of the light cycle. Quite similar rhythmic patterns were also seen in heart and breathing rates (Claridge *et al.*, 1973) and these metabolic rhythms correspond quite closely with the behaviour of river lampreys in their natural environment. For example, the greatest swimming activity would normally occur in the early months when the lampreys are making their passage through the estuary towards upstream spawning sites, while in January and February, when activity in the laboratory declined, they might be expected to have reached the areas where breeding occurs. Renewal of activity on a considerable scale in March would coincide with a period when nest-building and pre-spawning activity are in full swing and marked daytime activity appears. This change in activity patterns has

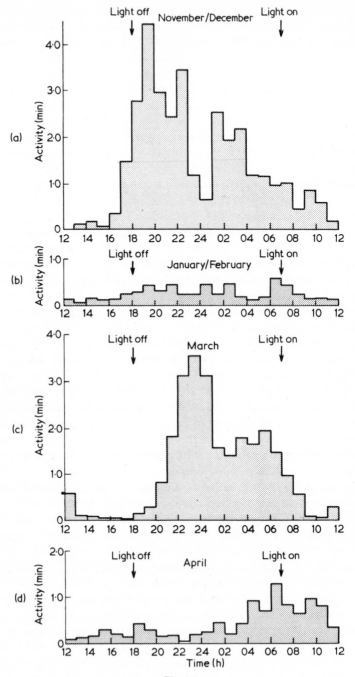

Fig. 4.3

been attributed to the accession of additional daytime activity rather than to the loss of the basic circadian rhythm of nocturnal activity (Sjöberg, 1977). The pineal, already known to be involved in diurnal pigmentary changes, may well be more widely implicated in these metabolic rhythms, although its role has not yet been investigated (Section 8.2.4).

4.3 Feeding

4.3.1 Adult lampreys

Extensive studies on the feeding habits of the landlocked sea lamprey of the Great Lakes Basin have shown that, in this species at least, the diet consists mainly of blood and that the tissues of the prey make only a very minor contribution to its diet. Laboratory investigations have shown that, when fed on rainbow trout, the blood ingested by this lamprey amounted on average to 11.6 % of the wet weight of the parasite and in some individuals was as high as nearly 30 % (Farmer *et al.*, 1975). The small amounts of muscle tissue that were ingested never exceed 2 % of the total quantity of blood consumed.

However, this picture of the sea lamprey as a blood-sucking ectoparasite does not apply universally to all species. For example, the gut of the river lamprey, *L. fluviatilis* has been found to contain muscle tissue, scales, spines, fragments of fins and ribs, pyloric caecae, swim bladder and fish eggs (Hardisty and Potter, 1971b). Similarly, the related species, *L. japonica* is said actively to attack fish in the manner of a true predator, gnawing out the tissues and internal organs (Nikol'skii, 1956). Reports of the feeding habits of the North American river lamprey, *L. ayresii* tell a similar story. In the gut of this species feeding in the Straits of Georgia, British Columbia, were found a variety of fish tissues, predominantly those of small herring, including parts of fins, scales, bones, muscle and fragments of gut. In laboratory tests, this species consumed the flesh of its prey, skeletonizing them completely within a few minutes (Beamish and Williams, 1976). On the other hand, the larger Pacific lamprey, *L. tridentata* probably feeds in much the same way as the sea lamprey, *P. marinus*, consuming only the superficial tissues of its victims leaving only a small hole through which it presumably draws off the blood. Less is known of the feeding habits of the smaller freshwater lampreys, but reports on the Danube lamprey, *E. danfordi* are of special interest, since they suggest that, like the hagfish, it may feed on dead and decomposing animal material (Chappuis, 1939). This species is said to collect around the effluent from slaughterhouses and to be attracted to bait of animal refuse (Grossu *et al.*, 1962).

In relatively confined areas and especially in lake systems, lampreys are

Fig. 4.3 Changes in the circadian activity rhythms of adult river lampreys, *L. fluviatilis* at various periods of the upstream migratory phase and maintained in a constant light-dark cycle of about 11–13 hours. From Claridge *et al.*, 1973.

capable of significant damage to fish populations. The recent history of sea lamprey depredations in the Great Lakes of North America is now widely known and for further details the reader may be referred to the account given by Smith (1971). This dwarf freshwater derivative of the anadromous sea lamprey of the North Atlantic, *Petromyzon marinus*, has apparently existed in Lake Ontario and other smaller lake systems, probably as far back as the last Glacial period, but was prevented by the Niagara Falls from entering the upper Great Lakes until the construction of the Welland canal in 1829. When these lampreys first penetrated through the canal is not known, but the first specimens were not recorded from Lake Erie until 1921 and although they have never flourished in this lake, by the middle 1930's they were already established in Lakes Huron and Michigan, finally appearing in Lake Superior somewhat later. Once in these upper Lakes the lampreys underwent a veritable population explosion, presumably due to a combination of favourable factors, including the abundance of suitable host fishes, the relative absence of predators and the existence in the Lakes' watersheds of ideal conditions for successful spawning and for the maintenance of their larval stages. The establishment of the sea lamprey was accompanied by a serious decline in the stocks of a number of fish species (Fig. 4.4). This was most important in the case of the lake trout, and led to the collapse of a flourishing commercial fishery. Although it is admitted that overfishing may have been a contributory factor to this decline, there can be no doubt of the destructive potential of the lamprey predator as a major factor in the dwindling populations of many of the more commercially valuable fish species. With the setting up of a Great Lakes Fishery Commission, control measures were instituted on an extensive scale and eventually with the development of selective larvicides, lamprey populations have been reduced to more acceptable levels. As an example of the extent of fish mortality attributed to the lamprey at the height of its invasion, it has been estimated that from 1954 to 1961 the lampreys of Lake Superior alone would have destroyed about 5800 million tons of fish (Smith, 1971).

Experiments described by Kleerekoper (1972) point to the importance of the olfactory sense in the location of the prey. Kept in dim light, feeding stage sea lampreys maintained for several days a diurnal rhythm of activity, but greatly increased movements occurred when water in which trout had been held, was introduced into the experimental apparatus, at periods when the animals would normally be relatively quiescent. These responses did not occur after the nasohypophysial opening had been blocked. If the olfactory stimulus was diffused throughout the whole body of water the movements of the lampreys was random in direction, but in a compartmentalized apparatus the lampreys moved towards the inlet through which the trout water was introduced, showing that it would be possible for these animals to locate and follow the source of an olfactory stimulus. Chemical isolation of the

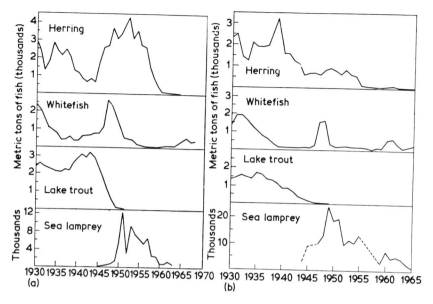

Fig. 4.4 Effects of the sea lamprey invasion of the Great Lakes on commercial fisheries. From Smith, 1971.

(a) Annual production (metric tons) of the lake trout, lake herring and whitefish in Lake Michigan from 1930–1968 together with the numbers of lampreys (in thousands) trapped at a weir on a spawning tributary to the Lake. (b) Annual production (metric tons) of lake trout, lake herring, and whitefish in Lake Huron from 1930–1965, together with the numbers of lampreys trapped on a spawning tributary.

components of the trout water showed that the effective odoriferous compounds were amines and that one of these, isoleucine methyl ester, was particularly effective not only in lampreys, but also in sharks and some teleosts.

The effectiveness of the lamprey's parasitic habits may be judged by the growth rates that it achieves and the rate at which it is able to accumulate its energy stores of lipid. For example over a parasitic phase of 2 or 3 years, the anadromous sea lamprey increases it body weight from 1–4 g to about 1000 g and its lipid content (as a percentage of wet weight) is increased about eightfold from 1.3% over 10% (Beamish, Potter and Thomas, 1979).

4.3.2 Hagfishes
Lacking the suction apparatus of a lamprey, the hagfish would be incapable of making a successful feeding attack on a mobile living fish and it is now generally accepted that they attack only dead or dying animals. Before the widespread adoption of trawling, hagfishes were indeed a menace to line

fishermen, attacking and totally destroying their bait within the space of one or two hours, as well as making inroads into the catch itself. For example, Cole (1913) quoted one example of a cod which, when hauled to the surface, had no fewer than 120 *Myxine* attached to it. As a result of his experience in trapping these animals. Cole asserted that they showed a distinct preference for recently dead or even dying fish, rather than for stale fish. Entering through the gills, it uses its slime secretions to block the movements of the operculum of a dying fish, only entering the body cavity when these movements had ceased. During its initial efforts at penetration, the hagfish is said to retain a vertical burrowing posture, all the time making vigorous swimming movements with the hinder part of the body. According to Cole, the liver is eaten first, then the gut and heart and finally the muscle tissues between the skin and backbone, working forwards from the posterior end. In underwater observations on *E. burgeri*, Fernholm (1974) found that an anchovy could be completely demolished within 12 minutes leaving behind only the skin and the backbone. Throughout their feeding, hagfishes show continuous swimming movements and in their efforts to tear off fragments of food they frequently resort to knotting behaviour (Fig. 4.8) to give them a better purchase. A number of observers have noted the secretion of slime around the food and although this may be a by-product of their strenuous activity, it could conceivably help to deter other competitive scavengers in the deep sea community from sharing the food resources of the hagfish.

The fact that hagfishes attack fish when these are used as bait does not of course, necessarily imply that this is their main source of food in their natural environment. As Strahan (1963) has pointed out, it seems unlikely that fish would normally be present in sufficient numbers to maintain the dense populations of hagfishes that are known to occur in certain areas. Some support for this view came from his analyses of gut and faeces. In one series, the gut contained only polychaete remains, but a larger number of faecal analyses made on *Myxine* showed parts of ribs, vertebrae and pectoral girdles of small herring, as well as hermit crabs, shrimps, priapuloids, remains of polychaetes, small lamellibranchs and gastropods. The fish and crustaceans were presumed to be dead when they were ingested, but Strahan believed that other organisms such as the small molluscs were probably ingested accidentally while the hagfishes were searching for food under the surface of the mud. Some experiments have also been carried out in the field involving the presentation of a variety of food materials. Underwater television observation showed that a polychaete and an anomuran were accepted by the hagfishes, but not a sea anemone. When herring was offered together with invertebrates, this was taken preferentially, as was decomposing rather than fresh fish.

Recent investigations have shown that the deep sea benthic community contains a surprisingly large variety of scavenging species and Jensen (1965) has remarked on the richness of the vertebrate and invertebrate fauna in the habitats of the hagfish off the San Diego coast, as revealed by bathyscaph observations. No doubt, within such competing communities, speed in the location and consumption of animal remains is at a premium. According to the stability-time hypothesis (Sanders, 1968), the long-term stability of the deep sea environment has encouraged the development of diverse communities of competing species which have reached an equilibrium through refined niche differentiation, particularly in regard to food specialization (Wolff, 1977). On the other hand, all our information tends to point to the hagfish as an indiscriminate feeder, taking from the sea bed any animal material that comes its way. As Dayton and Hessler (1972) have maintained, in order to survive, deep sea benthic species need to be flexible and opportunist in their choice of food, if they are to make full use of the available resources. This has therefore tended to blur a distinction between predators and deposit feeders, resulting in a community of 'croppers' which consume both live or dead and decaying organic materials. In the case of the hagfish, any preferences that it may appear to show are probably based on the ease with which a particular food can be located and the intensity of the olfactory stimulus that it generates. As a nocturnal feeder, depending entirely on its olfactory sense, the discovery of food, unless aided by currents, has been considered a random and inefficient process. This has been illustrated by aquarium experiments which have shown delays of 18–24 minutes between the introduction of bait and the first responses, such as the protrusion of the head from the burrow and the initiation of swimming. On the other hand, experiments in the field using intact and crushed fish as bait have shown that the responses may take place very rapidly when the olfactory stimulus is more intense. For example, when whole fish were used, the time that elapsed before the first appearance of the hagfishes was 6–16 minutes, whereas when crushed fish were presented, they appeared within half a minute (Foss, 1968).

Hagfishes are believed to feed only at infrequent intervals and, under aquarium conditions, are said to neglect food when this is offered more than once a week. However, this is compensated for by the voracity of their feeding and the amount they consume on a single occasion. Within minutes of the start of feeding, fragments of undigested material are passed from the anus as faecal tubes, surrounded by the mucous envelope secreted by the intestinal epithelium. Feeding normally occurs during their periods of nocturnal activity, although for a hungry hagfish, the olfactory stimulation may override the normal photokinetic responses. In underwater observations, *Myxine* has been seen to continue its search for bait even under the lighting used for underwater photography and once feeding began, it had to be touched before it reacted. A characteristic of laboratory-maintained hagfishes is their ability to survive for months without food. This is no doubt explained by their low metabolic rates (Section 5.2.3) and the extensive lipid deposits in the subcutaneous tissues and intestinal submucosa.

4.3.3 Larval lampreys

Although superficially resembling the protochordates in its filter feeding mechanisms, the presence in the ammocoete of a velum, capable of boosting the nutritive and respiratory currents, must have increased the rate of food intake and thus the body size that the ammocoete has been able to attain. While this feeding mechanism must put an upper limit on the size of food particle that can be accepted, it is doubtful whether it is otherwise selective as to the kind of organisms that are used as food (Hardisty and Potter, 1977). In general, the micro-organisms and detritus particles that have been found in the ammocoete gut are representative of those that occur in the surrounding water or on the surface of the silt substrate in the vicinity of the burrows (Moore and Beamish, 1973) (Fig. 4.5). Some of the Chlorophyta and diatom species may be more abundant in the water than in the gut, but this may be due to their tendency to form filaments too large to pass through the sieve-like mechanism of the buccal cirrhi. A similar selection based on size is also indicated for several genera of episammic diatoms. As might be expected, the size and composition of the diatoms found in the gut varies with the age of the ammocoete. Thus, the mean particle size in 0+ class of ammocoetes was found to be 35 μm compared with 100 μm in four-year-old animals. In sea lamprey ammocoetes, there were three times as many diatoms in the gut in the summer as compared with the winter. Surprisingly, at winter temperatures, up to 90 % of the diatoms were not completely digested and the surviving chloroplasts were capable of reproduction in suitable culture media, whereas at summer temperatures only 45 % were undigested. Consistent with these figures is the confirmation that the assimilation efficiency of the ammocoete is low and temperature-related (Moore and Potter, 1976a).

Although diatoms form an important part of the ammocoete diet, their resistance to extraction techniques has probably tended to exaggerate their importance. Although making up a relatively small proportion of all the organisms in the gut a wide variety of algal groups have been identified, including Chlorophyta, Chrysophyta, Cyanophyta and Euglenophyta. Other organisms identified in the gut include ciliates, rhizopods, rotifers, cladocerans, ostracods and copepods.

In addition to micro-organisms, ammocoetes ingest considerable quantities of detritus, more especially in winter and in streams where productivity is low. Indeed, in one English stream, algae accounted for only 0.01–0.2 % of the gut contents in winter and 0.1–0.3 % in summer. On the other hand, culture of ammocoetes in the laboratory showed that rates of growth on an algal diet are greater than on detritus and that both growth rates and condition factors are higher in streams of high productivity, where the gut has a greater algal content (Moore and Potter, 1976a,b).

Inevitably, the microphagous mode of feeding imposes an upper limit on the size that ammocoetes can attain at the end of a larval period varying in

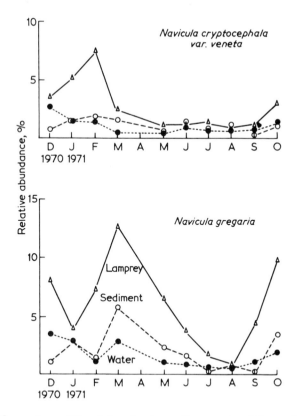

Fig. 4.5 Comparisons of the relative abundance of two species of diatoms in the water and in the sediments or the midgut of ammocoetes of the sea lamprey (*Petromyzon marinus*). From Moore and Beamish, 1973.

different species from 3–7 years or more, although even in those species with the longest larval phase, it is rare to find ammocoetes with body lengths greater than about 180 millimetres. Growth rates are higher in the early years of larval life and, in many species, there may be no further increase in length in the final year before metamorphosis. This period has been referred to as a 'rest period' or 'arrested growth phase' and is characterized by a conspicuous build-up in the lipid reserves in preparation for the long period of starvation that begins with the onset of metamorphosis and which (even in those species that feed parasitically after transformation) may continue for several months after its completion.

Since lampreys lack scales or other bony structures that might be used to determine age, this is usually estimated by identifying age classes from larval length-frequency distributions (Fig. 4.6). Where spawning is relatively

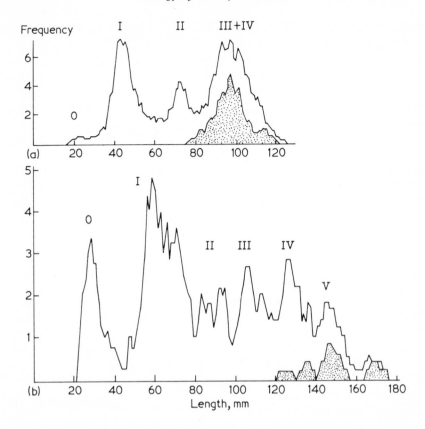

Fig. 4.6 The determination of age classes in ammocoete populations by the use of length-frequency distributions. From Hardisty and Potter, 1971 and Hardisty and Huggins, 1970.

(a) Length-frequency curves for a population of ammocoetes of the parasitic lamprey, *Lampetra fluviatilis* collected in the River Teme in July, just before the start of metamorphosis. The metamorphosing specimens of this species collected in the preceding year are shown by the shaded areas. At least three and probably, four year classes are believed to be present.

(b) Length-frequency curves for a population of ammocoetes of the non-parasitic lamprey, *Lampetra planeri* collected in the River Usk in August and September just after the start of metamorphosis. The shaded areas represent animals that had already begun to transform. At least five larval age classes are present (I – V) in addition to animals hatched in the spring of the same year (0).

restricted in duration and in streams where the larval population is more or less stable, each year class may be represented by a discrete peak in the curve. Thus, the number of age classes and hence the duration of the larval period can be determined. As judged by the proportionate height of the peaks representing successive age classes, it appears that annual mortality is low and relatively constant. In addition to the protection afforded by their concealed habits, it is thought that ammocoetes may be distasteful to predators such as eels (Pfeiffer and Fletcher, 1964) because of their skin secretions. These are produced from mucous cells and club and granular cells, elaborating proteins and peptides. As suggested by the expression 'a surfeit of lampreys' these animals have acquired a traditional reputation for producing digestive disturbances or even toxic effects and they are usually skinned or otherwise treated to remove the mucus before they are cooked. To some extent these unpleasant effects could be due to the presence of a biologically active peptide allied to, but not identical with, bradykinin that has been isolated from their skin secretions (Fischer and Albert, 1971).

4.4 Movements and swimming ability

In the laboratory, hagfishes exhibit long periods of inactivity, punctuated by only intermittent movements, but this may not necessarily reflect their habits in their natural environment, where nocturnal activity is more intense. *Myxine* is said to remain at times in a state of extreme passivity that has been referred to as 'sleep.' During these periods, its rate of velar pumping may fall from normal rates of 50–100 min^{-1} to only 25–30 contractions min^{-1}. These 'sleeping' hagfishes can be handled or even lifted out of the aquarium without any apparent response.

The eel-like swimming movements of lampreys and hagfishes are basically similar, involving lateral waves of contraction passing down the length of the body. Like the eel, the hagfish and, to a lesser extent, the lamprey show what has been termed 'freezing behaviour', in which the animal takes up a static S-shaped posture before swimming begins. Cyclostomes, like *Amphioxus*, show wide lateral oscillations of the head during forward swimming, whereas in the eel these oscillations are much more heavily damped in front of the animal's centre of mass, where the amplitude is usually at its minimum (Fig. 4.7). Behind the centre of mass, amplitude increases reaching a maximum at the tip of the tail, which makes the widest lateral excursions (Blight, 1977).

Because of the absence of a backward flexure in their myotomes, Nursall (1956) had incorrectly assumed that cyclostomes would be incapable of reverse swimming, involving the propagation of undulatory waves in a tail to head direction. On the contrary, hagfishes often swim backwards; a facility that they share with cephalochordates, eels and lampreys. In an intact hagfish, waves of undulation starting near the head and progressing backwards can be produced by stroking the tail, but after the spinal cord has been divided, these waves begin immediately behind the point of section. Stroking the gill region, on the other hand results in reverse waves, which after spinal section, begin

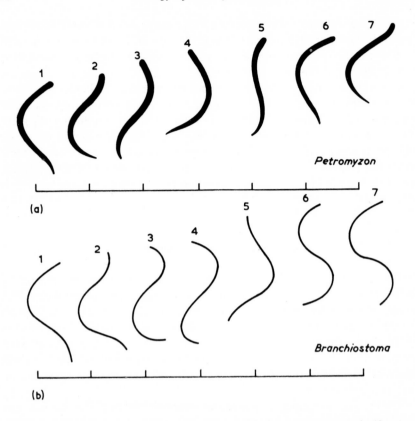

Fig. 4.7 Swimming movements of a 7 mm ammocoete of *Petromyzon* (a) and of a 45 mm
adult of *Branchiostoma* (b) traced from cine film records. Progress in a forward
direction is indicated by distance from the fixed baseline. The time intervals
between successive positions is about 16 ms in (a) and about 13 ms in (b). From
Blight, 1977.

just anterior to the point of operation. Thus, all segments of the cord are able
to initiate undulatory waves in either direction (Campbell, 1940).

In *Amphioxus*, a major factor in the production of either forward or
backward waves of contraction may be a longitudinal gradient in internal
bending resistance resulting from changes in the flexibility of the notochord
(Webb, 1973). Contractile elements like those of the cephalochordate
notochord have not been recognized in the hagfish, but the presence of
microtubules and membrane arrays surrounding the central vacuole of its
notochordal cells (Flood, 1969) at least raises the possibility that similar
mechanisms may exist. Lampreys, when faced by obstacles or by noxious
stimuli may retreat backwards for a short distance, before turning and

swimming away. However, as Blight (1977) points out, backward swimming in a lamprey is not achieved by a simple reversal of the processes involved in forward progression, since the head is held almost as straight as in forward swimming and does not therefore function as a substitute tail.

Hagfishes are described as graceful, elegant swimmers, but perhaps poor spatial orientation due to the degeneration of the eyes, may explain why they normally swim close to the bottom. Observations by divers have shown that they rarely swim more than 1.5 m from the sea bed and that they maintain this position by keeping the head above the horizontal, with the rest of the body sloping obliquely downwards. The fact that they often swim in an upside-down position could be related to their simple, toroidal labyrinth, which is said to be unable to monitor rolling movements as efficiently as pitch or yaw (Lowenstein and Thornhill 1970). On the other hand, Whiting (1972) has suggested that the ability of lampreys to maintain the dorsal side uppermost might be assisted by the fatty tissue of the fat column or 'protovertebral arch' above the nerve cord, acting as a buoyancy organ especially helpful to an animal without paired fins. However, while a similar body of adipose tissue is absent in the myxinoids, it is equally possible that it might be of adaptive value to the lamprey as a cushion above the nerve cord which (in the absence of complete vertebral arches), could offer some protection against the attacks of predators during periods when the animals are exposed in relatively shallow water in daylight (Potter *et al.*, 1978); a protection that would be less vital to a hagfish that normally lies completely buried during the daytime.

After destruction of the brain, the spinal hagfish completely loses its equilibratory reflexes and rolls over continuously whilst swimming, although in other respects its swimming movements are executed in a normal manner.

Because of the different arrangement of their parietal muscles, the body of the hagfish is much more flexible than that of a lamprey. An example of this flexibility is the way that they are able to perform complex contortions, described as 'figure of eight' and 'knotting' (Fig. 4.8). Both forms of behaviour are based on local responses to contact stimulation, seen in its simplest form in the type of movement called 'gliding'. This involves a lateral wave of contraction immediately behind the point of stimulation proceeding in a forward direction. An example of this type of movement is seen when a hagfish escapes tail first over the edge of a bucket. In the 'figure-of-eight' movement, the tail is first coiled so as to touch the body. This then glides over the body and the body under the tail, each contracting at or just behind the first points of contact. In 'knotting' behaviour, which may take one of two forms, the head may pass through the loop formed by the tail around the body, or in a second form described by Strahan, the tail is also passed through the first loop formed by the body. These responses may be used to escape from entanglement in slime or when the animals are held in the hand. They have also been seen in

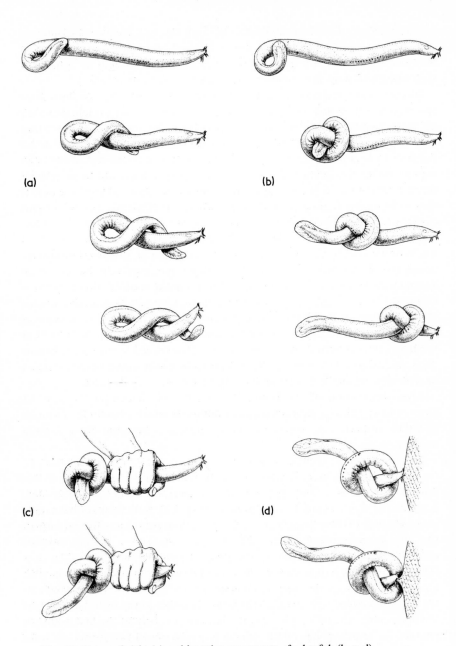

Fig. 4.8 Figure-of-eight (a) and knotting movements of a hagfish (b, c, d)
In (b) the hagfish can free itself from entanglement in slime, by rolling a knot from head to tail. In (c) it may use the movement to escape capture or in (d) to tear off its food. From Jensen, 1966.

animals attempting to tear off pieces of food when feeding in their natural environment.

4.4.1 Swimming speeds

Estimates of the rate of movement of sea lampreys migrating upstream have put their average rate of progress at about 5 cm s^{-1} or 0.18 km h^{-1} although it should be remembered that they are probably inactive for most of the daylight hours. Another estimate for the same species gave an average speed of 3.2 km h^{-1} against a slow current and 0.5 km h^{-1} in faster flowing water further upstream. Higher estimates have been given for the European river lamprey migrating into the rivers from the Gulf of Riga. Here the animals were said to travel up to 8–13 km in a single night and the average velocity in the early stages of the migration was estimated at 1–4 km h^{-1}. Indirect calculations based on the rate of depletion of energy reserves have put the average swimming speed of upstream migrating anadromous sea lampreys at about 6–7 cm s^{-1} or 200–250 m h^{-1} (Beamish, 1979). For large ammocoetes of the same lamprey, velocities of 0.11 and 0.26 m s^{-1} were recorded at 4–7°C and 0.45 m s^{-1} at temperatures of 20°C, but it is likely that speeds of this order would be maintained only for very brief periods. Laboratory experiments on feeding-stage sea lampreys showed that they were able to swim at their maximum speed of 35.5 cm s^{-1} for about 15 minutes, covering a distance of about 315 m (Fig. 4.9). On the other hand, the sexually mature migrant gave a vastly inferior performance and was able to sustain the same speed for only a minute, corresponding to a distance of 21.3 m (Beamish, 1974). This deterioration in swimming endurance might as the author points out be related to the cessation of feeding and the fact that at this stage the lamprey is relying on energy reserves in the form of lipids, built up during its parasitic phase.

Although there is no precise information of the swimming ability of hagfishes, observations made by skin divers indicate that they are capable of maximum speeds comparable with those of lampreys and that they are able to swim as fast as a diver is able to follow them (Fernholm, 1974). Strahan (1963) gives a maximum speed for *Myxine* of 25 cm s^{-1}, but other estimates both for this species and for *E. burgeri* have mentioned figures of the order of 1 m s^{-1}. This represents velocities of about three body lengths s^{-1}; considerably better than the performance of the sea lamprey.

Beamish has drawn attention to the fact that the swimming performance of the lamprey compares very unfavourably with that of teleosts. For example, the salmon is said to be able to maintain speeds about twice as great as the lamprey for periods up to at least an hour. The reason for this relatively poor swimming ability is obscure and there is certainly nothing in the respiratory or cardiovascular physiology of these animals that can be held responsible (Chapters 5 and 6). For this reason, Beamish considered that the answer might

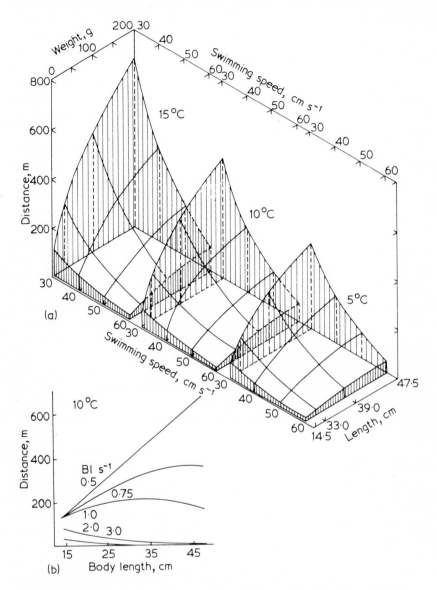

Fig. 4.9 The swimming performance of sea lampreys (*Petromyzon marinus*). From Beamish, 1974.

(a) Distance swum in relation to speed, body weight and temperature.

(b) Distance covered at different swimming speeds related to body length (Bl) Bl s^{-1} is compared for animals of varying lengths.

lie in hydrodynamic factors and perhaps in the relatively small caudal fin of the lamprey, which he considered to be poorly adapted to delivering a propulsive thrust. This seems all the more plausible, when it is borne in mind that the caudal fins of teleosts are said to be responsible for between 45 and 84% of the propulsive thrust (Bainbridge, 1963).

5

Respiration and feeding

5.1 Feeding and digestion

5.1.1 The feeding mechanisms

In both groups of cyclostomes the teeth are entirely horny epidermal structures. The grasping teeth of the oral disc of the lamprey have no counterpart in myxinoids, although the single median, recurved palatine tooth may help the hagfish to maintain its purchase on its food during the tearing action of its jaws. The biting and rasping teeth of the lamprey are carried on tooth plates above and below the mouth, as well as on the apex of the piston cartilage, whereas the grasping jaws of the hagfish are borne on a cartilaginous dental plate, which is everted during feeding, at the same time closing the jaws laterally (Fig. 1.4). Sliding in a groove on the upper surface of a fixed cartilaginous basal plate on the floor of the mouth, the U-shaped dental plate is moved forwards and backwards by the contraction of protractor and retractor muscles (Fig. 1.5b). As the dental plate is pulled over the edge of the basal plate the teeth are everted, and once the jaws come into contact with the food they are then rapidly retracted, snapping the teeth together so biting or tearing off pieces which are transferred to the mouth.

Although homologies with the hagfish feeding apparatus are obscure, the piston or tongue mechanism of the lamprey can be regarded as the functional equivalent of the dental plate (Fig. 1.5a). The long piston cartilage carries at its apex a transverse lingual lamina and above this a bilobed structure with longitudinal ridges bearing small denticles (Fig. 5.1). With the retraction of the tongue these teeth are brought together, exerting a cutting action on the tissues of the prey. The main retractor muscle is the long cardio-apicalis running from the head of the tongue to the pericardial cartilage at the rear of the branchial basket. The protractors arise mainly from the middle section of the piston cartilage, running forwards to attach to cartilages around the mouth, while shorter protractors from the head of the tongue are attached to the cornual cartilage. The complexity of the piston and pharyngeal pump

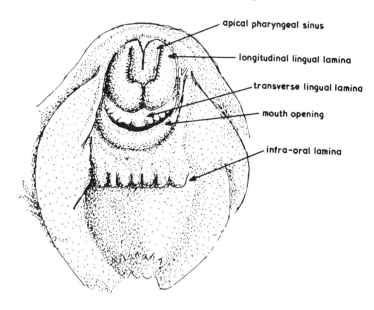

apical pharyngeal sinus

longitudinal lingual lamina

transverse lingual lamina

mouth opening

infra-oral lamina

Fig. 5.1 Dissection of the oral disc of a lamprey to show the arrangement of the rasping teeth. Redrawn from Lanzing, 1958.

mechanisms may be illustrated by the fact that these systems involve no fewer than twenty separate muscles. When feeding, the tongue rocks backwards and forwards, bringing into play the cutting actions of the bilobed head, as well as a rasping action of the teeth of the lingual lamina. Secretions of the salivary glands are poured out on to the wound, exerting cytolytic and anticoagulant effects on the tissues and body fluids of the host.

Attachment by the lamprey to its prey or to other surfaces, when it is not swimming, involve the development of reduced pressures within the buccal cavity. After the oral disc has been extended over the surface of attachment, its internal volume is reduced by contractions of the annular muscles (Fig. 5.2). The tongue is now retracted to seal off the oral passage behind, and the buccal cavity is once again enlarged to create the necessary vacuum. Water is then expelled from the pharynx through the water tube, largely by muscular compression of a dorsal diverticulum, the hydrosinus, thus reducing the pressure in this region also. Protraction of the tongue then allows some water to flow between pharynx and buccal cavity, equalizing the negative pressures in both compartments. In these processes, the velar valves play a vital role in isolating the branchial region from the pharynx and buccal cavities. Some idea of the suction pressure that lampreys can generate may be gained from the fact that when attached, even the largest lampreys can be suspended out of water under their own weight and a pressure of -120 cm H_2O has been measured

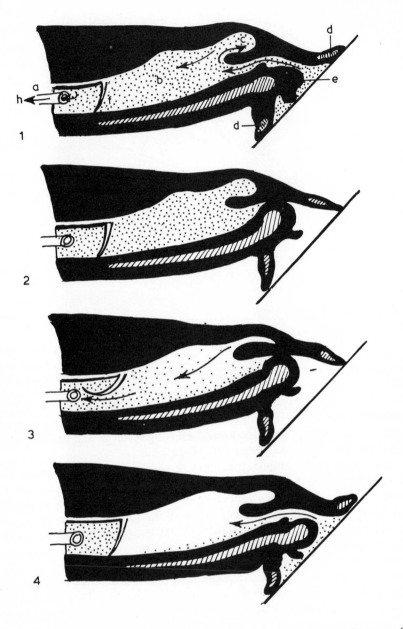

Fig. 5.2 The suction and feeding mechanism of the lamprey as seen in successive stages of its operation 1–4. Based on Reynolds, 1931.
a. oesophagus; b. pharynx; c. hydrosinus; d. annular muscle; e. tongue; f. velar tentacles; h. gill pouch.

when attempts were made to pull a lamprey off the surface to which it was attached (Gradwell, 1972). The effectiveness of the suction mechanism must be enhanced by the cirrhi forming a fringe round the edges of the oral disc (Fig. 1.3) and which are composed of masses of mucus-secreting cells (Lethbridge and Potter, in press).

5.1.2 Digestion

Unlike that of the lamprey, the hagfish intestine is not ciliated and the passage of food is assisted by the general movements of the body wall. In addition to mucous cells, the epithelium contains flask-shaped zymogen cells containing spherical acidophilic granules. Throughout its length the histological structure of the gut is said to be uniform and all regions are considered to be involved in both secretion and absorption (Adam, 1963b). Digestion and the evacuation of the faeces take place within a peritrophic membrane secreted by the epithelium, presumably protecting the mucosa against mechanical abrasion. This is likely to be particularly important in the hagfish where the proliferative capacity of the epithelium has been found to be exceptionally low in comparison with that of other vertebrates (Linna *et al.*, 1975). The intestine is also remarkable for its high lipid content, representing about 13 % of its wet weight (Spencer *et al.*, 1966).

Employing histochemical techniques, Adam (1963b) was able to detect an amylase with an optimum pH of 8–9 and also high lipase activity in the intestine of *Myxine*. In addition, a strong positive response was obtained with techniques used to detect leucoaminopeptidase activity. In common with those fish that are without a differentiated stomach, no HCl or peptic enzymes are produced in the digestive tract of cyclostomes. Nilsson and Fänge (1970) found some proteolytic activity at pH 4 in intestinal extracts from *Myxine*, but this was believed to be due to intracellular cathepsins, suggesting that during the evolution of the vertebrate stomach, the extracellular peptic enzymes may have been developed from this type of tissue enzyme. Among the hagfish enzymes that are active in the alkaline range of pH, these authors identified trypsin-like and chymotrypsin-like activity. Both enzymes occurred throughout the length of the intestine, although they were less prominent in the posterior third. This may be contrasted with conditions in the larval lamprey where the zymogen cells are concentrated to a marked degree at the extreme anterior end of the intestine. In the hagfish, Nilsson and Fänge also identified a leucoaminopeptidase and dipeptidase as well as carboxypeptidase that acts together with chymotrypsin. In other vertebrates these enzymes occur on the brush border membranes or within the mucosal cells and are concerned in the final degradation of proteins to amino acids.

Most of our information on the lamprey digestive tract comes from studies carried out on the ammocoete or on the migratory stages of the adult (Youson and Connelly, 1978; Hansen and Youson, 1978; Battle and Hayashida, 1965) and little is known of conditions in the adult feeding stages. In the ammocoete of *L. planeri*, zymogen cells with large acidophilic granules are confined to a short region at the anterior end of the intestine, adjacent to its junction with the oesophagus and similar cells are also found in the paired intestinal diverticula of the Southern hemisphere genus *Geotria*. This led Barrington (1972) to suggest that the concentration of zymogen cells in the anterior intestine and their aggregation in the diverticula of *Geotria* might parallel the early evolutionary history of the exocrine pancreas of the gnathostomes. Such a view would then place the lampreys closer to the higher vertebrates than the myxinoids, where these developments have not occurred. In the other Southern genus, *Mordacia*, where there is only a single diverticulum in place of the paired structures of *Geotria* (Fig. 5.3), no proteolytic activity has been detected in this region of the intestine, whereas in *L. planeri* Barrington was able to demonstrate a protease with an optimum pH of about 7.5, confined to the anterior region where the zymogen cells occurred. In *Mordacia*, proteolytic activity was present throughout the rest of the intestine but significantly, all the amylolytic and most of the lipolytic activity was restricted to the diverticulum (Strahan and Maclean, 1969).

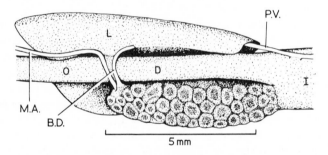

Fig. 5.3 Dorsal view of the intestinal diverticulum of an ammocoete of *Mordacia*. From Strahan and Maclean, 1969.
B.D. bile duct; D. diverticulum; I. intestine; L. liver; M.A. mesenteric artery; O. oesophagus; P.V. hepatic portal vein.

In their ultrastructure, the secretory cells of the lamprey intestine resemble those of the hagfish and their high tryptophan content presumably reflects the presence of trypsin and chymotrypsin (Luppa, 1964; Luppa and Ermisch, 1967). It has been suggested that they may differ from pancreatic acinar cells in having both absorptive and secretory functions, perhaps in this respect representing an early stage in the evolution of the zymogen cell, before these two functions had become separated in distinct and differentiated cell types. During metamorphosis, the zymogen, mucous and ciliated cells of the anterior intestine are largely replaced by ion-transport cells (Youson and Connelly,

1978; Hansen and Youson, 1978). In contrast to the posterior regions, this part of the adult gut has been regarded as a storage rather than an absorptive area, perhaps foreshadowing the evolution of the vertebrate stomach.

Feeding lampreys are remarkable for the amounts of food that they consume, although it is possible that the results obtained in laboratory experiments may be biased by the fact that food fishes are ready to hand, whereas under natural conditions a considerable time may be spent in the location of prey. In one series of experiments, Farmer *et al.*, (1975) established that young feeding sea lampreys, *P. marinus*, when presented with trout whose blood cells had been labelled with radioactive chromium (^{51}Cr), fed at rates varying from 2.9–29.8 % of their wet body weight a day. Such rates of feeding are greater than in most teleost species, perhaps reflecting the ease with which a blood diet can be assimilated. This, together with its high protein content must also contribute to the high conversion efficiency of the lamprey which, in these experiments, was found to vary from 10.6–56.9 % compared with an upper limit of about 30 % measured in teleosts. As might be expected, faecal losses were very low, representing only about 3 % of the energy content of the food, as against figures of about 10 % reported in some fish species.

The contrast between the high efficiency rates of the adult lamprey and the low rates recorded in the ammocoete (Moore and Potter, 1976b) has been linked with the changes that take place at metamorphosis in the histology of the gut and its circulatory system. These include an increase in the absorptive surface through the development of longitudinal folds (Fig. 2.3), improvements in the arterial supply, the development of vascular couples in which the venous and arterial blood flows in opposite directions through the intestinal wall, the disappearance of muscle fibres separating the blood vessels of the lamina propria from the epithelium and finally, the development of a more efficient venous return through the hepatic portal system (Percy and Potter, 1979).

5.2 Respiration

One of the more interesting aspects of cyclostome physiology is the existence within the group of three quite distinct methods of gill ventilation. That of the ammocoete has been regarded as the most primitive and, as in the filter feeding protochordates, the pharynx is undivided and the unidirectional water current serves both for feeding and respiration. With the development of macrophagous feeding in the agnathans, it became necessary to separate these functions, but this division has been achieved in quite different ways in the hagfishes and the lampreys; the myxinoids making use of the nasohypophysial tract to convey the water current to the posteriorly placed gills; the adult lamprey developing a separate water tube connected to the internal gill openings, making it possible to use pumping, tidal mechanisms of gill ventilation that can be operated even when the animal is engaged by its sucker.

5.2.1 The respiratory current

In the hagfish, the respiratory current is produced by the pulsations of the complex scroll-like velar folds, operated by extrinsic muscles attached to the neurocranium and axial skeleton and inserted on the cartilages that support them (Fig. 5.4). These folds are housed in a velar chamber developed from the nasohypophysial tract. When at rest, the folds are rolled upwards towards the

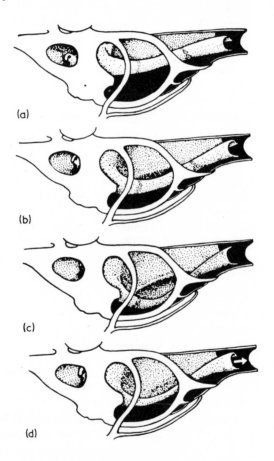

Fig. 5.4 The mode of operation of the velum in the hagfish as seen in a cut-away view into the velar chamber. From Strahan, 1963.
(a) resting position (b) velar scroll beginning to unroll (c) velar scroll fully unrolled (d) scroll beginning to move dorsally (arrow indicates direction of water flow).

dorsal side of the chamber, subsequently unrolling ventrally and laterally. This has the effect of scooping up the water and deflecting it backwards towards the pharynx, while at the same time, the reduction in pressure within the velar chamber draws in fresh water through the nostril (Johansen and Strahan, 1963).

Although most authors believe that the velum of lampreys and myxinoids are homologous, this view is not without its difficulties and although a comparable structure may have been present in the fossil agnathans, we have no direct evidence to support this belief. The ammocoete velum consists of two simple flap-like paddles, but these are arranged vertically rather than horizontally as in the hagfish (Fig. 1.6). Here its movements are produced by intrinsic muscles and it serves not only as a respiratory pump, but also acts as a valve preventing the reflux of water from the pharynx into the mouth during the expiratory phase of the respiratory cycle. Neither in the hagfish nor in the ammocoete does the velum act alone in producing the respiratory current. In the hagfish, the activity of the velum is assisted by peristaltic contractions of the muscular gill pouches and their afferent and efferent ducts. In the ammocoete, the current is augmented by alternate contractions and expansions of the branchial basket; the inhalatory phase being assisted by the passive elastic recoil of the cartilaginous framework of the pharyngeal wall.

Given the engagement of the lamprey mouth and sucker during feeding and the forward position of the gills, the tidal method of ventilation represents the only possible alternative to the solution adopted by the myxinoids. In the lamprey, the velum has completely lost its water-pumping functions and has been reduced to a small flap-like valve, preventing the entrance of food into the water tube and playing a role in the maintenance of suction. As in the ammocoete, the water current is produced by alternate changes in the volume of the branchial basket; contraction of the branchial constrictors and the diagonal muscles reducing its volume and elastic recoil bringing about its subsequent enlargement (Randall, 1972). The direction of the water flow is controlled by two sets of valves in the external gill openings; namely an outer ectal valve and an inner pair of ental valves. Water flow from the water tube into the pharynx can be controlled by the velar valves. Even when lampreys are swimming, the tidal pumping mechanism is thought to be maintained, although it has been shown that water flow through the mouth and out of the gills is also possible under certain circumstances.

5.2.2 Morphological and functional aspects of cyclostome gills

In spite of the evolutionary distance separating the cyclostomes from the teleosts and the fundamental differences in the mechanics of gill ventilation in the two groups, there are some remarkable parallels in the gross and microscopic structure of the gills and their functional correlates. These resemblances have been shown to extend to the relation between filament lengths and numbers and body weight, the spacing and areas of the secondary lamellae and the total gill areas (Lewis and Potter, 1976; Lewis, 1976).

Amongst the factors that influence the rate of gaseous exchange, gill areas are of obvious importance, although it should be remembered that since the gills are exposed to osmotic or ionic stresses, gill areas may represent a compromise between respiratory requirements and osmotic limitations (Randall, 1970). In the ammocoete of *L. fluviatilis*, gill areas have been estimated at between 1462–2717 mm^2 g^{-1} and similar values have been recorded for the adult (1402–2337 mm^2 g^{-1}) (Lewis and Potter, 1976). Both of these values come well within the upper end of the range for teleosts and are comparable with those of the most active species (Hughes and Morgan, 1973).

In the fine structure of the gills and the water-blood pathway there are further parallels with gnathostome fishes. Except in the marginal and basal regions the surface of the gill lamella is composed of two cell layers (Fig. 5.5); an outer layer of platelet cells whose surfaces are covered with microvilli and below them, the basal cells. These rest on a basement membrane containing collagen fibres, in places interdigitating into the plasma membranes of the

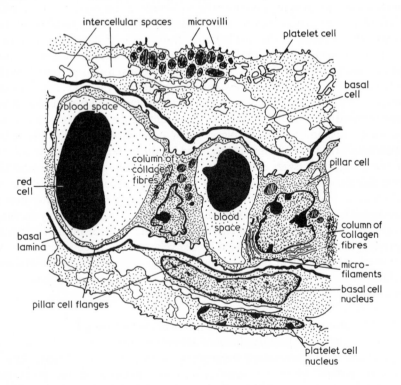

Fig. 5.5 Section of the gill lamella of an ammocoete. × 3600. Drawn from an electron micrograph by Youson and Freeman, 1976.

pillar cells. to form the collagen columns that support the lamellae and maintain the arterial blood pressure. The pillar cells have cytoplasmic extensions – pillar cell flanges which meet those of adjacent cells to enclose the blood spaces. Estimates of the thickness of the water-blood barrier have given values of 4.57 μm and 1.21 μm in an ammocoete and adult of *L. planeri* of similar body weight (Lewis, 1976).

As might be expected from the close parallels in their gill morphology, the lamprey gill appears to be quite as efficient as that of a gnathostome fish as a medium for gaseous exchange. This efficiency has been measured by Lewis (1976) from the diffusing capacity D_t (Hughes, 1972) calculated from the expression $D_t = K.A/t$ where:

K is Krogh's (1919) permeation constant (0.00015 ml cm^2 mm Hg^{-1} min^{-1});

A the total gill area; and

t the thickness of the water-blood barrier.

When expressed in terms of body weight, values of D_t for adult *L. planeri* and *P. marinus* were 1.84 kg^{-1} and 0.51 kg^{-1}; quite similar to those of teleosts (Hughes and Morgan, 1973).

In larval lampreys, the ventilation of the gills is adjusted to varying oxygen demands by alterations in the frequency of pumping or in the volume of water passing over the gills at each contraction (stroke volume). At 20°C the mean ventilation rate (volume of water passing over the gills in unit time) was found to be 0.60 ml min^{-1}g^{-1} and the maximum rate for any animal was 3.3 ml min^{-1}g^{-1} (Rovainen and Schieber, 1975). Various stimuli were effective in increasing ventilation volumes, but after exercise, mechanical obstruction of the water current, low oxygen tensions or increased carbon dioxide concentrations, the stroke volumes were raised to a greater extent than the ventilation frequencies. Stroke volumes varied from 4–25 μl g^{-1} and these, like the ventilation volumes, lie within the normal ranges for fishes i.e. 2–31 μl g^{-1} and 0.09–13 ml g^{-1} min^{-1}. For an adult lamprey, *L. (Entosphenus) tridentata*, ventilation volumes varied from 0.25–10.5 ml g^{-1} min^{-1} (Johansen *et al.*, 1973); values which again are very similar to those of teleosts. These comparisons indicate that, in spite of the vast differences in the ventilatory mechanisms of ammocoetes, adult lampreys and fishes, each appears to be equally effective in delivering water to the gills.

In teleosts, the rate of gaseous exchange between water and blood can be regulated by changes in the flow pattern of the blood in the secondary lamellae of the gills and, since microfilaments are present within the pillar cells of lamprey gills (Fig. 5.5) a similar form of control may also exist in these animals (Youson and Freeman, 1976).

The amount of oxygen that passes into the blood in the gills will depend on the surface area of the gills, the diffusion distance and the oxygen gradient between blood and water. The efficiency with which the animal is able to utilize the oxygen contained in the respiratory water can be assessed by calculating the extraction rates, represented by the percentage difference in the oxygen content of the inspired and the exhaled water. For larval stages of *P. marinus* this has been found to be about 43 % (Rovainen and Schieber, 1975) compared with values of 10–28.4 % in the adult lamprey, *L. (Entosphenus) tridentata* (Johansen *et al.*, 1973). These apparently higher extraction rates in the ammocoete are in line with the general picture in fish, where low extraction rates are characteristic of active species living in running water, while more sluggish species living in still water tend to show higher rates.

Because of its habit of feeding with its head and branchial region frequently burrowed deeply into the carcasses of fish, it has been suggested that cutaneous respiration may be a significant factor in the oxygen uptake of the hagfish. This is borne out by the fact that *Myxine* can survive for a considerable period after its nasal opening has been obstructed. In the case of the lamprey, *L. fluviatilis*, it has been estimated that up to 18 % of its oxygen uptake may take place through the skin (Korolewa, 1964) and these animals are certainly capable of surviving for several days out of water at temperatures below 10°C. On the other hand, survival is reduced to a matter of only a few hours if the gills are covered (Czopek, 1970), indicating that when the animal is out of water, respiration may still be taking place through the moist gill surfaces. Cutaneous respiration may be significant in small ammocoetes which, once they have settled down under laboratory conditions at low temperatures and relatively high oxygen tensions, may cease to show either velar or branchial contractions.

5.2.3 Oxygen consumption

Both larval and adult lampreys lend themselves to measurements of standard oxygen consumption, defined as the nearest practicable approach to the basal metabolic rate. Suitable conditions can be provided for measurements on ammocoetes by employing an artificial substrate of glass beads into which they will burrow and settle down quietly. Adults can be maintained within glass tubes to which they will attach themselves and remain motionless for long periods.

The importance of allowing the ammocoete to burrow has been demonstrated by comparing the oxygen consumption of burrowed and unburrowed animals (Potter and Rogers, 1972). The unburrowed ammocoetes took longer to reach a constant level of oxygen consumption than the burrowed larvae and this level was always higher than for animals that had been provided with a substrate (Fig. 5.6). Measurements of the oxygen consumption of burrowed ammocoetes show little variation between species and by vertebrate standards the rates are low, although greater than the rates of oxygen consumption of the protochordate *Amphioxus*, whose burrowing habits are not very dissimilar to those of an ammocoete.

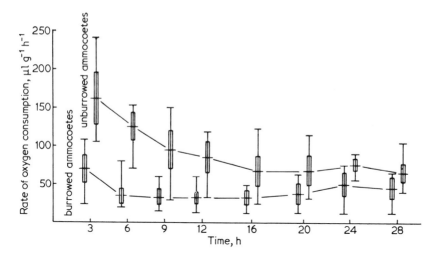

Fig. 5.6 The oxygen consumption of burrowed and unburrowed ammocoetes under light conditions. From Potter and Rogers, 1972.

The contrast in the metabolism of larval and adult lampreys may be illustrated by following the changes in oxygen consumption that take place during metamorphosis in the closely related parasitic and non-parasitic species, *L. fluviatilis* and *L. planeri* (Lewis and Potter, 1977). Before metamorphosis, the mean larval rates of oxygen consumption at $10°C$ were 30 $\mu g\ g^{-1}h^{-1}$ for *planeri* and 43 $\mu g\ g^{-1}h^{-1}$ for *fluviatilis* rising at the completion of transformation to 74 $\mu g\ g^{-1}h^{-1}$ and 89 $\mu g\ g^{-1}h^{-1}$ in the young adults (macrophthalmia), representing a doubling of the rate over that of the ammocoete stage (Fig. 5.7). In addition, with the eruption of the eyes, a circadian rhythm of oxygen consumption develops, involving an increase in the rate during the dark phase. In the case of the non-parasitic *planeri*, which almost immediately enters the phase of sexual maturation, sex differences in oxygen consumption begin to appear, similar to those observed in the sexually mature *fluviatilis*. Curiously, the rates of oxygen consumption (expressed in terms of body weight) are remarkably similar in adults of the two species, in spite of the vast differences in their body weights. Thus, for adult males and females of *fluviatilis* with a mean body weight of 45 g, the rates were 90 and 50 $\mu g\ g^{-1}h^{-1}$ compared to 108 and 65 $\mu g\ g^{-1}h^{-1}$ for male and female *planeri*, whose mean body weight was only 2.5 g. For a wide variety of animal groups, the relationship between metabolic rates and body weights can be expressed by $M = aW^b$, where M is the metabolic rate, W the body weight, a a constant of proportionality and b the exponent (Wilson, 1972). For most species that have been studied the value of b lies between 0.7–0.8, but although

Biology of the Cyclostomes

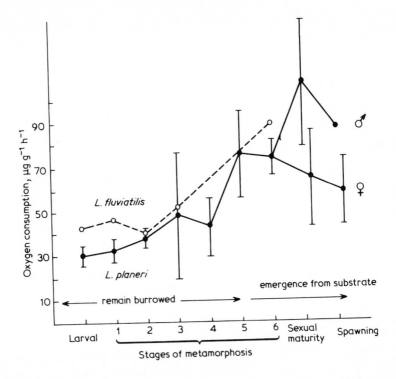

Fig. 5.7 Increase in the oxygen consumption of non-parasitic and parasitic lampreys during metamorphosis. After Lewis and Potter, 1977.

this applies to the ammocoete, the value for the adult lamprey is close to unity.

The limited information on the metabolic rates of hagfishes is interesting in so far as it suggests that there may be pronounced species differences in oxygen consumption related to their ecology and more especially to the depths of water they normally frequent. Laboratory experiments on the comparatively shallow water species *E. stouti* (section 4.1.1) indicated low oxygen consumption rates of $12-15\,\mu g\,g^{-1}h^{-1}$ for animals acclimated at temperatures of $4°$ and $10°C$, resting quietly on the bottom (Munz and Morris, 1965). More recently, a remote-controlled underwater vehicle (monitored by television cameras), has been used in conjunction with a respirometer, to measure oxygen consumption in a number of benthic and abyssal animals, including the hagfish, *E. deani*, on the sea bed under natural conditions (Smith and Hessler, 1974; Smith, 1978). Measurements on this hagfish at a depth of 1230 m and a temperature of $3.5°C$ in the San Diego Trough gave an average value of only $3.1\,\mu g\,g^{-1}h^{-1}$ over a period of 13 h and during this time, no significant changes in the rate of oxygen uptake were observed. In the same area, a very similar respiration rate was also recorded in a deep sea benthic teleost, the rattail, *Coryphaenoides aerolepis.* Moreover, later measurements on an abyssopelagic rattail, *C. armatus* at depths of 2700–3600 m and a temperature of 3°C in the North West Atlantic again produced very low values of 4.9–5.3 μg g^{-1}h^{-1} (Smith, 1978). Such low metabolic rates are now believed to be characteristic for deep water species and have been related to an environment of low temperatures and high pressures, where opportunities for feeding are infrequent and availability of food is low.

Even making allowances for the rather low temperatures at which these measurements were made, the metabolic rates of the hagfishes are in striking contrast to lampreys of a similar size range. They may be compared with oxygen consumption rates at 5°C of 15 μg g^{-1}h^{-1} for adult river lampreys (*L. fluviatilis*) in winter rising to about 50 μg g^{-1}h^{-1} at spawning time and a figure of 53 μg g^{-1}h^{-1} for the larger sea lamprey, *P. marinus* at the same temperature (Claridge and Potter, 1975; Beamish, 1973).

The extent to which the rates of oxygen consumption of the lampreys can be raised during activity may be illustrated by the experiments carried out by Beamish (1973) on the effects of swimming activity in the sea lamprey, *P. marinus*. For example, in the case of feeding stage adults with mean body weights of 52 g, he recorded mean oxygen consumption rates of 475 μg g^{-1}h^{-1} in animals swimming for 20–60 minutes at speeds of 30–40 cm s^{-1} at 10°C. Taking the upper values for oxygen uptake of swimming lampreys, these represented active rates some seven times higher than the resting rates; increases that are of the same order of magnitude as those recorded in the highly active trout, whose oxygen uptake during exercise may be from four to eight times higher than its resting values.

5.2.4 Responses to oxygen depletion

To a far greater extent than the myxinoids, both larval and adult lampreys must be able to cope with changes in their external environment, particularly changes in temperature and in the oxygen content of the water. Larval lampreys do not normally live in stagnant water, but the currents over the burrows may be very slow and both oxygen tensions and water temperatures will show quite large seasonal variations. Naturally, the degree of oxygen depletion that the ammocoete is able to tolerate will be dependent on temperature. For example, at 5°C ammocoetes of *I. hubbsi* were able to survive for at least 4 days at oxygen tensions of only 7–10 mm Hg, whereas at a temperature of 22.5°C, survival for a similar period was only possible at much higher oxygen tensions of 19–21 mm Hg (Potter *et al.*, 1970).

When exposed to low oxygen or high carbon dioxide concentrations, the ammocoete displays a number of behavioural adaptations, tending either to minimize the oxygen deficiency or to move the animal into a more favourable environment (Section 4.2.1). Physiological adaptations include increases in ventilatory frequency or stroke volume. The increase in the volume of water passing over the gills must be capable of compensating for the decreased oxygen content of the water as well as for the additional energy expenditure involved in the greater respiratory effort. Under conditions of severe oxygen deficiency, ammocoetes may show pumping rates up to 200 beats min^{-1}, but at tensions approaching the lethal levels, the rate of pumping again begins to decline.

Adult lampreys show a remarkable ability to maintain, or even increase their oxygen uptake with decreased oxygen concentrations in the water (Fig. 5.8). Starting from air saturation values of 100%, *L. fluviatilis* increased its rate of oxygen uptake at 9.5°C from 38 $\mu g\, g^{-1}h^{-1}$ to 52 $\mu g\, g^{-1}h^{-1}$ at 30% air saturation (Claridge and Potter, 1975).

In adult lampreys, both ventilatory frequencies and heart rates increase as oxygen tensions are reduced and in the case of *L. fluviatilis* the former may reach levels some three times higher than those of resting animals (Fig. 5.8). At air saturations of less than 20%, hyperventilation becomes much more marked and at this stage the animals begin to show restless, continuous swimming activity, often leaping clear of the water. Similar avoidance responses are observed in fish when oxygen concentrations approach the lethal levels. Under conditions of hypoxia, teleosts and elasmobranchs develop a slowing of the heart beat (bradycardia) attributed to increased vagal tone and in trout this is known to be accompanied by an increased stroke volume, thus maintaining a relatively constant cardiac output (Shelton, 1970). This is in marked contrast to the condition in the lamprey, where the heart rate increases slightly under reduced oxygen tensions and in this connection it should be noted that vagal stimulation of the lamprey heart or the administration of acetylcholine, results in an acceleration of the heart rate (Section 6.4.3).

The increased oxygen uptake of the lamprey under hypoxic conditions is almost certainly associated with the additional energy expenditure involved in increased gill ventilation. In fish generally, the metabolic cost of ventilation has been put at about 30% of the standard rates (Randall, 1972), but this will almost certainly vary widely with the habits and environments of different species. For example, this energy expenditure will tend to be higher for fish living in still water, where more active pumping is required, and in these cases, ventilation volumes will tend to be low and oxygen extraction rates high. For species living in fast-flowing water, the energy costs will tend to be lower, but larger volumes of water will be passed over the gills, from which comparatively less oxygen will be extracted (Randall, 1970). This would be advantageous in so far as it would keep oxygen tensions high over the entire length of the secondary lamellae. It has usually been assumed that the tidal respiration of lampreys would be more costly in terms of energy than the mechanisms

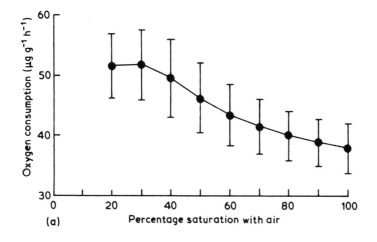

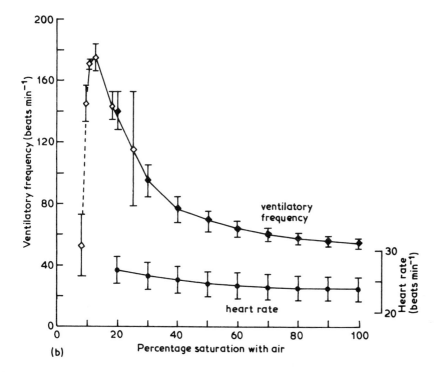

Fig. 5.8 Physiological responses of an adult lamprey, *Lampetra fluviatilis* to reduced
oxygen tensions. From Claridge and Potter, 1975.
(a) Changes in oxygen consumption. (b) Changes in ventilatory frequency and
heart rates.

employed by fish, but this has been questioned on the grounds that the passive elastic recoil during the expiratory phase of the lamprey cycle might well compensate for any general inefficiency in tidal breathing (Shelton, 1970).

6

The heart and circulatory system

6.1 Vascular structures

In its basic plan, the circulatory system of the cyclostome conforms to the characteristic craniate pattern. In the pharyngeal region, the ventral aorta gives rise to a series of aortic arches supplying the gills, of which there are eight in the lamprey; the first – the hyoid – serving the anterior hemibranch of the most anterior gill pouch. The remaining afferent arteries divide to supply the anterior and posterior hemibranchs of adjacent gill pouches, and a similar pattern is seen in the efferent arteries, joining the dorsal aorta below the notochord (Fig. 6.1). In myxinoids, the number of arterial arches varies with the number of gill pouches and unlike those of lampreys or other craniates, each arch supplies the hemibranchs of the same pouch. This also applies to the efferent vessels, which join two lateral aortae above the pharynx. These unite behind the branchial region to form the median dorsal aorta, which is continued forwards between the lateral vessels. The homologies of these vessels are by no means certain, but Holmgren (1946) considers that in its embryonic development, the dorsal aorta of *Myxine* is closer to the gnathostome pattern than that of the lamprey.

The venous system shows the typical craniate pattern of paired anterior and posterior cardinal veins, the latter formed by the bifurcation of the caudal vein. The anterior cardinals of the hagfish are asymmetrical, the left emptying into the sinus venosus independently, while the right cardinal joins a median inferior jugular draining the posterior tongue muscles. The proximal part of the right anterior cardinal constitutes an anterior portal vessel, derived from the right Cuvierian duct and opening into the contractile portal heart (Fig. 6.2). In the lamprey both anterior cardinals enter the single right duct of Cuvier, independently of the inferior jugular. In both groups an hepatic portal system is present, but neither possess a renal portal system. In the myxinoids, the portal heart receives blood from the branchial region through the anterior portal vessel and from the gut through the intestinal vein, transmitting it to the

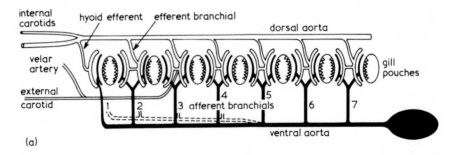

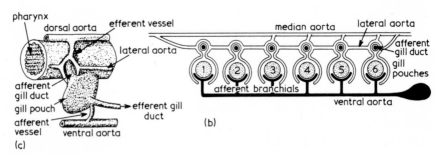

Fig. 6.1 The branchial circulation of a lamprey (a) and a hagfish (b). Details of the gill
pouch and its vascular relations are shown in (c).

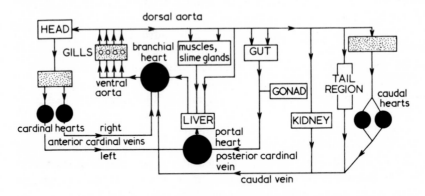

Fig. 6.2. Diagrammatic representation of the circulatory system of a hagfish. Shaded
areas represent sinuses and black circles, contractile structures. Based on Jensen,
1966.

liver through a common portal vein. In the lamprey, the portal system is
concerned solely with blood from the gut conveyed to the liver via the
mesenteric vein.

A characteristic of the cyclostome circulation is the presence of an extensive system of sinuses or blood lacunae, communicating on one side with the arterial system and on the other with the veins. This open type of circulation is particularly well developed in the hagfishes, where the blood pressures are low and the maintenance of the circulation is aided by 'accessory hearts', supplementing the pumping activity of the branchial or systemic heart (Fig. 6.2). In addition, the general movements of the body of the hagfish make a significant contribution to the circulation by exerting pressure on the sinuses, thus forcing blood into the veins. In the lamprey, where the sinuses are well developed in the branchial region, they play an important part in the mechanics of respiratory movements. In the myxinoids, the sinus system is particularly important in the head regions, where apart from the skin and brain, there are no true veins and these lacunae replace the capillary bed (Johansen, 1963).

These conditions have been compared to the primitive type of circulation seen in many invertebrates and in the cephalochordates, where peristalsis of the blood vessels and contractions of the body wall musculature are important factors in circulatory dynamics. Further primitive features have also been described in the fine structure of the capillaries of the neurohypophysis and thyroid of the hagfish (Casley-Smith and Casley-Smith, 1975), where open connections have been described between the endothelial cells. These openings, measuring from 20–50 μm (even in some cases up to 1.0 mm), would allow macromolecules to be taken up into the blood from the surrounding tissues. Similar open junctions have been described in *Amphioxus*, but are otherwise characteristic of the lymphatic vessels of higher vertebrates. True lymphatics are thought to appear for the first time in elasmobranchs and teleosts, where presumably, the higher blood pressures would have required closed capillary junctions. Such forms have also developed alternative mechanisms in the shape of capillary fenestrations. These structures have been found neither in *Amphioxus*, nor in the hagfish material examined by Casley-Smith, but they are known to occur in the capillaries of elasmobranchs. Thus in the evolution of these vascular structures, the myxinoids appear to stand between the cephalochordates on the one hand and the fishes on the other.

6.2 Blood and blood cells

In keeping with their extensive systems of blood sinuses, cyclostomes are characterized by large blood volumes and extracellular water compartments; the volume of the extracellular water being of the same order as in the elasmobranchs and distinctly greater than in teleosts (Table 6.1). Correlated with their more extensive lacunar system, blood volumes in hagfishes are larger than in lampreys.

Both in hagfish and lamprey, the protein concentrations of the plasma are

Table 6.1 Total body water, extracellular fluid and blood volumes of cyclostomes (expressed as percentages of body weight). Data from Robertson, 1974.

	Total body water	Extracellular fluid	Blood
Hagfishes	72.3–74.6	27.0	18.7
Lampreys	75.6–86.8	23.9–24.4	8.0–8.5

higher than in elasmobranchs or teleosts (Larssen *et al.*, 1976; Drilhon, 1968). In these fishes concentrations are said to range from 2.5–3.5 %, but they may reach over 7 % in the sea lamprey during its parasitic phase, falling in the non-feeding migrant stage to reach minimum values (2.1–3.8 %) in the sexually mature animal (Webster and Pollara, 1970). In contrast to mammalian sera where the dominant component is the albumin fraction, an analogous fraction is either absent or represented only by a trace in *Myxine* serum (Manwell, 1963) and in the lamprey, *L. fluviatilis* forms only a minor component of the electrophoretic patterns (Drilhon *et al.*, 1968). In all cyclostomes, as in fishes, the major serum constituents have mobilities analogous to the α and β globulins of mammals (Rall *et al.*, 1961). In higher vertebrates, these form lipoprotein complexes and their dominance in the cyclostomes implies a similar capacity to bind and transport lipid. In fact, a lipid-rich fraction has been isolated from the β_2 globulin zone of the female sea lamprey (Drilhon, 1968). Lipoproteins are also abundant in the serum of the hagfish (Manwell, 1963; Thoenes and Hildemann, 1970) and in the latter, the high density fraction is said to be unique amongst vertebrates in the absence of cholesterol esters; a condition that is shared only with invertebrates (Lee and Puppione, 1974). Lampreys differ from gnathostomes in that vitamin A is transported in association with these plasma lipoproteins, rather than being bound to specific retinol-binding proteins (Shidoji and Muto, 1977). Serum fractions with mobilities analogous with mammalian gamma globulins are also present in cyclostome sera, although these do not show immunological activity (Boffa *et al.*, 1967; Thoenes and Hildemann, 1970). These same serum fractions, however contain transferrins; single-chained proteins with molecular weights of 75–80 000, able to bind iron at two binding sites, suggesting that they may have been produced as a result of gene duplication (Drilhon *et al.*, 1968; Webster and Pollara, 1969; Boffa and Drilhon, 1970; Aisen *et al.*, 1972). At least in the higher vertebrates, transferrins may show bactericidal activity, but this has not been demonstrated in the cyclostomes.

The blood cells of lampreys fall into the same broad categories as those of higher vertebrates, although some doubt still remains on the functional and structural correspondence of certain elements of the leucocyte series. Among these, granulocytes, eosinophils and neutrophils can be distinguished and macrophages are present, perhaps corresponding to the monocytes described by earlier workers and apparently derived from large lymphocytes (Percy and Potter, 1976). The cells of the granulocyte and erythrocyte series originate in small lymphocytes, which these authors regard as stem cells; a view that differs radically from that proposed by several previous authors who had traced the origins of the haemopoietic stem cells to fixed reticular cells in the haemopoietic tissues.

Initially, at the time when the prolarva is hatching, the formation of blood cells occurs in the blood islands around the remaining yolk mass, but as the yolk is absorbed and the intestine opens, haemopoiesis is taken over by the typhlosole. By the time the ammocoete has reached lengths of about 20 mm, haemopoiesis begins in the intertubular and adipose tissues of the kidney fold and together with the intestine, these remain the active sites throughout the remainder of larval life. With the approach of metamorphosis, these larval blood-forming tissues undergo involution and their functions are taken over by the fat column above the spinal cord, which remains the active haemo-poietic tissues of the adult lamprey until the end of its spawning run, when it too undergoes massive degeneration (Percy and Potter, 1977). This change in haemopoietic sites at metamorphosis coincides with a switch from larval to adult haemoglobins, although it is not known if these are contained in distinct erythrocyte populations or are solely produced in separate haemopoietic sites. The nucleated red cells of cyclostomes are exceptionally large and those of the lamprey have a volume some eight times that of the human red cell, containing about six times as much haemoglobin (Riggs, 1972).

Less is known of the developmental history of the blood cells of myxinoids, but it now appears that the main areas of blood cell formation are the intestinal submucosa and the central mass of the pronephros; the first of these tissues making up as much as 10 % of the whole intestinal volume (Fänge, 1973). The implication of these sites in haemopoiesis is supported by the intense activity in these tissues of delta-aminolevulinic acid dehydrase (ALA-D); a key enzyme in porphyrin synthesis (Olsson, 1973). Large numbers of mononucleate, lymphocyte-like cells are present in the blood of *Myxine*, but these are said to be difficult to distinguish from spindle cells, thrombocytes, monocytes or haemocytoblasts described by other authors. The mitotic activity of erythroblasts in the circulating blood has been confirmed by their incorporation of H^3-labelled thymidine, but strong RNA synthesis is indicated in spindle cells and some lymphocyte-like blast cells by their intense uptake of H^3-labelled uridine (Fänge and Edström, 1973). The identification of cells functionally equivalent to the lymphocytes of higher vertebrates is made more difficult by the fact that these, unlike the lymphocytes of the lamprey, are relatively insensitive to radiation treatment (Good et al., 1972; Finstad et al., 1969). Macrophages are present in the peritoneal fluid and have been activated by thioglycollate treatment (Thoenes and Hildemann, 1970). Ultrastructural studies on the granulocytes of the intestinal wall of *Myxine* (Östberg et al., 1976) suggests that these animals, unlike the lamprey, may

possess only a single type of granulocyte with heterophilic granules and
equivalent to the neutrophil of the higher vertebrates.

The oxygen capacity of cyclostome blood appears to be lower than that of
active teleosts and for the lamprey, *L. (Entosphenus) tridentata* this has been
put at 9.15 vol. %. Haematocrits, often used as an indication of oxygen
capacity, vary considerably throughout the life cycle of the lamprey and show
a dramatic fall during and after spawning, when the haemopoietic activity of
the fat column is declining. For the ammocoete, values of 24.7 and 28.7 % have
been recorded, but in adult lampreys records vary from 21.5–64 %, coming
within the range encountered in active teleosts and higher than the haematocrits of lungfishes or elasmobranchs. In spite of this being a time of intense activity, the spawning period is characterized by a massive destruction
of blood cells in the liver and kidney, accompanying the final involution of the
haemopoietic tissue of the fat column. As a result, haematocrits decline
sharply and the blood of some spawning or spent lampreys may become
virtually colourless. Meanwhile, bile pigments derived from the breakdown of
the haemoglobin may be responsible for the change in liver colour from
orange to green that occurs in some animals (Section 7.3.4). In the sea
lamprey, *P. marinus* the skin of the sexually mature animal changes from a
greyish metallic blue to orange and this also has been attributed to the
accumulation of biliverdin (Kott, 1973). Haemoglobin concentration in the
lamprey ($5.8–13.0$ g 100 ml^{-1}) are comparable with the range in active
teleosts ($7–17$ %), but reports for several species of hagfish suggest that these
animals may have distinctly lower haematocrits ($20–25$ %) and lower
haemoglobin concentrations (4.6 g 100 ml^{-1}) than adult lampreys.

6.3 Cyclostome haemoglobins

6.3.1 Physiological properties

In the structure and properties of their haemoglobins, the cyclostomes have
been regarded as the most primitive of vertebrates. Whereas the haemoglobins
of the higher vertebrates are composed of four chains $\alpha_2 \beta_2$ with four haem
oxygen binding groups and molecular weights of about 68 000, the haemoglobins of lampreys and hagfishes are monomeric, with only one haem group
to each molecule and their molecular weights are only about one quarter those
of the higher vertebrates. Not surprisingly, in view of their monomeric
character, cyclostome haemoglobins, and more especially those of hagfishes,
show at best only slight indications of the allosteric effects that have developed
in the multi-chained haemoglobins of the higher vertebrates, including co-
operativity between the haem binding sites and the modifying effects of pH
and CO_2 (Bohr effects).

In relation to the oxygen transport functions of the blood and the
conditions under which oxygen is taken up by the respiratory organs and

unloaded to the tissues, the important factors are defined by the oxygen dissociation curves. Among these are the shape of the curve for percentage oxygen saturation (y) plotted against partial oxygen pressure (p) (Fig. 6.4), the oxygen affinity of the haemoglobin (partial oxygen pressure at 50% saturation – P_{50}), the effects of pH and CO_2, and temperature dependence and allosteric effects, involving substances such as the organic phosphates which may modify the oxygen equilibrium. When plotted as log $y/100-y$ against log p, the data fall approximately on a straight line (except at extremes of high and low saturations) and the slope of the line n or sigmoid coefficient, expresses the degree of interaction between haem binding sites or the degree of co-operativity between the sub-units, such as occurs in tetrameric vertebrate haemoglobins, where oxygenation of one haem group favours the oxygenation of another. At values for n close to unity, as in the monomeric haemoglobin or myoglobin, the dissociation curve has the form of a rectangular hyperbola, but above values of about 1.6 the curve becomes recognizably sigmoidal in shape. This has important consequences in the greater efficiency with which the haemoglobin takes up or unloads oxygen to the tissues. Thus a shift of the sigmoid curve to the right and a decrease in oxygen affinity, means that at a given pO_2 it will deliver more oxygen to the tissues than in the case of the hyperbolic curve of the monomer and is better adapted to unloading its oxygen over a wider and higher range of oxygen tensions. At the same time, it has been emphasized that from the standpoint of oxygen transport, it is not the shape of the curve in itself that is significant, but rather the influence of this shape on the actual amounts of oxygen that are transported. The higher the range of oxygen tensions over which unloading occurs, the greater will be the gradient between blood and tissues and the more efficient the delivery. Thus, with or without co-operativity, the further the curve is shifted to the right, the lower the oxygen affinity and the greater the volumes of oxygen that can be delivered (Fig. 6.3).

Although they share the monomeric character of invertebrate globins, the physiological properties of the cyclostome haemoglobins (and more especially those of the lampreys) are unique in showing a significant degree of co-operativity ($n > 1.0$) and distinct Bohr effects. The latter is of great importance in coupling oxygen and CO_2 transport; the CO_2 produced by tissue metabolism favouring the delivery of oxygen, while oxygenation in the respiratory organs promotes the removal of the CO_2. Compared to the monomeric invertebrate globins or the myoglobin of vertebrates which have $n = 1.0$ and high oxygen affinities, the cyclostomes may be regarded as transitional to the heterotetrameric haemoglobins of higher vertebrates, where n is usually greater than 2.5 and oxygen affinities are low. Although their possession of a monomeric haemoglobin might be thought to exclude haem interactions, the existence in lampreys (and to a lesser extent in hagfishes) of haemoglobins which show values of n in excess of unity,

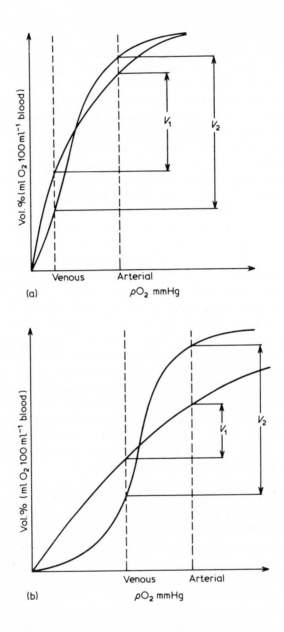

Fig. 6.3 The influence of co-operativity on the volumes of oxygen delivered to the tissues
in haemoglobins of high (a) and low (b) oxygen affinity. After Coates, 1975.
V_1 volume of oxygen delivered without co-operativity.
V_2 volume delivered by a haemoglobin showing co-operativity.

especially in concentrated solutions or at lower pH, indicates some degree of co-operativity. These properties are explained by the tendency for the monomeric globin to aggregate to form homodimers and temporary tetramers (Behlke and Scheler, 1970; Andersen and Gibson, 1971). The equilibrium between these aggregated sub-units and the monomers is dependent on oxygen tension; deoxygenation or reduced pH favouring aggregation, whereas oxygenation promotes the dissociation of the sub-units. In the higher vertebrates, the oxygen affinity of the haemoglobin is controlled by organic phosphates whose binding is dependent on pH, thus increasing the Bohr effect. Neither in lampreys nor in hagfishes do these compounds affect the oxygen equilibria and this form of regulation must therefore have been evolved after the divergence of agnathans and gnathostomes.

Information now exists on the oxygen equilibria of a variety of lamprey species, both adult and ammocoete and on the haemoglobins of three species of myxinoids (Table 6.2). In drawing conclusions from data of this type, considerable care needs to be exercised because of the variety of techniques that have been used and the differences in experimental conditions, particularly in regard to temperature, haemoglobin concentrations or pH. All these factors may significantly modify the oxygen equilibria. For example, in solutions oxygen affinities are usually higher, Bohr effects reduced or abolished and haem interaction tends to be increased. The physiological properties of the haemoglobin are also modified by the intracellular environment of the erythrocyte, so that interpretations in terms of the whole animal should preferably be based on whole blood rather than on haemolysates.

Among the more significant points to emerge from Table 6.2 are the following:

(1) In general hagfish haemoglobins have higher oxygen affinities than the haemoglobins of adult lampreys.
(2) Amongst lamprey haemoglobins, those of the ammocoete always show considerably higher oxygen affinities than those of the adult. (Fig. 6.4)
(3) In lampreys there appears to be a broad correlation between the adult oxygen affinities and the body size and migratory habits of the different species. This may be illustrated by the series *P. marinus*, *L. tridentata*, *L. fluviatilis*, *I. unicuspis* and the non-parasitic *L. lamottenii*. Here the P_{50} ranges from about 17–20 mm Hg in the large, wide-ranging *P. marinus* or *L. tridentata* to 2.0–8.0 mm Hg in the dwarf and non-migratory brook lamprey.
(4) The haemoglobins of the three hagfish species show distinct levels of oxygen affinity. The lowest values are those of *E. burgeri*, which vary from 6–25 mm Hg depending on pH and concentration. In *M. glutinosa* two independent reports have given values of 4.2 and 9.0 mm Hg, but the highest oxygen affinity is almost certainly that of *E. stouti* (2.4 mm Hg).

Table 6.2 Some physiological properties of cyclostome haemoglobins.

	n	Bohr effects $\log P_{50} \log pH$	Oxygen affinity P_{50} mm Hg	Conditions	Authors
Myxinoids					
E. stouti	0.9–1.1	−0.2	1.6–2.2 (pH 7.5–6.2) 2.0–4.0	haemolysates 0.2 % Hb, 20°C	Li et al., 1972
	1.0	None		haemolysates and cell suspensions 5–15°C	Manwell, 1963
E. burgeri	1.02–1.2 (pH 7.8–7.0)		8.6–11.4 (pH 7.8–7.0)	haemolysate 0.59 mM, 22°C	Bannai et al., 1972
M. glutinosa	1.04	−0.07	4.2 (pH 7.3)	haemolysates 0.1 mM, 20°C	Bauer et al., 1975
	1.0 (pH 7.0)	Slight	9.0 (pH 7.0)	haemolysates 3.5 % Hb, 21–25°C	Manwell, 1963*
Lampreys					
P. marinus adult	1.2–1.4 (pH 7.3)	−0.7	18.0	haemolysates 4.0 % Hb and cell suspensions 25°C	Manwell, 1963*
ammocoete	1.9	−0.7	4.0	cell suspensions 25°C	Manwell, 1963*
L. fluviatilis adult	1.21–2.32	−0.22	10.7 (pH 7.75)	cell suspensions 10°C electrolytic technique	Bird et al., 1977
ammocoete	1.02–1.49	−0.25	1.9 (pH 7.75)	electrolytic technique	Bird et al., 1977
L. tridentata	1.9	−0.41	17.0	haemolysates	Johansen et al., 1973
L. japonicus	1.54	−	17.0	haemolysates. 0.3 mM Hb, 20°C	Dohi et al., 1973

ammocoete	1.8–1.9	Slight	10.0 (pH 7.7)	erythrocyte suspensions 15.5°C	Potter et al., 1970
L. lamottenii	1.8–1.9	—	9.0	erythrocyte suspensions 15.5°C	Manwell, 1963.

* In some cases these figures are estimates based on the author's text-figures.

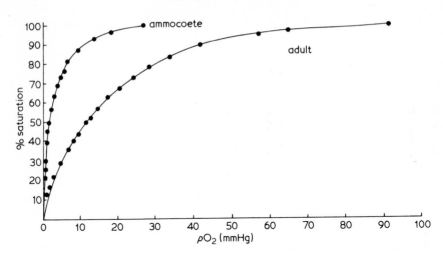

Fig. 6.4 Oxygen dissociation curves for the haemoglobins of ammocoetes and adults of the non-parasitic lamprey, *L. planeri*. From Bird, Lutz and Potter, 1976.

(5) As judged by the values for *n*, haem interactions are much more marked in lampreys than in hagfish haemoglobins.

(6) Bohr effects in the hagfishes are slight or absent, but appear to reach significant levels in lampreys.

In an aquatic vertebrate the haemoglobin must be adapted to combining with oxygen at the gills over the range of partial pressures that exist in the water of its environment, while at the same time being able to unload its oxygen at tissue pressures compatible with their metabolic rates. Hence, the more active species tend to have haemoglobins with low oxygen affinities and large Bohr effects, while the converse tends to apply in more sluggish species living in water relatively depleted of oxygen. Sigmoidal oxygen dissociation curves, which tend to be more pronounced in the active species, can be regarded as a compromise between the high oxygen affinity needed for loading in the gills and the lower affinity that would favour the delivery of oxygen to the tissues. For example, the carp, a more sluggish teleost living in still water with lower oxygen tensions, has a high affinity haemoglobin with a P_{50} of 4.0 mm Hg, whereas the active trout has a low affinity haemoglobin with a P_{50} of 38 mm Hg. It has also been maintained that Bohr effects tend to be smaller in species that live in environments where CO_2 tensions are variable.

Within the cyclostomes there are similar ecological parallels between the low metabolic rates, burrowing habits and environment of the hagfishes and larval lampreys, which are reflected in the comparatively high oxygen affinities of their haemoglobins (Fig. 6.4). In the case of the myxinoids, Manwell (1963) linked their slight or non-existent Bohr effects with what he termed their 'saprophytic' mode of feeding. This may involve the hagfish burrowing into the carcasses of animals, where they may be confronted by high or variable CO_2 tensions. Much more significant Bohr effects have been recorded in the haemoglobins of larval lampreys, but although these animals spend most of their time beneath the mud surface, the tip of the oral hood usually projects from the mouth of the burrow and the respiratory current is therefore drawn direct from the water flowing above it.

In keeping with their much higher rates of oxygen consumption and highly active life, the haemoglobins of adult lampreys have much lower oxygen affinities, pronounced Bohr effects and values of *n* well in excess of unity. Indeed, the substitution of adult for larval haemoglobins at metamorphosis is a graphic illustration of the modification of the properties of haemoglobins in response to changes in ecological requirements (Fig. 6.4). Similar changes occur in other vertebrates, for example the change from foetal to adult haemoglobins in mammals or the transition from the larval haemoglobins of the tadpole to those of the adult frog. In lampreys, the larval and adult haemoglobins differ in their electrophoretic patterns (Adinolfi *et al.*, 1959; Uthe and Tsuyuki, 1967; Potter and Nicol, 1968), but it is not yet known whether there are corresponding differences in their primary structure.

Among the hagfish species there are significant differences in the properties of their haemoglobins, matched by differences in their primary structure

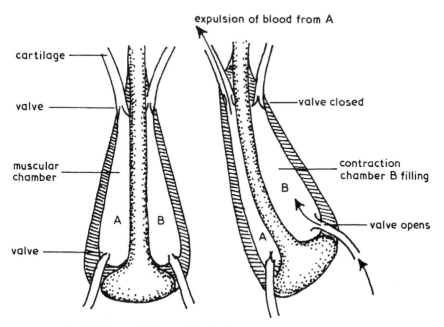

Fig. 6.5 The caudal heart of the hagfish. Redrawn from Jensen, 1965.

(Section 13.1 and Tables 6.2, 6.3). For example, the haemoglobin of *E. burgeri* which has the lowest oxygen affinity of the three species, also shows the most marked tendency for the aggregation of sub-units ($n = 1.2$). So little is known of the habits and environments of myxinoids that attempts to relate the properties of the haemoglobins of different species to their ecology must be highly speculative. Nevertheless, it may be recalled that *E. burgeri* is a species that inhabits shallow water and undergoes a seasonal migration; habits that could well be related to the distinctive characteristics of its haemoglobins.

In relation to the conditions under which blood gas transport occurs, the little information that we have on internal oxygen tensions points to an interesting contrast between the lampreys and hagfishes. In *Eptatretus*, Manwell found that the arterial blood was never more than 50% saturated, while the venous blood was almost completely deoxygenated. From the oxygen dissociation curve it was clear therefore, that blood gas transport would take place at very low oxygen tensions, normally less than 10 mm Hg. This would locate oxygen transfer on the steepest part of the hyperbolic curve. On the other hand, in *L. tridentata*, arterial blood saturations range from 94–100% and arterial tensions from 58–77 mm Hg. Venous blood had saturation values of 75% and oxygen tensions of 20–40 mm Hg (Johansen *et al.*, 1973). Such differences must obviously have important consequences for the relative volumes of oxygen transferred to the tissue in the two groups of cyclostomes.

6.3.2 Multiple haemoglobins

In common with many invertebrates, fish, amphibia and reptiles, the blood of cyclostomes usually contains a mixture of several haemoglobin species, differing in physico-chemical characteristics and often in amino acid composition (Chapter 13). The properties of the blood may therefore be a result of the interaction of the various components and many of the more recent biochemical studies on cyclostome haemoglobins have been concerned with the characteristics of these isolated haemoglobin species.

In the hagfish, *E. burgeri* four main components $F^1–F^4$ have been isolated (Bannai *et al.*, 1972) and yet a further component was identified in some animals. F^1 and F^2 are present in proportions of 17% and 13% and are virtually identical in their amino acid composition. In this respect, they differ from F^3 and F^4 which together make up about 35% of the haemoglobin. F^1 and F^2 have similar oxygen affinities, which are lower than those of F^3 and F^4 (Table 6.3). None of the isolated components showed a significant degree of haem interaction and when the various components were combined these interactions were only observed between F^3 and F^4. This combination produced higher oxygen affinities than any other component alone and had an n value of 1.31. From the sedimentation data it was inferred that this combination formed tetramers, which would therefore be hybrid molecules. Only F^3 showed any tendency to aggregate by itself. Unlike the haemoglobins of lampreys, which hardly aggregate at all in the oxygenated state, this did occur in *E. burgeri* at high concentrations, although in the deoxygenated state it occurs more readily at lower concentrations. Because of the aggregation behaviour of F^3 and F^4, it was suggested by the authors that these components might be comparable with the α and β chains of higher vertebrates, but we should be very cautious in drawing far-reaching conclusions from this type of experiment, where the conditions may be very far removed from those that exist *in vivo*.

In the hagfish, *E. stouti*, electrophoretic patterns have disclosed the presence of 2–3 major and at least two minor haemoglobin components (Ohno and Morrison, 1966; Li *et al.*, 1972) each differing in amino acid composition. As in *E. burgeri*, the sedimentation data indicate some aggregation in the oxygenated state. All these components differ in electrophoretic mobility, but have identical oxygen affinities (Table 6.3).

The three main haemoglobin components of *Myxine* (Paleus and Liljeqvist, 1972) have distinct oxygen affinities, but the value of *n* is close to unity and does not suggest the existence of haem interactions (Table 6.3). In whole haemolysates, the addition of CO_2 increases the small Bohr effect and also the value of *n*, indicating some aggregation of sub-units which is absent in the separate components (Bauer *et al.*, 1975).

Among the lampreys, no less than six haemoglobin species have been recognized in *P. marinus* (Rumen and Love, 1963), (Table 6.3) differing in sedimentation properties, isoelectric points, oxygen affinities, *n* values and in primary structure. Of these, fractions I and IV differ in oxygen equilibria from III, V and VI. In the deoxygenated state, IV, V and VI aggregate to form homo- and heterotetramers, but II does not do so and I and III are intermediate in character. In *L. fluviatilis* on the other hand, only two major components are present (Allison *et al.*, 1960) and these are identical in their primary structure, except that in one the N-terminal residue is blocked, while in the other it is free. In the closely related *L. planeri* also, two components can be recognized in electrophoretic patterns (Adolfini *et al.*, 1959) and it seems quite likely that these may have the same kind of relationship to one another as the components of *L. fluviatilis*. Two components are present in *L. (Entosphenus) japonicus* in proportions of 92.3% and 7–8% with different electrophoretic properties, but their amino acid composition seems to be identical to the haemoglobins of *L. fluviatilis* (Dohi *et al.*, 1973).

These examples of haemoglobin heterogeneity in the cyclostomes raise the much debated problem of its possible biological significance. While it may be natural to assume that the existence of these multiple haemoglobins in the lower vertebrates has some respiratory significance, perhaps as an adaptation to variations in ambient conditions, this would be alien to the strictly neutralist view, which maintains that a majority of duplicate or redundant gene copies have no particular adaptive significance, but permit the accumu-

Table 6.3 Some properties of the isolated haemoglobin components of cyclostomes. Data from Riggs 1970; Paleus *et al.*, 1971; Bannai *et al.*, 1972; Li *et al.*, 1972; Bauer *et al.*, 1975.

| | Hagfishes | | | | | | | | | | Lamprey | | | | | |
| | *Eptatretus burgeri* | | | | *Eptatretus stouti* | | | *Myxine glutinosa* | | | *Petromyzon marinus* | | | | | |
	F1	F2	F3	F4	A	B	C	I	II	III	1	2	3	4	5	6
$S_{20,w}$ deoxy	2.05	1.84	2.44	1.91	2.05	2.05	2.05	—	—	—	2.7	1.0	3.5	4.5	4.5	4.5
g/100g	17	13	35	35	—	—	—	27	33	40*	10.3	7.8	3.5	28.2	40.0	10.3
P_{50} mm Hg	20	20	6.7	4.8	3.16	2.69	2.69	3.7	2.3	6.3	2.52	—	6.61	9.55	6.76	6.32
n	1.0	1.01	1.09	1.05	1.0	1.03	1.02	1.03	1.03	1.03	1.1	—	1.4	1.55	1.4.	1.3
pl	—	—	—	—	—	—	—	4.5–5.0	6.3–7.2	8.1–8.8	4.5	4.3	5.4	4.7	5.8	6.0

* These *Myxine* components accounted for only 52% of the whole haemolysate.

lation of random mutations, free from the rigorous restraints of natural selection. However, the fact that haemoglobin heterogeneity is so widespread in aquatic poikilotherms does seem to indicate that it may be related to the greater environmental variability that confronts these animals, particularly in relation to oxygen tensions and temperature (Riggs, 1970). On the other hand, it has been pointed out that Antarctic fish, living in a relatively stable environment, also exhibit multiple haemoglobins (Grigg, 1974). Referring to the situation in the hagfish *Myxine glutinosa*, Bauer *et al.*, (1975) suggest that a rise in ambient temperature might favour the increased synthesis of Hb II, whose higher oxygen affinity could then counteract the impaired oxygen uptake due to the high heat of oxygenation. In the case of *E. stouti*, where the oxygen equilibria of the major components are almost identical and therefore would be neutral to selection, it has been emphasized that each has a different isoelectric point. Because of the Donnan equilibrium, the pH within the red cell would therefore vary with the changes in the proportions of the haemoglobins and, since a small Bohr effect is present, these could result in changes in oxygen affinity that might be of some adaptive value (Li *et al.*, 1972). This could only apply however, if the various haemoglobin species are present within the same erythrocyte.

6.4 Cyclostome hearts and vascular physiology

Relative to their body weight the cyclostomes have remarkably large hearts (Table 6.4). The heart ratio for the lamprey is greater than for any other vertebrate group, with the exception of mammals, and even the somewhat smaller heart ratio of *Myxine* is conspicuously greater than that of the teleosts. As a general rule, vertebrate heart ratios are broadly correlated with the activity and metabolic rates of the various species and this also applies within the life cycle of the lamprey, where in keeping with its lower metabolic rate and more sedentary habits, the heart ratio of the ammocoete is only about a half

Table 6.4 Vertebrate heart ratios. Data from Hesse, 1921; Poupa and Oštádal, 1969; Fänge, 1972; Claridge and Potter, 1974.

	Heart weight/Body weight × *100*
Lampreys	
ammocoetes	0.09–0.1
adults	0.25–0.41
Hagfish	0.16
Teleosts	
active species	0.09
sluggish species	0.54
Mammals	0.64

that of an adult lamprey (Claridge and Potter, 1974). Although the absence in the cyclostome heart of a coronary circulation might be expected to limit their capacity for sustained work at heavy loads, the characteristically spongy texture of the myocardium probably assists in the interchange between blood and muscle.

A noticeable feature of cyclostome hearts is the presence of large numbers of chromaffin cells (Section 11.3) and the high concentrations of catecholamines. In lampreys, the sinus venosus has high concentrations of both adrenalin and noradrenalin, whereas the atrium and ventricle contain mostly adrenalin. On the other hand, in *Myxine*, adrenalin predominates in the ventricle, but noradrenalin in the atrium and portal heart (Fänge, 1972). The hagfish heart is the only vertebrate heart to be completely aneural, whereas the lamprey heart is innervated by branches from the vagus. No evidence has been found for the innervation of the chromaffin cell system and although cholinergic fibres have been traced to neuromuscular endings on the muscle fibres of the sinus venosus, none have been identified either in the atrium or the ventricle (Beringer and Hadek, 1972).

In addition to the main systemic or branchial heart, hagfishes have several accessory contractile structures (cardinal, portal and caudal hearts) which tend to work independently of the systemic heart and which have different inherent rhythms of contraction (Fig. 1.2). Unlike the other accessory hearts, the portal heart is a true heart with muscular tissue of the same type as that of the systemic heart. It has been suggested that one of the main functions of this organ may be to assist in overcoming the capillary resistance of the liver (Fänge *et al.*, 1973). Its contractions begin in the proximal part of the supraintestinal vein, which has therefore been regarded as a pacemaker region. At temperatures of about 20°C, its frequency is said to average 47 beats min^{-1} falling to 24 beats min^{-1} at 10° C. The pressures that it generates are said to vary from 3–6 mm Hg (Johansen, 1960) and its output has been estimated at 0.1–0.3 ml min^{-1}. The caudal heart is a quite unique structure, located below the notochord and pumping blood from the large subcutaneous sinuses of the tail region into the caudal vein (Fig. 6.5). It consists of a median cartilage with two chambers, one on each side, communicating separately with the caudal vein and provided with valves, preventing reflux from the vein. Valves also occur at the openings from the heart chambers into the sinuses. The filling and emptying of the heart chambers is brought about by extrinsic muscles, attached at one end to the median cartilage and at the other to a posterior cartilaginous knob. By alternate contractions of the muscles of either side, the chambers are alternately compressed and dilated. Histologically, the muscles are of the skeletal type and are innervated from the spinal cord.

6.4.1 Circulatory dynamics
What fragmentary information we have on the intravascular pressures of the

cyclostomes, indicates that these are much lower in the hagfishes than in lampreys. This helps to explain the necessity for accessory hearts in the myxinoids. Under resting conditions systolic pressures in the systemic heart of the hagfish vary from about 1.0–8 mm Hg falling to zero at diastole (Johansen, 1963; Chapman *et al.*, 1963). Nevertheless, at maximum load with increased venous pressures, the heart is capable of systolic pressures of 20–30 mm Hg. When the aortic valves are open, the pressure drop in the ventral aorta is negligible, but the gill capillaries impose a considerable resistance between the dorsal and ventral aortae. As a result, pressures measured in the dorsal aorta were only about 3–5 mm Hg and the pulse pressure was of the order of 2 mm Hg. The average pressure drop in the gill capillaries is said to be about 2–4 mm Hg. In *Myxine*, Johansen detected a distinct pulse in the dorsal aortic pressures, which he attributed to the rhythmic contractions of the gill pouches, although this was not observed by Chapman *et al.* (1963) in *Eptatretus*. No measurements of intracardial pressures have been made on lampreys, but pressures of 18–34 mm Hg have been recorded in the dorsal aorta (Johansen *et al.*, 1973). Making allowance for a pressure drop in the gill capillaries of the same order as in teleosts (40–50 %), this would suggest that the pressures in the ventral aorta and ventricle may be within a range of 30–60 mm Hg at systole. The recorded pressure in the dorsal aorta is similar to that of teleosts and very much higher than in the hagfish. Furthermore, pulse pressures in the lamprey are said to be about 9 mm compared to only 2 mm in the myxinoid. Although there are no records of vascular pressures in the ammocoete, the relatively small size of its heart suggests that these would be lower than in the adult lamprey. In addition, the extensive changes that have been described in the larval circulation at the time of metamorphosis, including improvements in arterial supply and venous return would almost certainly lead to a more efficient circulation (Percy and Potter, 1979).

In elasmobranchs and lungfishes where the heart is contained within a non-compliant pericardium, venous pressures around the heart are usually negative, favouring the aspiration of the venous blood on its return to the heart. Negative pressures of -0.7 to -3.0 mm Hg have also been recorded in the lamprey, which also possesses a rigid closed pericardium.

Heart rates in the systemic heart of the hagfish are said to vary from 18 beats min^{-1} at 11°C to 26 at 18°C, whereas in the isolated heart the rates may be restricted to 5 min^{-1} at 8°C rising to 12 beats min^{-1} at 15°C and to 24 at 20°C. Frequencies tend to be higher in the portal heart and in intact animals, rates of 24 min^{-1} at 11°C and 30 beats min^{-1} at 18°C have been recorded. The caudal heart tends to beat intermittently and irregularly, but when it is contracting it may show a rate as high as 88 beats min^{-1} at 16°C (Chapman *et al.*, 1963). Being no larger than a pinhead, this structure can make only a local and minor impact on the circulation and no obvious ill effects are said to follow its removal.

In line with their higher metabolic rates, the heart rates of lampreys are considerably higher than those of the hagfish. Frequencies of 40–60 min^{-1} have been recorded for the isolated heart of *L. fluviatilis* and in anaesthetized or decapitated specimens of *L. tridentata*, rates rose from 25 min^{-1} at 5°C to 110–130 beats min^{-1} at 25°C. In *L. fluviatilis*, the heart rates of intact animals showed a steady increase throughout the migratory period and at a constant temperature of 16°C increased from 33 beats min^{-1} in December to nearly 50 beats min^{-1} in March. Parallel with these increases in heart rate there were similar increases in ventilatory frequency and standard oxygen consumption (Claridge and Potter, 1975). Judging from the limited information obtained on *L. tridentata*, the cardiac output of the lamprey heart in resting animals (32 ml kg^{-1} min^{-1}) comes well within the range recorded in teleosts (5–100 ml kg^{-1} min^{-1}) or elasmobranchs (21–25 ml kg^{-1} min^{-1}).

6.4.2 Circulatory control

In keeping with the habits of an animal that may spend most of its life in almost total quiescence, punctuated by short bursts of vigorous activity, the systemic heart of the hagfish shows exceptionally low systolic pressures. In higher vertebrates, cholinergic and adrenergic cardiac nerves regulate the heart in response to varying demands, either through changes in the heart rate (chronotropic effects) or by alterations in the force of its contractions (inotropic effects). Except for embryonic hearts, the systemic heart of the myxinoids is unique in the absence of these cardiac nerves, but it is still able to adjust its activity to meet environmental demands or sudden increases in its locomotory activity.

In the resting condition or when the heart is isolated, it is purely myogenic with its own inherent rhythmic activity. This is attributable to the spontaneous firing of pacemaker cells, widely dispersed throughout the cardiac tissue. If a ligature is tied between atrium and ventricle so as to effectively separate the two chambers, the ventricle shows a decreased rate of contraction. Thus, each chamber has its own pacemaker activity, but while this is of similar magnitude in the atrium and sinus venosus, it appears to be weaker in the ventricle (Bloom *et al.*, 1963). Electrical stimulation of this chamber has no apparent effect on its rhythm, but the atrium and sinus venosus show a considerable increase in the frequency of their contractions. Characteristic pacemaker potentials have been recorded by Arlock (1975) from cells in the atrium of the systemic heart of *Myxine*, but not from the ventricle. These show a depolarization phase of about 25 mV, following a slow depolarization, and during subsequent repolarization the potentials reached a higher value of about 30 mV.

The immediate stimulus that is responsible for raising the output of the hagfish heart is an increase in the backflow of venous blood (Fig. 6.6). The resulting distention causes an acceleration in heart rate and an increase in the

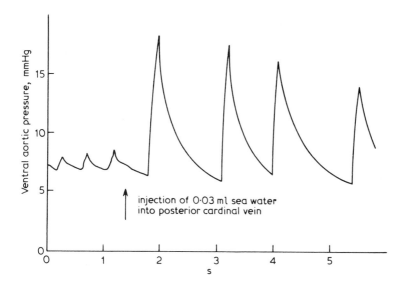

Fig. 6.6 The response of the hagfish heart to increased venous inflow. Redrawn from Chapman *et al.*, 1963.

force of its contractions, obeying Starling's Law of the Heart. This ability of cardiac pacemakers to alter their rate of spontaneous firing in response to changes in tension or pressure is widespread in vertebrates and has been referred to as 'intrinsic rate regulation'. This could be regarded as a primitive condition, phylogenetically predating the development of neural control systems (Jensen, 1969).

Although the lamprey heart has a vagal nerve supply, a similar type of intrinsic rate regulation is still present (Jensen, 1969) and the heart rate is sensitive to moderate distention of the ventricle. The quiescent heart of *P. marinus* responds to an increase in the intraventricular pressure from 0–15 mm Hg by raising its rate of contraction from 24–44 beats min^{-1}, subsequently returning towards normal as the pressure decreases.

In response to stimulation, the portal heart of the hagfish behaves in a similar way to that of the systemic heart, showing acceleration with weak electrical stimulation and inhibition when the stimulation is more intense. Like the systemic heart, it also responds to increases in venous inflow or to slight mechanical stimulation by an increase in the rate of contraction and in stroke volume. Pacemaker potentials have also been observed in the portal heart and the pacemaker region is apparently located in the proximal part of the portal vein where this enters the heart. The caudal heart on the other hand, is under reflex control from the spinal cord and its contractions disappear when the latter is destroyed. It is also said to stop beating after intense mechanical stimulation of the skin surface or during vigorous movements of the body. Responses to postural changes have been observed as when the tail of the hagfish is raised above the level of its head, resulting in increased rates of contraction.

6.4.3 The pharmacology of cyclostome hearts

In the response of the heart to drugs, not only are there significant differences between the lamprey and the hagfish, but these responses are quite distinct from those seen in the hearts of fishes and higher vertebrates. For example, the aneural hagfish heart shows complete insensitivity to acetylcholine, whereas the heart of the lamprey responds to this substance, or to vagal stimulation, by increases in heart rate (positive chronotropy) and a diminution in the amplitude of its contractions (negative inotropy). This may be contrasted with the situation in teleosts and elasmobranchs, where vagal stimulation or the application of acetylcholine slows the heart rate (negative chronotropy) and reduces the amplitude of the contractions. Moreover, at least in some teleost species, there is evidence that the vagus may also contain some adrenergic excitatory fibres as well as the cholinergic inhibitory fibres (Randall, 1970).

Atropine (which blocks acetylcholine receptors of the muscarine type such as those of the vertebrate heart) increases the heart rate in fish, but in the case of the lamprey heart, the accelerating effects of acetylcholine are blocked by curare and not by atropine. This indicates that the acetylcholine receptors are of the nicotinic type. In fact, the application of nicotine mimics the positive chronotropic and negative inotropic effects of acetylcholine itself, whereas muscarine or pilocarpine are without effect. Thus the lamprey heart resembles that of a fish in its cholinergic nerve supply, but differs in the type of acetylcholine receptors.

The high concentrations of catecholamines and the large numbers of chromaffin cells in the cyclostome heart have been referred to earlier. The lamprey heart is particularly sensitive to reserpine, which depletes catecholamine stores and limits the uptake of these compounds. After treatment with reserpine, the heart stops in diastole, although its contractions can be restored by the application of acetylcholine. Large doses of noradrenalin exert positive inotropic and chronotropic effects on the heart of the lamprey, acting *via β*-adrenergic receptors, but these effects are never as pronounced as those observed with acetylcholine. However, even in high concentrations, catecholamines are without any effect on the hagfish heart. At the same time, reserpine treatment may reduce the normal response to increased venous inflow and this response can to a certain extent be restored with the subsequent application of catecholamines. As is the case in the lamprey, large doses of reserpine stop the hagfish heart in diastole, but in this animal, contractions can then be restarted by catecholamines and not, as in the lamprey by acetylcholine. It has been suggested that acetylcholine may act indirectly on the lamprey heart through liberating catecholamines from the chromaffin cells, but the relative insensitivity of the heart to these compounds presents some difficulties for this interpretation (Lukomskaya and Michelson, 1972). On the other hand in the isolated ventricle of the ammocoete, colchicine (known to inhibit the release of hormones and neurotransmitters, including catecholamines) has been found to inhibit the chronotropic effects of acetylcholine (Lignon, 1975).

From the hagfish heart, Jensen (1963) isolated a highly active cardioactive agent which he called eptatretin. This is apparently peculiar to the hagfish and is believed to be an unstable amine, but not a catecholamine. Although its role in cardioregulation is unknown, Jensen suggested that it might facilitate the repolarization of the pacemaker potentials to the threshold levels required for their spontaneous discharge.

6.4.4 Cardiovascular responses of lampreys during activity and under conditions of oxygen depletion

In the lamprey, *L. tridentata*, Johansen *et al.* (1973) have been able to analyse some of the cardiovascular changes that occur during exercise or in hypoxia, (Table 6.5). With decreasing oxygen concentrations, both arterial and venous oxygen tensions declined and the differences between them were reduced, as was the blood/water oxygen gradient. This reduction in blood oxygen tensions and particularly the venous tension, means that these are now positioned on the steepest part of the oxygen dissociation curve and oxygen utilization (expressed as the percentage difference in arterial and venous oxygen tensions) increased some three- or four-fold. A very significant factor in the resistance to hypoxia is the high degree of venous saturation. This is said to act as a reservoir compensating for the reduced availability of oxygen in the water.

During activity, the venous oxygen tensions decrease to a greater extent than the arterial tensions (Table 6.5), but because of their positions relative to the oxygen dissociation curve and the relatively high oxygen affinity of the haemoglobin, the reduction in arterial tension has little effect on arterial saturation. As a result, oxygen utilization is approximately doubled.

Nothing in our fragmentary knowledge of respiratory or cardiovascular physiology in the lampreys seems to explain their relatively poor swimming performance and lack of staying ability (Section 4.4.1). Structurally, the gills are very similar to those of teleosts and their areas are comparable with those of the most active fishes. The tidal respiration of the adult lamprey might be more energy costly, but as yet we have no evidence to support this view. Neither does the relative size of the heart or its functional capacities encourage the belief that cardiovascular factors are likely to limit the activities of these animals. Perhaps, as Beamish (1974) has suggested, the answer may lie more in the direction of hydrodynamic factors or even in some biochemical or physiological characteristics of the parietal muscles. Attempts to compare the

Table 6.5 Cardiovascular changes in resting and active lampreys (*L. tridentata*) and under reduced oxygen tensions. Data taken from Johansen *et al.*, 1973 and representing mean values. For the hypoxia experiments, the figures under (a) are for the start of the experiment and under (b) for the same animals at the conclusion.

	Water oxygen tension mm Hg	Arterial oxygen tension mm Hg	Venous oxygen tension mm Hg	Blood/ water gradient	% oxygen saturation		% oxygen utilization
					arterial	venous	
Resting	114	65.1	33.5	49.0	97–100	75	24.8
Active	105	49.1	13.5	57.8	–	–	58.7
Hypoxia (a)	115	75.0	47.0	40.0	96	87	8.9
(b)	32	13.0	8.8	19.0	37	23.5	30.1

energy costs of swimming in lampreys and teleosts have been complicated by uncertainty over the distances actually travelled by the migrating sea lamprey (Beamish, 1979), but assuming that they follow a direct route, the energy cost has been calculated as $1.31-1.48$ cal $g^{-1}km^{-1}$. This would be much higher than the figure of $0.33-0.42$ cal $g^{-1}km^{-1}$ reported for the eel, using rather similar swimming mechanics (Schmidt-Nielsen, 1972).

7

The skeleton and the muscular system

7.1 The cranial skeleton

7.1.1 The neurocranium

In comparison with the difficulties that we face when attempting to relate the splanchnocranium of the cyclostomes to that of the gnathostomes, the neurocranium presents relatively few problems of interpretation and, in its broad outlines, this structure conforms to a pattern that is readily recognizable in the skull of higher vertebrates, consisting essentially of the sense capsules, parachordals and trabeculae (Fig. 7.1). In its simplest form, in the ammocoete, the brain is supported ventrally by a cartilaginous framework attached to the cranial extension of the notochord, with paired parachordals continuing forwards as the 'trabeculae'. The latter diverge again in the midline to form a hypophysial fenestra at the site of the pituitary and where the trabeculae and parachordals meet, there is a small, basitrabecular process. The ear capsule is united to the parachordals, but the small olfactory capsule is connected to the membranes covering the brain only by fibrous tissue. In the course of metamorphosis, the trabeculae extend to form lateral cranial walls and the auditory capsules are united by a cartilaginous bridge (tectum synoticum). The anterior part of the hypophysial fenestra is reduced by the development of an intertrabecular plate and the developing eye becomes supported by a massive sub-ocular arch formed from an extension of the basitrabecular process (Fig. 7.2)

In the myxinoids, the trabeculae arise from the front of the otic capsule rather than directly from the parachordals. Below the pituitary there is a sub-hypophysial plate, connected anteriorly to the nasal capsule. The latter is a hemicylindrical framework of longitudinal rods, joined in front and behind by half hoops of cartilage. Peculiar to the hagfishes is an elongated cylindrical framework supporting the nasohypophysial canal and below this is a massive sub-nasal cartilage. The homologies of these specialized structures remains obscure and their inclusion in the neurocranium is quite arbitrary.

118

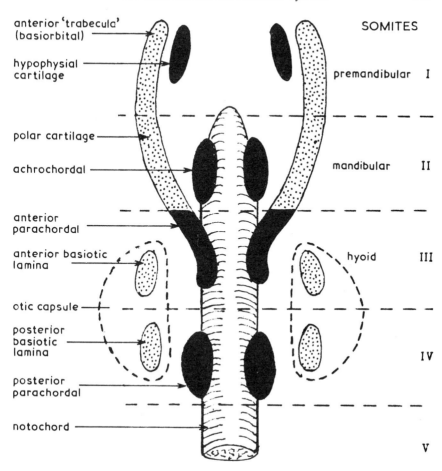

anterior 'trabecula' (basiorbital)

hypophysial cartilage

polar cartilage

achrochordal

anterior parachordal

anterior basiotic lamina

otic capsule

posterior basiotic lamina

posterior parachordal

notochord

SOMITES

premandibular   I

mandibular   II

hyoid   III

IV

V

Fig. 7.1 Mesodermal elements of the lamprey cranium and their embryonic sources (After Bjerring, 1977).
The numbering of the somites does not follow Bjerring's scheme which recognizes a terminal segment anterior to the premandibular.
Myotomic derivatives indicated by dotted areas; sclerotomic elements in black.

The precise homologies of the cyclostome trabeculae have been disputed, but it now seems probable that at least a part of these structures may correspond to the gnathostome polar cartilages, lying between the cranial end of the parachordals and the trabeculae (Section 3.2). An alternative view is that the true trabeculae of the lamprey may be represented by the transverse commissure linking the cranial ends of the 'trabeculae' and which are also present in the myxinoid embryo. This interpretation is strengthened by the claim that at least in the lamprey, this commissure is ectomesenchymal in origin. Jarvik (1964) believes that as in the gnathostomes, the cyclostome

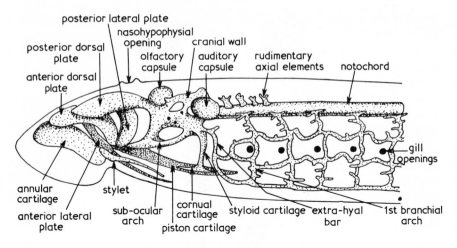

Fig. 7.2 Skeleton of the head and branchial region of a lamprey.

visceral arches were originally made up of three separate elements, which in the cephalaspids had fused to one another and to the cranium to form a continuous unit. The medial part of the mandibular arch (infrapharyngeal) still lay in its original position and would correspond to the trabeculae. True trabeculae, in the sense of paired longitudinal rods would therefore have been absent. If the condition in the cyclostomes has followed that of the cephalaspids, Jarvik's view would support the homology of the trabecular commissure with the gnathostome trabeculae.

7.1.2 The splanchnocranium

In the ammocoete, the head skeleton is for the most part composed of a unique tissue – mucocartilage – derived from mesectoderm and presumably of neural crest origin. In early embryos, this tissue forms a continuous layer covering the head region, later separating into dorsal and ventral areas around the mouth and pharynx. The mucocartilage of the ammocoete, although more localized, forms a kind of supporting collar around the oral hood, vestibule and buccal cavity giving these areas the necessary elasticity to act antagonistically to the muscles of these regions.

During metamorphosis, the cartilaginous skeleton of the adult lamprey develops in several ways. In some areas, the mucocartilage of the ammocoete breaks down and is replaced, sometimes quite independently, by hyaline cartilage. In other cases, larval mucocartilage may be transformed directly into adult cartilage. Finally, some adult cartilages are neoformations developed from embryonic blastemas.

During metamorphosis, the major modifications take place in the buccal region involving the elongation of the prenasal region and the transformation of the oral hood into the suctorial funnel of the adult. The latter is supported by an annular cartilage, carrying slender stylets (Fig. 7.2) serving as attachments for muscles that are able to alter the position of the funnel relative to the body axis. The roof of the funnel is supported by an anterior dorsal plate which can be moved vertically to assist in creating the necessary suction pressure for attachment. Behind this are anterior lateral cartilages, whose compression occludes the passage between the oral funnel and the buccal cavity. A similar function can also be attributed to the large posterior dorsal cartilage, joined behind to the trabecular commissure and forming the roof of the mouth cavity. The movements of these dorsal plates relative to one another are also involved in the changes in the volume of the hydrosinus that plays such an important part in the suction mechanism (Section 5.1.1 and Fig. 5.2). A further pair of posterior lateral cartilages, attached by elastic tissue to the subocular arch, are concerned in lateral compression of the buccal cavity.

The skeleton of the branchial region remains basically similar to that of the ammocoete. The most important changes are the anchorage of the whole structure to the cranial skeleton through the union between the dorsal longitudinal bars and the parachordals. At its anterior end, the branchial skeleton now shows an additional arch, passing forwards and upwards from the first branchial arch to unite with the cranium at the point where the velar arch (styloid cartilages) join the sub-ocular arch (Fig. 7.2).

With a few exceptions, all the cartilages of the adult myxinoid skull are fused to a form a continuous structures (Fig. 7.3). From the front of the 'trabeculae' extends the dorsal longitudinal palatine bar, joined to the auditory capsule behind, and separated from it by a fenestra through which the facial nerve (VII) emerges. At their cranial ends these longitudinal plates are joined by the palatine commissure, carrying the palatine tooth. This palatine bar may correspond to the subocular arch of the lamprey, where it

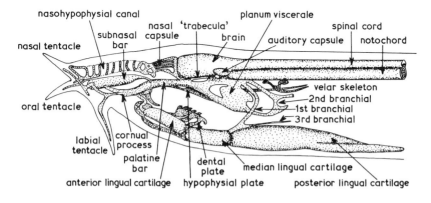

Fig. 7.3 Head cartilages of a hagfish. Based on Cole, 1905.

forms the floor of the orbit, and the subocular fenestra of the latter would then have its counterpart in the fenestra between the palatine bar and 'trabeculae' of the hagfish, both of which carry the main roots of the trigeminal (V). The complex 'tongue' skeleton is attached to the first branchial arch through the posterior end of the median lingual cartilage and in front of the latter are two pairs of anterior lingual cartilages, one medial and two lateral. These have been compared with the medial, apical and lateral apical cartilages of the lamprey tongue skeleton (Holmgren, 1946), while the paired middle and single posterior segment of the hagfish apparatus have been likened to the anterior 'head' and the main body of the lamprey piston cartilage. Jarvik (1965) on the other hand has discounted a strict homology between the two structures, although he allows the possibility that both may have been derived from some basal elements of the visceral arch system of the ancestral vertebrate.

In the hagfish the posterior and functional gill sacs have no branchial skeleton, other than cartilaginous rings around their external openings. The first so-called branchial arch attached above to the otic capsule and below to the median lingual cartilage has been compared to the first visceral arch of the lamprey, both lying outside the corresponding nerve and therefore referred to as extrabranchials. The second arch of the hagfish is united to the first and joins the posterior angle of the velar plate (planum viscerale) supporting the walls of the buccal cavity in front of the velar folds. The third rudimentary arch emerges from the posterior end of the median lingual plate.

Interpretations of the cyclostome skull have been based mainly on extrapolation from conditions in the gnathostomes and this has led many authors to search for the same visceral arch elements that are found in the skulls of the jawed vertebrates. The rationale for this approach is at least open to question. Whether the fused and continuous branchial skeleton of the cyclostome or the cephalaspid is a primitive or as Jarvik has suggested, a secondary condition is difficult to decide. What is clear is that the buccal regions of ammocoetes, lampreys and hagfishes have been adapted to quite different functional requirements and as Damas (1958) has stressed, the search for mandibular or even premandibular arch elements in the cyclostome cranium can hardly be justified in a vertebrate that never possessed true jaws. It may be legitimate and convenient to describe the skeletal components of particular regions of the head as premandibular, mandibular or hyoidean, but speculation on their precise homologies with the visceral arch elements of gnathostomes is hardly likely to be fruitful.

7.2 Phylogenetic considerations

While there can be no doubt that many of the peculiar features of the cyclostome splanchnocranium have arisen as adaptations to their highly specialized mode of feeding, some of the difficulties in detecting a common organization plan in the cyclostome cranium may be due to the extensive evolutionary regression that has probably occurred. In lampreys, the presence of the rudiments of an axial skeleton in the form of vestigial cartilages around the notochord, seems to offer clear evidence of such secondary reduction. Skeletal regression is a well-established phenomenon in the evolutionary history of several major vertebrate groups (Romer, 1968) and Orvig (1968) believes that the Devonian cephalaspids also show evidence of a reduction in the bony exoskeleton. This trend is believed to have involved a reduction in the middle vascular layer of the corium in the trunk region behind the cephalic shield. During its metamorphosis, the corium of the lamprey also exhibits three zones; the middle of which is a vascular layer. Behind the head this layer is absent, suggesting that the naked condition of the petromyzonids may represent the final stage in an evolutionary reduction that had already begun in its middle Devonian cephalaspidomorph ancestors.

In its distribution and development, the unique mucocartilage of the ammocoete has sometimes been regarded as a relic of a continuous endoskeleton of the cephalaspid type. Such comparisons are really more appropriate to the mesectoderm that precedes ontogenetically the differentiation of the mucocartilage, and which is presumably derived from the neural crest. As Damas has shown, this tissue does in fact, form a continuous envelope, covering at its maximum extent, the cephalic and branchial regions. In later stages, its distribution becomes more restricted through the increased development of other structures, including the musculature. The relationship between the mucocartilage of the ammocoete and the adult cartilages are still far from clear. Earlier work had suggested that mucocartilage was entirely converted into true cartilage, but other authors have maintained that cartilage cells never arise directly from the tissue of the mucocartilage. The truth perhaps may lie between these two extremes. Certainly, the basal cartilages of the skull and branchial regions are formed before the embryonic mesenchyme has differentiated into mucocartilage. What is more, adult cartilage appears during metamorphosis, in places where no mucocartilage exists and at some sites, cartilage may apparently be formed within the degenerating larval muscles. Johnels (1948) considered mucocartilage to be a specialized larval tissue, developed secondarily to cartilage in an evolutionary sense and thus should not be regarded as an ontogenetic stage in the formation of true cartilage. Phylogenetically therefore, mucocartilage would be of more recent origin than cartilage and in this respect the adult lamprey would be more primitive than its larval stage. Like the rudiments of the paired eyes, the larval mucocartilage could thus be thought of as a kind of persistent embryonic primordium, whose further development has been suspended (Section 12.3). This view is re-inforced by the relatively late histological differentiation of the mucocartilage, which for a long period, retains the characteristics of an embryonic mesenchyme. It is also consistent with the histochemical development of the tissue. Throughout the whole of the larval period, the intercellular matrix is characterized by the presence of hyaluronic acid typical of a loose connective tissue. In older ammocoetes, chondroitin sulphates, characteristic of cartilage matrix (Section 13.8.1) begin to appear in the tissue, although this is not yet reflected by changes in its morphological character. Significantly, as the animals approach metamorphosis, there is an increase in the synthesis of chondroitin sulphates, showing that the cells are beginning to differentiate in the direction of cartilage (Mangia and Palladini, 1970).

Apart from the question of the distribution of mesectoderm or mucocartilage, there are a number of other indications that the cartilaginous skeleton of the present day cyclostomes may be a relic of what was at one time, a more extensive exoskeleton. The great variability that is seen in the detailed morphology of the prebranchial arches in myxinoids, or in the branchial basket and vertebral rudiments of lampreys, is consistent with the variability that is characteristic of many vestigial structures no longer subject to selective pressures. In lampreys, particularly during metamorphosis, there are tendencies towards aberrant cartilage development, involving in some cases, reduction of existing cartilages and in other cases, their transformation into connective tissues (Johnels, 1948). In addition, at some sites, blastemas, otherwise identical to those which produce cartilage, may fail to do so and subsequently regress.

The existence in the petromyzonid ancestry of a massive bony cranium of the cephalaspid type has been questioned because of its implications for the musculature of the head (Damas, 1944). As in the gnathostomes, the presence of an immoveable occipital region of the skull might be expected to have led to the disappearance of the myotomes of the post-otic region. However, in the head of the lamprey the myotomal regions of the 4th and 5th head segments remain functional, projecting forwards over the dorsal surface of the head, and it is partly from these post-otic myotomes that the unique corneal accommodatory eye muscles are developed (Section 1.8.1). If, as Janvier (1975) believes, these muscles were absent in the cephalaspid eye, this would tend to support the view that the corresponding myotomes had already disappeared from the occipital region. Thus, as Damas (1944) has pointed out, the derivation of petromyzonids from cephalaspid-like ancestors would appear to be contrary to Dollo's law, since it would involve the reappearance in phylogeny of structures that had previously been suppressed. On the other hand, because of the entirely different position of the branchial region and the cranium in lampreys and cephalaspids, the relationships of the myotomes and the skull may also have been quite different. The absence of the forward extension of the epibranchial parts of the myotomes in the cephalaspids need not necessarily imply that the corresponding myotomes had disappeared and their present projection over the head of the lamprey could have been a secondary development (see Section 3.3).

In the pteraspids, the existence of a phylogenetic trend towards skeletal regression is even more problematical and the naked condition of the myxinoids may well be a primitive character (Spjeldnaes, 1968), derived from an ancestral form (which may or may not have been related to the pteraspidorphs) which had not, at that evolutionary stage, acquired a rigid exoskeleton. In relation to the internal factors that may govern calcification, Moss (1968) has referred to the view that there may be a critical level of ionic strength in the body fluids, beyond which calcification does not occur. Living agnathans (presumably referring more specifically to the hagfishes) are said to have higher ionic strengths than either teleosts or elasmobranchs, which might have something to do with the absence of hard tissues in myxinoids, although in this case it would be necessary to assume lower levels of ionic strength in the ancestral and fossil agnathans with their calcified tissues.

Attempts have been made to identify in myxinoid embryos, a structure resembling the head shield of the pteraspids, but these are far from convincing. On the other hand, it does seem likely that the cartilaginous endoskeleton of the hagfish has undergone extensive modification and regression. This is obviously true of the visceral arches. Holmgren (1946) who accepted the relationship of the myxinoids to the pteraspidomorphs, explained the caudal position of the functional gill sacs as a result of the intercalation of myotomes between the anterior end of the branchial region and the skull, rather than to their backward migration. Such an extension of this 'neck region' could not have occurred in a pteraspid-like ancestor, where this part of the body was covered by a bony head shield. The backward movement of the gills and the extension of the prebranchial zone could therefore only have taken place in such ancestral forms if the dermal skeleton had already regressed.

7.3 The muscular tissues

The somatic musculature of the cyclostomes shows a primitive and complete segmentation, extending in a continuous series of myotomes from the tip of the tail to the head. The parietal muscles of the lamprey extend over the dorsal surface of the head as far forwards as the posterior dorsal plates and, in the hagfish, to the area of the nasal capsules. In lampreys, where the branchial region is placed immediately behind the head, the continuity of the myotomes is interrupted by the gill ports, which split up the myotomes into epibranchial and hypobranchial sections. This is not the case in myxinoids, even in those species with separate gill openings, and the gill ports are placed below the ventral margins of the parietal muscles. The intervening region is occupied by the complex longitudinal musculature of the tongue mechanism. In the hypobranchial zone of the lamprey the segmentation no longer corresponds with that of the epibranchial region above. In front of the first gill pore and under the eye, these muscles are connected indirectly to the annular cartilage through a sub-ocular muscle, controlling the side to side and vertical movements of the head. The small corneal accommodatory muscle attached to the outer sclerotic coat of the eyeball is also derived in part from the epibranchial sections of post-otic myotomes.

Unlike the parietal muscles of the lamprey that meet in the mid-ventral line, those of the hagfish end ventro-laterally at the level of the slime glands and the ventral surface is covered by a layer of unsegmented oblique muscles and segmental rectus muscles; an arrangement that helps to explain their ability to coil and contort their bodies, as well as the extreme extensibility demanded by their voracious feeding habits.

Contrasting with the lamprey, all the head muscles of the hagfish belong to the visceral division, originating from branchial mesoderm and not from the myotomes. Accordingly they are innervated by cranial nerves V–X. In both groups, the muscles that operate the complex feeding mechanisms are controlled by the trigeminal complex. Other important visceral muscles are those of the velum and branchial regions.

7.3.1 The organization of the myotomes

Cyclostome myotomes may be described as $\gtrless$ -shaped with one forward and two backwardly directed flexures. In addition, each myotome is directed obliquely backwards from the medial to the lateral surface, so that successive units are telescoped into one another and a single transverse section will pass through a number of successive myotomes (Fig. 7.4). Within each myotome, the muscle fibres run parallel to the long axis of the body and are attached at either end to the connective tissue septa dividing one myotome from the next. Acting antagonistically to the lateral contractions of the myotomes is the elasticity of the notochord, whose mechanical properties are due to the turgidity of the large vacuolated cells that make up its central core.

Each myotome is composed of a series of horizontal compartments, extending from the lateral to the medial surfaces. As in other lower vertebrates, two types of fibres can be readily distinguished within each compartment; transparent and colourless, centrally placed fibres and more superficial, opaque, brownish parietal fibres. These two types of fibre have quite distinct cytological, biochemical and physiological properties which have led to their characterization as white, fast or twitch fibres and slow, red non-twitch fibres. The superficial location of the slow fibres is common to a majority of fish species (Bone, 1966) in which they are generally found immediately below the skin.

The approximately rectangular muscle compartments of the hagfish consist of from 2–4 layers of fast fibres, stretching horizontally from the skin surface to the medial border of the myotome; the slow fibres forming a single layer over the ventral and lateral surfaces of each compartment (Fig. 7.5). In lampreys, each compartment is usually made up of four layers of fast fibres,

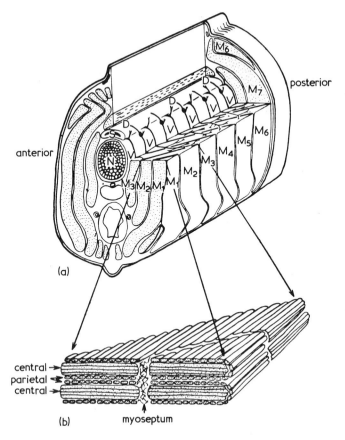

Fig. 7.4 Arrangement and composition of lamprey myotomes. From Peters and Mackay, 1961.
D. dorsal nerve roots; M. myotome; N.notochord; S. spinal cord; V. ventral nerve roots.

50–60 μm in diameter, devoid of a basement membrane and each consisting of some 30–40 individual fibres. Unlike the hagfish, the fast fibres are surrounded on all except the medial side by a layer of slow fibres with a diameter of about 65 μm and a common basement membrane. The larger diameters of the slow fibres is a general feature and seems to be even more marked in hagfish muscles.

Within the chordates, Flood (1973) has described a phyletic trend in the organization of the axial musculature (Fig. 7.5). In the cephalochordates, the myotomes are made up of a continuous system of muscle plates with no compartmentalization, although Flood suggests that differences in mitochondrial content of the myofibrils may be an indication of incipient differentiation

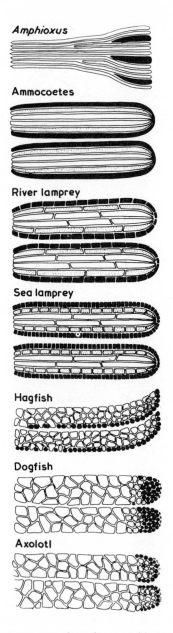

Fig. 7.5 The muscle compartments of *Amphioxus* and lower vertebrates. From Flood, 1973.
Red, slow fibres shown in black; intermediate fibres, shaded; and white, fast fibres unshaded.

towards the muscle types of higher forms. In the larval lamprey, each lamella in the compartment is regarded as a cellular unit and division into separate fibres only occurs in the adult lamprey. Compartmentalization is also seen in fishes and urodeles, but in these groups the slow fibres are more restricted in their distribution and tend to be confined to the lateral surfaces.

Compared to conditions in parasitic lampreys, the muscle compartments of the non-parasitic *L. lamottenii* show some interesting features (Teravainen, 1971). Here, the muscle compartments contain only three, rather than the usual 4–5 layers of central fibres that are characteristic of the parasitic lampreys. Moreover, the middle member of these central fibres extends as a continuous sheet from medial to skin surface, recalling the undivided muscle plates of the ammocoete (Fig. 7.6). This failure on the part of an adult brook lamprey fully to develop characteristics normally associated with the adult form, has been noted in a number of other morphological and physiological features and has been interpreted as evidence of paedomorphic trends.

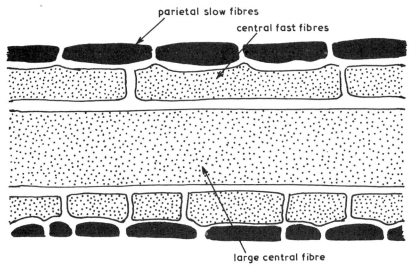

Fig. 7.6 Diagrammatic transverse section through the parietal muscle of a non-parasitic lamprey, *L. lamottenii* Redrawn from a micrograph by Teravainen, 1971.

7.3.2 Cytological differentiation of the muscle fibres

In the parietal muscles of hagfishes and lampreys, the cytological differentiation of fast and slow fibres follows a similar pattern. In *Myxine*, the fast white fibres, apart from their larger diameters, have sparse mitochondria and a low content of lipid and oxidative enzymes (Korneliussen and Nicolaysen, 1973). The arrangement of the myofibrils is said to conform to the type described as 'Feldstruktur' in which the fibrils are continuously and evenly distributed within the fibres. The myofibrils are separated by straight Z discs and distinct M bands are present. The smaller slow fibres have a much higher content of lipid and mitochondria. The myofibrils branch and anastomose to a greater extent than in the fast fibres and are surrounded by a conspicuous capillary network. Among cyclostome parietal fibres, a third and intermediate type has been described (Lie, 1973). These have diameters between those of the typical slow and fast fibres, but with greater numbers of mitochondria than the latter and with less densely packed myofibrils. However, these do not branch and anastomose like those of the typical slow fibres. Of the visceral muscles, the longitudinal tongue retractor consists entirely of fibres which, from their lipid and glycogen content, as well as in their fibrillar structure have been classified as white fibres. On the other hand, two other visceral muscles — the craniovelaris and spinovelaris — are entirely composed of red fibres, but of a quite distinctive type (Nicolaysen, 1966). These have very small diameters, few and small lipid droplets, but abundant mitochondria and glycogen deposits. Other distinctive features are the absence of M-lines and the presence of numbers of myosatellite cells. These satellite cells are thought to be involved in muscle regeneration and growth and their presence could be significant in a muscle on whose continuous and rhythmic contractions rests the main responsibility for the maintenance of the respiratory current.

In lampreys, the smaller slow fibres are said to have maximum diameters of about 70 μm compared to 164 μm for the fast fibres. They also have fewer mitochondria and fat droplets. The fast fibres have no basement membrane, except at their tendon ends along the medial surface (Teravainen, 1971). As in *Myxine*, an intermediate type of fibre has been distinguished in the parietal muscles, lying immediately adjacent to the single layer of slow fibres (Lie, 1973). Among the visceral muscles of lampreys, a distinctive type of 'myotube' fibre occurs in the muscles of the velum and in the branchial constrictors. In this type of muscle, the myofibrils are concentrated around the periphery, surrounding a central core of sarcoplasm containing the nucleus (Ciaccio and Dell'Agata, 1973). These fibres have been compared to the nodal tissue of the mammalian myocardium and although their physiological characteristics have not been investigated in detail, their peculiar structural features could well be related to the rhythmical activity involved in respiratory movements. Another visceral muscle, the tongue retractor of the adult lamprey is considered to be transitional between typical fast and slow fibres in its diameter, arrangement of myofibrils, lipid content and mode of innervation (Fominykh, 1970). In addition to their distinctive morphological and biochemical characteristics, vertebrate fast and slow fibres are now known to be distinguished by their relative content of fast or slow myosin isoenzymes. Differentiation of the two types of muscle apparently develops ontogenetically through changes in the biosynthesis of these myosins, probably under the influence of their distinctive motor innervation (Gauthier *et al.*, 1978).

7.3.3 Innervation

The fast and slow fibres of the cyclostomes differ fundamentally in their modes of innervation. Fast, central fibres are innervated at their ends by the plate-like terminals of motor nerves and are activated by propagated action potentials. The slow, parietal fibres on the other hand, have a distributed innervation by motor nerves which pass along the length of the fibre, branching to form 'bouton-like' terminals fitting into deep invaginations on the surface of the muscle and activating it through junction potentials. At least in the case of the lamprey, it has now been established that the axons innervating the two types of somatic muscle come from distinct fast and slow motoneurones in the spinal cord (Section 8.1.2).

The fast fibres of lampreys have nerve endings near the myotendinous junctions at both septal ends of the fibre (Peters and Mackay, 1961; Nakao, 1976). At the same time only those fibres closest to the abdominal surface have a direct motor innervation – the rest, like vertebrate smooth muscle, being electrically coupled to innervated fibres (Teravainen, 1971). With the exception of cardiac muscle, this indirect stimulation is a unique feature for a striated muscle. Its morphological basis may be the absence of a common basement membrane, the close packing of the individual fibres and the presence of the desmosome-like junctions described by Jasper (1967), which may enable the entire muscle lamella to function as a single unit.

In place of the double-ended innervation of the lamprey central fibres, the fast muscle fibres of the hagfish are innervated at only one of their septal ends by motor nerves ending in a typical motor end plate (Jansen and Andersen, 1963; Alnaes *et al.*, 1964; Korneliussen, 1973). In both hagfish and lamprey, the same motor axon may innervate corresponding fast fibres in neighbouring myotomes, thus co-ordinating contractions over a wider area. This is also true of the slow fibres of the hagfish, although in this case they receive a motor nerve supply from both ends.

Of the other muscle types that have been distinguished, the tongue retractor of the lamprey (regarded as intermediate in structure and lipid content between fast and slow fibres) has no multiple innervation like that of a typical slow fibre (Fominykh, 1970). However, in spite of its monosynaptic innervation and activation by conducted action potentials, sensitivity to acetylcholine has been demonstrated over the entire length of the fibre, in this respect resembling the slow fibre and suggesting that this represents a primitive and transitional condition from which the differentiated innervation patterns of fast and slow fibres might have evolved (Rozhkova, 1972). Like those of the lamprey heart, these cholinoreceptors are of the nicotinic rather than the muscarinic type.

In the parietal muscles of the hagfish, the intermediate type of fibre that has been distinguished by morphological criteria, resembles the typical slow fibres in being innervated at only one end by plate-like terminals (Korneliussen,

1973). Of the visceral muscles that have been investigated, the longitudinal lingual muscle consists solely of fibres of the fast type with single axon innervation, whereas the spinovelaris and craniovelaris fibres are all of the slow type with distributed synaptic sites. According to Nicolaysen (1966) these are about 10 μm apart, whereas in the slow parietal muscles there are said to be about 28 synaptic sites to each fibre and these were estimated as being about 100 μm apart (Alnaes *et al.*, 1964). Korneliussen (1973) found some evidence for monoaminergic innervation of the red muscle of the craniovelaris, although it is possible that the catecholamines are involved in controlling the metabolism of the lipid and glycogen stores.

7.3.4 Functional comparisons of fast and slow muscles

As implied in the descriptive terminology applied to them, the most obvious physiological characteristics of the two main muscle types are the speed and duration of their contractions (Table 7.1). These differences in mechanical response can be demonstrated by indirect stimulation of hagfish muscle tissue containing both types of fibre (Fig. 7.7). Under these conditions, a weak stimulus elicits a twitch response of short duration, but as the intensity of stimulation is increased, a slow component appears, increasing in amplitude as more and more fibres are brought into play. With repeated stimulation, the fast component is eventually blocked and the slow component isolated. This has a peak contraction time of about 120 ms and a duration of about 500 ms, compared to a peak contraction time of 60–80 ms and a duration of 150 ms for the fast fibres (Andersen *et al.*, 1963; Jansen and Andersen, 1963). In the visceral muscles of the hagfish, peak contraction times in the longitudinal lingual muscle (consisting solely of fast fibres) are about 60 ms, while in the slow velar muscles the corresponding figure is about 125 ms (Nicolaysen, 1966).

Table 7.1 Some physiological and electrical characteristics of fast and slow muscle fibres in the cyclostomes. Data from Alnaes *et al.*, 1964; Teravainen, 1971; Teravainen and Rovainen, 1971b.

| | FAST FIBRES | | SLOW FIBRES | |
	Lamprey	**Hagfish**	**Lamprey**	**Hagfish**
Activation	Propagated action potentials		Junction potentials	
Response	All or none		Graded contractions	
Tetanic fusion frequency	13.2s^{-1}	50s^{-1}*	7.4s^{-1}	35s^{-1}*
Contraction time, ms	200	150	300	500
Peak contraction time, ms	–	60–80	–	120
Membrane potential, mV	77–88	75	64–75	46
Membrane capacity, μF cm^2		4.0		4.0
Membrane resistance 10^3 Ohms cm^2	5.0	5.0	12–31	17.0
Time constant ms	4.4–6.1	21	61–123	72

*These figures for tetanic fusion frequencies for hagfish refer to the visceral muscles of *Myxine*.

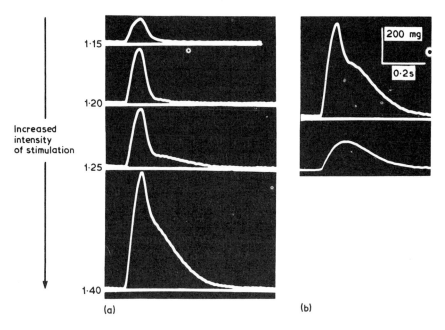

Fig. 7.7 Contractions of fast and slow fibres of the hagfish. From Andersen, Jansen and Loyning, 1963.

In (a) a slow component appears as the intensity of stimulation is increased. In (b) the fast component is blocked by repeated stimulation, thus isolating the slow component.

When stimulated, the fast fibres respond in an all-or-none fashion, with the appearance of an overshooting action potential and the minimum duration of stimulus required to produce this response is proportional to the time constant (Table 7.1). In the slow fibres, on the other hand, there is a graded response, varying with the extent of the membrane depolarization. However, in *Myxine*, the summation of junction potentials is not very effective in raising the amplitude of the response and the ceiling of depolarization is reached quite early in the train of impulses (Jansen and Andersen, 1963). The fast fibres require a higher frequency of stimulation for tetanic fusion and similar differences have been observed in the visceral muscles of *Myxine* (Table. 7.1).

Electrical constants for the two types of fibre reflect the way that membrane changes resulting from the release of transmitter at the motor terminals, are propagated through the fibre membrane. For example, high membrane resistance or capacitance may be taken as an indication of relatively low ionic permeabilities, while the latter is also responsible for time effects. Thus, the time constant may be expressed as the time taken for the electronic pulse to decay to about 37 % of its maximum value. In their absolute values, the high membrane resistances of the cyclostome slow fibres are of the same order as those observed in amphibians and reptiles (Prosser, 1973) and in cyclostomes they are some two to three times higher than in the fast fibres. In *Myxine*, membrane capacitance is apparently similar in both types of fibre, but in lampreys, as in the frog they are about twice as high in the slow fibres as in the fast type.

In both groups of cyclostomes, the resting membrane potentials of the fast fibres are considerably larger than in the slow fibres, although the difference seems to be more marked in the case of the hagfish (Table 7.1). Commenting on the high specific resistance and low resting potentials of the slow fibres of *Myxine*, Alnaes *et al.*, (1964) drew attention to the exceptionally high sodium concentrations in the muscle fibres of the hagfish, pointing out that if this reflected an unusually high sodium and a relatively low potassium conductance, these factors might explain their electrical characteristics. On the other hand, it may be noted that these conditions are not present in the lamprey, where the membrane resistance of the slow fibres may be even higher than it is in the hagfish. In an examination of sodium and potassium concentrations in muscles of various types in a number of marine vertebrates and invertebrates, Nesterov (1972) has claimed that the more rapidly contracting and active muscle fibres contain lower sodium and higher potassium concentrations. For example, in three teleost species, the potassium concentrations and the Na/K ratio were always much higher in the deeper white muscle than in the more superficially placed and presumably red fibres. Unfortunately, nothing is at present known of ionic distributions in the central and parietal muscles of the cyclostome.

Electrophysiological studies have revealed the existence in fast and slow fibres of miniature potentials, believed to represent the release of quanta of transmitter at the motor terminals (Fig. 7.8). In fast fibres of the hagfish these potentials are observed only close to one end of the fibre and never at both ends. Moreover, with increasing distances from the end of the fibre, their amplitude decreases in accordance with the electrical properties of the membrane. On the other hand, the miniature junction potentials of the slow fibres are observed along the entire length of the fibre and show a varying time course. This picture is of course, consistent with the innervation of the slow fibres by multiple, distributed endings and with the presence in the fast fibres of a single motor end plate at one end of the fibre.

7.3.5 The biological significance of fast and slow muscles

As Bone (1966) showed in his experiments with spinal dogfish, the higher glycogen stores of the red muscle are not depleted by long periods of slow swimming or even after the animals have been exhausted by short bursts of violent activity. On the other hand, continuous slow exercise substantially

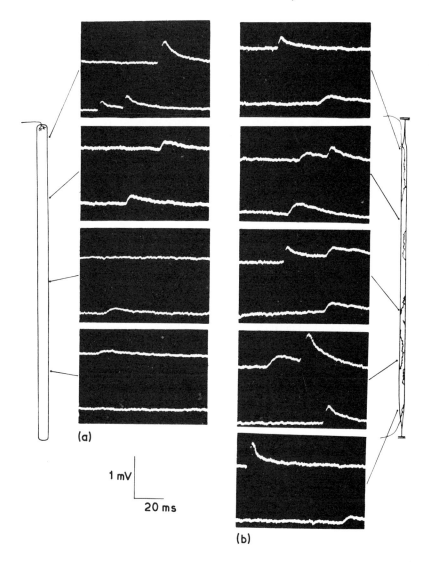

Fig. 7.8 Distribution of miniature potentials in fast fibres (a) and slow fibres (b) of *Myxine*. From Jansen and Andersen, 1963.
     The intracellular recording sites are indicated on the drawings of the fibre at the side of each diagram.

reduced their lipid content. Depletion of the glycogen of the white muscle did occur in violent exercise, but these carbohydrate stores were not affected by long periods of slow swimming. These results are clearly consistent with the histological characteristics of the two types of muscle. In their biochemical constitution, location in the myotomes and structural features, the slow red fibres are obviously well adapted to the aerobic metabolism of lipids. This is illustrated by their numerous, large mitochondria (containing oxidative enzymes) and their high lipid content, while their superficial position and abundant capillaries would favour the delivery of oxygen. Conversely, the small and scanty mitochondria, poor vascularization and low lipid content of the fast, white fibres is in keeping with their role in more violent activity, employing the anaerobic metabolism of glycogen as their energy source. In this case, the duration of their activity would be limited by the accumulation of lactic acid and the rate at which oxygen can be delivered to the muscle by the respiratory and cardiovascular systems. As an example, after forced swimming for 20 minutes, lactic acid concentrations in lamprey blood have been found to increase sevenfold (from 0.32 to 2.3 mM) and in strips of parietal muscle after 5 minutes of electrical stimulation, lactate concentrations increased from a resting level of 30 μmol g^{-1} wet weight to 58 μmol (Phillips and Hurd, 1977).

As Bone has emphasized, the existence in fish of two distinct motor systems, powered by different metabolic processes and with separate innervation, explains on the one hand, the rapidity with which these animals are fatigued by sudden bursts of violent activity and on the other hand, their ability to sustain slow swimming speeds over great distances. However, it appears that these two motor systems are not always completely separate and that in some of the higher teleosts, the fast fibres are also active during continuous swimming at speeds well below the maximum the fish is able to sustain for long periods (Bone, 1975). These differences appear to be related to the type of innervation of the fast fibres. Whereas, in the dogfish, the herring and the cyclostomes these fibres are focally innervated at one end, in the gadoids and in the goldfish or carp which appear to use their fast fibres at cruising speeds, they are multiply innervated.

As outlined earlier (Section 4.4), the lampreys are notable for their relatively poor performance in slow sustained swimming and are able to maintain their maximum speeds only over short distances. Attempts to correlate the locomotory abilities of different fish species with the distribution of the two types of muscle have not been entirely convincing (Boddeke *et al.*, 1959) and, at present, a satisfactory histological or physiological characterization of 'sprinters' or 'stayers' is hardly practicable. This is not surprising when we consider the complexity of the factors that are involved in swimming performance. In addition to the characteristics of the muscular system, these would include the cardiovascular and respiratory systems and hydrodynamic factors. Among the latter, mucus secretions play an important part in aquatic animals by reducing frictional drag and, in some teleosts, a 25% dilution of mucus has been found to reduce friction by 50–60%. On the other hand, in the hagfish, in spite of its prodigious slime producing capabilities, this factor appears to be less important and it has been reported that mucus from *E. stouti* only produced a 12% reduction in friction when compared to a sea water standard (Hoyt, 1975).

Reference has already been made (Section 4.4.1) to the sharp decline in the swimming endurance of the sea lamprey at sexual maturity. A possible clue to this inability to sustain the maximum swimming speeds of which its white musculature are capable, may lie in a reduction in the capacity of the liver to synthesize glycogen and glucose from the lactic acid, produced as a result of sustained muscular effort. During the upstream migration, degenerative changes occur in the liver (Autori and Bertolini, 1965; Bertolini, 1965; Sterling *et al.*, 1967), which are associated with starvation rather than sexual maturation. These changes are manifested by an alteration in liver colour from brown to green, as a result of the accumulation of bile pigments, resulting from haemoglobin degradation (Sawyer and Roth, 1954). Significantly, the green liver of the river lamprey, *L. fluviatilis* has been found to possess only half the gluconeogenic capacity of the brown liver as gauged from the relative activities of two enzymes (phosphoenolpyruvate carboxylkinase and pyruvate carboxylase) involved in the early stages of gluconeogenic pathways from lactate (Phillips and Hurd, 1977).

8

The nervous system

8.1 The spinal cord

Although some progress has been made towards a functional analysis of the cyclostome spinal cord, this task has been made more difficult by virtue of its peculiar characteristics which have impeded direct comparisons with the distribution of the cells and fibres in the spinal cord of higher vertebrates. The unique ribbon-like shape of the cord (said to be shared only by *Latimeria*) is almost certainly a secondary acquisition, since a more usual cylindrical form is present in earlier ontogenetic stages. Whiting (1972) considers that this flattened shape may have been due to the downward expansion of the dorsal fat column; a structure that he believes may already have been developed by heterostracans (Section 4.4). In the absence of medullated nerve fibres there is no sharp division between white and grey matter, although the cells form a broad central mass surrounded by the fibres. Distinct dorsal and ventral horns are also missing and the dorsal and ventral spinal nerve roots of the lamprey are unique in failing to unite. In myxinoids, where the nerve roots do unite, these junctions lie far outside the spinal cord in the region of the parietal muscles.

8.1.1 The reticulospinal system
In both groups of cyclostomes, the principal descending fibre pathways consist of large unmyelinated axons in the ventral and lateral columns (Fig. 8.1). These arise from cell bodies located in the medial regions of the brain stem and constitute the reticulospinal suprasegmental motor control system (Fig. 8.2). In keeping with their large diameters, the rate of conduction in these axons ($6-10 \text{ m s}^{-1}$) is higher than in any other nerve fibre (Shapovalov 1972). In the lamprey, about 50 of these reticular neurones have been identified in the brain stem, varying considerably in size (Rovainen *et al.*, 1973). Some, larger than the usual somatic motor neurones, are more conspicuous and have been referred to as Muller cells, but there have been differences in the number of

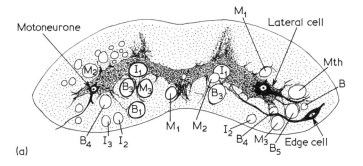

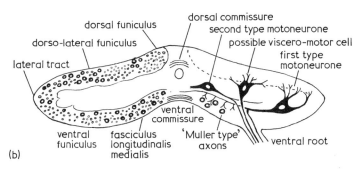

Fig. 8.1 The cyclostome spinal cord.
(a) Spinal cord of a lamprey showing some of the identified neurones and fibres. Drawn from the data of Rovainen *et al.*, 1973.
I, Muller fibres of the isthmic group; B, Muller fibres of the bulbar group; M, Muller fibres of the mesencephalon; Mth. Mauthner fibres.
(b) Fibre tracts and neurones of the spinal cord of *Myxine*. Redrawn from Bone, 1963.

such cells that have been recognized by different authors. Muller cells are arranged bilaterally in pairs and are constant in their location. Leaving aside those cells on which there has been disagreement, at least 7 pairs have been identified within the brain stem. In *Myxine*, none of the axons in the medial longitudinal bundle (fasciculus longitudinalis medialis) can readily be distinguished from the others by size nor can they be homologized with the Muller cells of the lamprey. For this reason, Bone (1963) preferred to speak only of 'Muller-type' axons, considering true Muller cells as specialized elements of a reticular through conducting system to be absent in hagfishes. In addition, unlike the Muller fibres of the lamprey, the Muller type axons of *Myxine* appear to cross in the ventral commissure to descend in the medial longitudinal bundle of the opposite side (Fig. 8.1).

Apart from the Muller neurones, Rovainen *et al.* (1973) have identified two pairs of Mauthner neurones in the lamprey brain stem, one of which is regarded

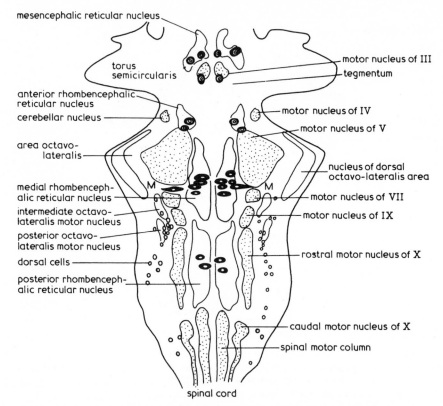

mesencephalic reticular nucleus

torus
semicircularis

anterior rhombencephalic
reticular nucleus

cerebellar nucleus

area octavo-
lateralis

medial rhombenceph-
alic reticular nucleus

intermediate octavo-
lateralis motor nucleus

posterior octavo-
lateralis motor nucleus

dorsal cells

posterior rhombenceph-
alic reticular nucleus

motor nucleus of III

tegmentum

motor nucleus of IV

motor nucleus of V

nucleus of dorsal
octavo-lateralis area

motor nucleus of VII

motor nucleus of IX

rostral motor nucleus of X

caudal motor nucleus of X

spinal motor column

spinal cord

Fig. 8.2 Topological chart of the brain stem of the lamprey, *Lampetra fluviatilis* showing
the positions of the main cell masses. Redrawn from Nieuwenhuys, 1972.
Periventricular cell masses are shaded; unshaded areas indicate reticular and
migrated nuclei; positions of individual reticular and Mauthner neurones are
shown in solid black.

as the homologue of the Mauthner neurones of fish and amphibia. These cells,
which appear to be absent in the hagfish, are distinguished from the Muller
neurones by their proximity to the acoustic nerve (VIII) and the fact that their
axons cross before descending on the opposite side of the cord. Throughout
their passage down the cord, the Muller axons establish synaptic contacts with
the dendrites of giant interneurones (Fig. 8.3) which show both chemical and
electrical coupling (Ringham, 1975; Christensen, 1976). In addition they form
monosynaptic connections with spinal motoneurones (Batueva and Shapovalov, 1974). Thus, like the giant command systems of invertebrates, they
connect the brain directly to the motor system and provide a rapid pathway
for the execution of predatory or escape reactions. Such frantic and
indiscriminate escape responses are seen when resting lampreys are suddenly
disturbed and have been attributed to prolonged 'afterdischarges' that have
been observed in Muller cells after experimental stimulation of the cranial

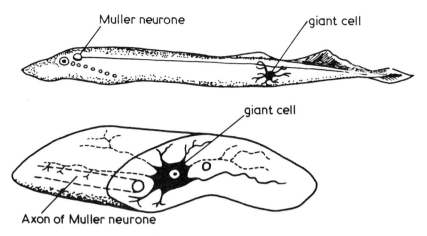

Fig. 8.3 Position of giant interneurones of the lamprey and their relationship to the Muller fibres. Redrawn from Christensen, 1976.

nerves (Wickelgren, 1977a,b). Unlike the invertebrate systems or the Mauthner neurones of teleosts which give all-or-none responses to single shocks, the Muller and Mauthner fibres of lampreys require high frequency stimulation and their responses may either habituate or continue with steady stimulation.

Because of its relationship to the brain stem and the ascending somatic sensory tracts, the reticular system is able to link sensory inputs from various sources to the motor units. Through their long dendritic connections these cells receive convergent inputs from a variety of sensory systems, including afferent impulses from both distance receptors and the somatic sensory systems. For example, stimulation of the optic nerves of the lamprey results in excitatory or inhibitory post synaptic potentials (EPSP's and IPSP's) in the reticular elements, although this does not occur in ammocoetes before the retina and optic tectum have differentiated. Rovainen (1967) considered that the relationship of the reticular system to secondary sensory polysynaptic pathways was indicated by the long latencies of responses to cranial nerve stimulation, but Wickelgren (1977a,b) has found evidence that different kinds of sensory activity in cranial nerves (olfactory, optic vestibular and trigeminal) may be relayed more directly through independent short latency pathways.

The direct relationship of the Mauthner neurones to the VIII nerve suggests that they may be important in the maintenance of equilibrium. In teleosts, these neurones play a part in the development of effective larval swimming movements, especially in relation to the functions of the tail and it is significant that these cells may be missing in species in which the tail is reduced (Whiting, 1957). The apparent lack of differentiated Mauthner cells in hagfishes could be related to the reduction of the acousticolateralis system and in this connection it may be recalled that these animals often swim upside down.

It has been suggested (Whiting, 1957), that the Mauthner neurones developed phylogenetically from a system of a few large and distinctive neurones in the ancestral vertebrates, carrying out sensory, motor and correlating or co-ordinating functions. This condition is still represented in primitive chordates and in the early ontogenetic stages of vertebrates by the Rohon-Beard and Rhode cells of *Amphioxus* and the Muller cells of the lower vertebrates.

8.1.2 Cell types of the spinal cord
According to Rovainen (1974a), each segment of the lamprey spinal cord contains about 500 neurones on each side, of which the majority are small cells less than 15 μm in diameter, whose functions have not yet been investigated. Of those that have been studied, the following types are recognized and most of them have their counterparts in the spinal cord of the hagfish.

(1) Somatic motoneurones, innervating slow and fast muscle fibres, of which there are about 40 pairs in each segment.
(2) Edge cells or 'Randzellen' numbering about 20 per segment.
(3) Lateral cells − not more than one per segment.
(4) Giant interneurones − not more than one per segment.
(5) Dorsal cells − about two per segment.

(a) *Fast and slow motoneurones*
In the spinal cord of *Myxine*, two types of primary motoneurone were described by Bone (1963). The first more numerous cell, often situated at the lateral borders of the grey matter, was believed to represent a somatic moto-neurone innervating the fast muscle fibres of the myotomes. A second less abundant type, situated medially to the point of emergence of the ventral root of the spinal nerves, was thought to supply the slow myotomal fibres. From both cell types processes extend towards the ventral commissure and in the case of the 'fast motoneurones' these divide into ascending and descending branches on the opposite side of the cord. In addition, branches are also given off to the medial longitudinal fibre bundle and especially to the ventral area occupied by 'Muller type' fibres (Fig. 8.1).

In the lamprey, the primary motoneurones are among a group of trans-versely orientated disc-like cells, immediately lateral to the Muller fibres and dorsal cells (Fig. 8.1). Intracellular stimulation of these cells results in one-to-one contractions in the muscles of the same side, occasionally spreading over two or three segments and representing 5−12 myotomal sub-units (Ter-avainen and Rovainen, 1971a). Contrary to Bone's suggestion in regard to *Myxine*, the slow motoneurones of the lamprey are believed to be the more abundant type, representing from two-thirds to four-fifths of the total number of motoneurones in each segments. Both types show similar membrane potentials (60–80 MV), but differences in input resistance and time constants may reflect variations in cell size and in the degree of dendritic arborization. In fact, the fast motoneurones are somewhat larger than the slow type, with diameters of 15–25 × 45–70 μm compared to 20–25 × 25–30 μm, and in addition they show more extensive dendritic branching. Unlike the moto-neurones of *Myxine*, the processes of the lamprey motoneurones do not extend to the opposite side of the cord, although some dendrites approach the mid-line in the region occupied by the Muller fibres.

(b) *Edge cells or 'Randzellen'*
This type of spinal interneurone is about 20–50 μm in diameter and located in the lateral tracts of the lamprey cord (Fig. 8.1). Their axons extend forwards, in some cases towards the opposite side of the cord and appear to have their counterparts in the lateral arcuate cells of *Myxine*, which have decussating axons or a long axon on one side with dendrites ending on the limiting membrane. Because of their relationship to the peripheral dendritic network of the somatic motoneurones, Bone suggested that they may be involved in the reciprocal excitation and inhibition of motoneurones on either side of the cord during swimming movements. A somewhat similar conclusion was reached by Teravainen and Rovainen (1971b) who observed activity in the edge cells of the lamprey during myotomal reflexes when they inhibited the motoneurones of the opposite side. Some of the edge cells in the lamprey were excited by intracellular stimulation of one of the Muller axons (Fig. 8.4), but no antidromic action potentials could be elicited in edge cells from more posterior regions of the cord.

(c) *Lateral cells*
These are amongst the largest cells in the lamprey cord with transversely directed dendrites confined to the same side (Fig. 8.1). Their long axons extend towards the tail and since lateral cells behind the brain have been stimulated in the tail region, these axons must in some cases extend to 8–10 cm in length. Intracellular stimulation of the lateral cells failed to elicit synaptic potentials in giant interneurones, edge cells or other lateral cells and indeed for the most part in motoneurones. However, they were found to inhibit certain unidentified medially placed cells some of which may have been motoneurones. Rovainen (1974a) suggested that these inhibited cells may be connected with local contractions in adjacent myotomes and that the function of the lateral cells may be to bring about the relaxation of myotomes behind them, in this way contributing to the production of undulatory waves of contraction during swimming. Lateral cells were consistently excited by Muller axons (Rovainen, 1974b).

(d) *Giant interneurones*
In *Myxine*, these cells are located in the lateral or medio-lateral areas of the grey matter; several pairs being alternately arranged in each segment. Their

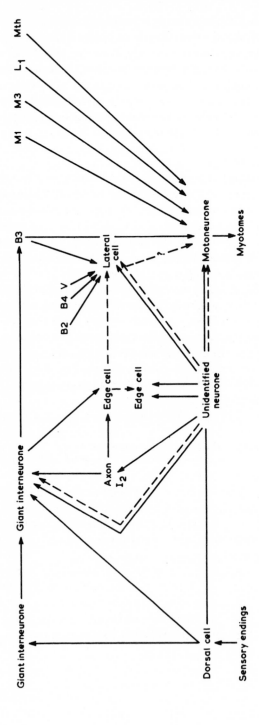

Fig. 8.4 Synaptic connections between neurones in the spinal cord of the lamprey. From Rovainen, 1974b.

Solid lines indicate excitatory interactions; dashed lines, inhibitory. That the same cell B3 excited motoneurones and in turn was excited by giant interneurones was not tested experimentally by Rovainen. The diagram should not be taken to imply that individual edge cells or unidentified cells make all the connections that are shown.

axons join the ventral commissure and run in the median longitudinal bundle or the ventral funiculus. Like the giant Rhode co-ordinating neurones of *Amphioxus*, with which they have been homologised, the axons of posteriorly placed cells ascend the cord, while the more anterior axons descend caudally.

The giant interneurones of lampreys (or large internuncials of Whiting), are located in the lateral grey of the posterior third of the cord (Fig. 8.3) making synaptic contact with a pair of Muller axons and sending dendrites into the dorsal or dorso-lateral fibre columns, where they would be in a position to intercept sensory axons (Rovainen, 1974a). Their axons ascend towards the brain, but although it has been possible in only one case (Rovainen *et al.*, 1973) to trace them into the brain stem, the fact that stimulation produced excitatory post-synaptic potentials in a contra-lateral Muller cell shows that they must reach at least as far as the medulla. Regarded as higher order sensory neurones, these cells have not produced movements of the body or synaptic potentials in motoneurones when stimulated intracellularly. On the other hand, they are readily excited by mechanical stimulation of the tail or skin surface (Teravainen and Rovainen, 1971b) and the fact that they fire together suggests that they may play a part in swimming movements (Martin *et al.*, 1970). Dorsal cells when stimulated, elicit both monosynaptic and polysynaptic EPSP's in the giant interneurones (Martin and Wickelgren, 1971), whose role has been defined as part of a convergent multispecific sensory system extending towards the brain (Rovainen 1974a). The arrangement of the giant interneurones in the hagfish could allow them to play a part in co-ordinating the contractions of successive groups of motoneurones. This, together with the cranio-caudal directions taken by their axons, has led to a suggestion that they might be implicated in rapid responses in either direction of the cord following stimulation of trunk or tail, like as those observed in the knotting movements of the hagfish when it is escaping from slime (Bone, 1963).

(e) *Dorsal cells*
On either side of the mid-line of the lamprey cord are two rows of dorsal cells or Hinterzellen, whose fibres emerge through dorsal roots to terminate in the skin. Most of these cells appear to have anterior processes extending at least as far forwards as the branchial region and after antidromic stimulation of their rostrally projecting axons, Martin and Bowsher (1977) found that in three-quarters of the neurones they investigated, these axons reached the isthmic region of the brain stem. Measurements of conduction velocities showed a decrease in the rostral direction, suggesting the existence of numerous collaterals in the spinal cord of brain stem. These cells, which have been homologised with the Rohon-Beard cells of earlier ontogenetic stages of lower vertebrates, have not been identified in myxinoids, although Kuhlenbeck (1975) has suggested that they may have their counterparts in certain comparatively large cells dorsal to the central canal. Dorsal cells are regarded

as first order sensory cells in which action potentials are produced by stimulation of their sensory endings and not by synaptic activation of the cell body via other elements in the central nervous system (Martin and Wickelgren, 1971; Birnberger and Rovainen, 1971). Individual dorsal cells are responsible for responses to touch, pressure or nociceptive stimuli and their receptive fields are located in the ventral or ventro-lateral body surfaces. Although differing in some electrophysiological properties, the trigeminal ganglion cells are also regarded as first order sensory neurones equivalent to dorsal cells and responding to the same range of sensory modalities, but with their receptive fields on the surface of the head (Matthews and Wickelgren, 1978). Dorsal cells have been shown to be involved in reflex movements of the dorsal fin after the skin of the same side of the trunk has been stroked and similar fin movement follows intracellular stimulation, in some cases on the opposite side of the body. Stimulation of single dorsal cells does not produce a response in single motoneurones and an appropriate response probably requires the simultaneous activity of a number of these cells.

8.1.3 Reflex behaviour and neuronal interrelationships
The application of electrophysiological techniques has thrown some light on the interrelationships of some of the larger neurones in the spinal cord of the lamprey (Fig. 8.4) but we are still unable to form a clear picture of the neural mechanisms involved in normal swimming behaviour. Local or segmental reflexes which require the presence of intact dorsal roots, appear to involve only the slow motor units and in experimental investigations, this has generally been true also of the long reflexes that continue rostrally after destruction of the dorsal roots (Teravainen and Rovainen 1971b). Activity in fast muscle fibres has been noted only after intense mechanical stimulation as when the tail is pinched or pricked. These differences in the reflex behaviour of the two motor systems suggest that the slow units are more active and have a lower threshold. Stimulation of dorsal cells has produced myotomal movements and both inhibitory postsynaptic potentials (IPSP) and excitatory postsynaptic potentials (EPSP) in fast and slow motoneurones, probably via interneurones, although no response occurred in single motoneurones after intracellular stimulation of a single dorsal cell. In simple fin reflexes, only two groups of neurones – dorsal cells and fin motoneurones – have been implicated.

Intracellular stimulation of neurones of the reticular system has shown that motoneurones are excited directly by certain Muller cells (M^1, M^4) in the mesencephalon and further neurones in the isthmic region (I_1), as well as by the larger Mauthner cell (Fig. 8.4). On the other hand, Muller neurones in the rhombencephalon excited lateral cells and only one activated motoneurones (Rovainen, 1974b). During mechanical stimulation, this latter cell was excited by a giant interneurone. Swimming responses have been observed in an intact lamprey after stimulation of a Muller cell of the rhombencephalic group and in another case, stimulation at a frequency of 100 s^{-1}, resulted in short contractions on the same side of the body, although no undulatory waves were produced. Presumably, normal swimming requires the simultaneous activity of larger numbers of neurones, but as decapitated lampreys and hagfishes show swimming behaviour after intense mechanical stimulation (Section 4.4), many of these cells must be intraspinal. Relevant to these wider effects is the fact that in addition to their monosynaptic connections with motoneurones, Muller axons may act through other relay pathways and that many other thinner descending fibres as well as interneurones and dorsal root afferents may provide synaptic inputs to motoneurones (Batueva and Shapovalov, 1977a,b). That the integrity of the reticular neurones is not essential to the execution of normal swimming reflexes is also demonstrated by experiments involving complete severance of the spinal cord (Selzer, 1978). Under these conditions, ammocoetes regain apparently normal co-ordination of their swimming movements after several weeks, although the large reticulospinal axons degenerate beyond the zone of transection. Functional recovery has been attributed to the short distance sprouting of axons of giant interneurones or dorsal cells, which may establish polysynaptic connections with unspecified interneurones across the severed cord.

The Mauthner cell of the lamprey produces composite EPSP's in a majority of the myotomal motor neurones (Rovainen, 1974b) and it seems likely that, as in teleosts, these cells may be involved in rapid escape reactions. In fish, a 'startle response' can be elicited by acoustic or vibrational stimuli, exciting the Mauthner cells through the acoustic nerve. This in turn activates the motoneurones over a larger area, producing a sudden, rapid flexing of the whole body to one side, followed by a less extensive and slower return flip in the opposite direction (Eaton *et al.*, 1977). Thus, in teleosts, the significance of the Mauthner system lies in its ability to activate directly a large part of the myotomal system and the same effect can be produced experimentally by antidromic stimulation of the Mauthner neurones or by electrical stimulation of the VIII nerve (Zottoli, 1977). In contrast to lampreys, where responses to vibrational stimuli are located within the labyrinth (Lowenstein, 1970) and could therefore activate the Mauthner neurones via the acoustic nerve, the labyrinth of hagfishes does not respond in this way to mechanical stimuli (Lowenstein and Thornhill, 1970; Lowenstein, 1973). This might well account for the absence in the hagfish of a differentiated Mauthner cell system. In *Myxine*, vibrational responses have been detected from a skin area below the degenerate eye and although it has been suggested that this might correspond to the rudimentary lateral line area of *Eptatretus*, it should be noted that it is innervated by the trigeminal (V).

8.2 The brain

In the linear arrangement of the fore, mid and hind brain, the lamprey (or more appropriately the ammocoete) has been claimed to show the closest approach to the archetypal vertebrate pattern and on Halstead's (1973 a, b) interpretation, a similar arrangement was also present in the heterostracans (Section 3.2.3). Moreover, except for the presence of a bilobed cerebellum, the cephalaspid brain has been held to show marked parallels to that of the lamprey (Section 3.2.2). At the same time, in the adult cyclostomes, these simpler brain patterns have been considerably distorted by the changes that occur during embryonic or larval development. For example, in the adult lamprey, the disposition of the telencephalon in relation to the diencephalon is affected by the development of the sucker and olfactory organ (Fig. 8.10) and in the hagfish, the dorso-ventral flattening of the brain has been related to its development within a rigid shelled egg. In addition, the relative development of various brain regions in the hagfish must reflect the impoverishment of certain sensory systems and its greater reliance on others. Some of its peculiar features, such as the reduction of the ventricular cavities and the absence of choroid plexuses, are secondary developments arising in the course of its embryonic life.

8.2.1 The medulla

According to classical interpretations of the topography of the central nervous system, the walls of the neural tube have been divided into longitudinal functional zones – a dorsal and primarily sensory alar plate area separated by a groove, the sulcus limitans, from a ventral, primarily motor, basal plate area. In further elaborations of this pattern, the alar plate has been further sub-divided into dorsal somatic sensory and ventral viscero-sensory zones and the basal plate into medial somatic motor and lateral viscero-motor

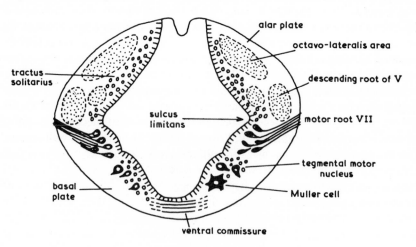

Fig. 8.5 Distribution of longitudinal functional systems in the alar and basal plate areas of an ammocoete medulla. Redrawn from Kuhlenbeck, 1968.

areas (Fig. 8.5). In its segmental arrangement of branchial nerves, dorsal roots and spinal nerves, the cyclostome medulla retains much of this primitive segmental pattern of the spinal cord, but superimposed upon it are intersegmental correlating mechanisms provided by the longitudinal functional systems. In his topographical analysis of the brain stem of the lamprey, Nieuwenhuys (1972) finds that the medial somatic motor and lateral viscero-motor columns are clearly recognizable in the rhombencephalon; the former consisting of caudal motor nuclei belonging to the most anterior spinal nerve roots, together with the three motor co-ordinating centres (tegmental motor nuclei) of the reticular formation (Fig. 8.2). The lateral viscero-motor column consists of the five efferent nuclei of the so-called branchial nerves – cranial nerves V, VII and IX, together with the rostral and caudal nuclei of X. On the other hand, although afferent fibres representing the general and special somatic sensory and visceral sensory divisions reached the alar plate area, Nieuwenhuys concluded that the centres where these fibres terminate do not show the typical dorso-ventral pattern of the longitudinal functional zones. As is the case with the dorsal spinal nerve roots, the afferent fibres of the branchial nerves divide into ascending and descending bundles after they enter the medulla.

In the lamprey, the special somatic sensory division is represented by the so-called octavo-lateralis or acoustico-static area, occupying virtually the entire rostral alar plate region. This consists of three longitudinally arranged cell masses; the first two representing the end stations of lateral line nerves and the third those of the acoustic nerve. From these centres arise arcuate fibres which cross to form an ascending tract – the lateral lemniscus – passing to the tectum or cerebellum. In addition, other secondary fibres descending from the vestibular nucleus may join the longitudinal medial bundle to reach the spinal cord. The Mauthner neurones, considered by some to have arisen within the vestibular nucleus, are displaced in a medial direction towards the tegmental nucleus, although their axons descend in the lateral tracts, rather than in the main reticular bundle.

The lateral line system, represented in lampreys by a longitudinal row of neuromasts as well as tracts around the snout and eye regions, is virtually absent in the hagfish, which must rely to a greater extent on cutaneous touch or chemoreceptors, situated on the head region and especially on the tentacles. Correlated with this reduction of the lateral line system and vestibular organs and its greater dependence on cutaneous receptors, the myxinoid medulla shows some modifications from that of the lamprey (Fig. 8.6). The main afferent pathways from the peripheral nerves enter the anterior projecting horns of the medulla on either side of the mesencephalon and here the dominant component is the general sensory division of V. The enormous size of the general cutaneous system, especially the sensory V, is one of the more conspicuous features of the myxinoid medulla and compared to this, the acoustico-lateralis and visceral divisions are insignificant. Although the

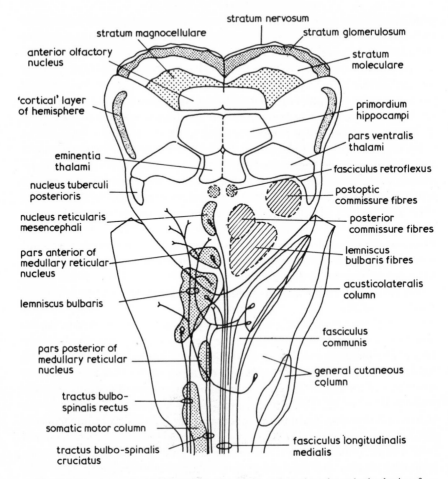

Fig. 8.6 Diagrammatic representation of a horizontal section through the brain of *Myxine*. From Bone, 1963.

primary sensory fibres are said to be arranged in functional systems, their boundaries are not well-defined. The reduced acoustico-lateralis area lies between the trigeminal and visceral afferent areas. The latter is made up of some fibres from VII, but mainly those from the IX/X complex. Arcuate fibres from the various sensory areas constitute a conspicuous ventral commissure forming an ascending lemniscus, where they are joined by other fibres reaching as far forwards as the tectum of the mesencephalon.

The motor nuclei of the branchial nerves are situated at the sides of the medulla. In the ventral somatic motor column, the nucleus of VI is absent and the caudal area belongs to the first of the spinal nerve roots. Medial to the viscero-motor column are the diffuse nuclei of the reticular motor system containing large multipolar neurones up to 75–80 μm in diameter. The

'Muller type' axons of these cells are less conspicuous than in lampreys and although some originate in larger 'Muller type' cells, others are said to come from smaller reticular elements. As some of these reticular axons cross before joining the median longitudinal bundle, it has been maintained by some authors that they represent a Mauthner system. Another line of argument has been that certain crossed bulbar-spinal axons which arise in the acoustico-lateralis area should be regarded as the homologues of the Mauthner neurones of lampreys and gnathostomes.

Some of the viscero-motor functions of the lamprey medulla have been investigated experimentally in relation to the control of respiratory movements (Rovainen, 1974c; Rovainen and Schieber, 1975; Homma, 1975). These studies have shown that the muscles of the ammocoete velum are controlled from the V motor nucleus, although a few velar motoneurones may also be located in the nucleus of VII. In both adult and ammocoete, the muscles of the branchial basket, including those of the gill pouches and ectal valves, are controlled from the motor nuclei, of IX and X. Using isolated preparations of the brain, velum and branchial region of the ammocoete, Homma was able to show that during typical breathing activity, the potentials recorded from the surface of the velum preceded those of the branchial basket by about 117 ms and were of longer duration – averaging 248 ms against 130 ms. During this respiratory activity, periodic bursts of spike discharges were recorded from the motor nucleus of V and these were synchronized with the discharges from the velum. Similar periodic discharges were also recorded from the motor nucleus of X, but these were of greater amplitude, shorter in duration and with longer latencies than the discharges from the velar musculature. Moreover, these bursts in the X nucleus were always preceded by activity in the velum. Transection of the brain stem behind the nucleus of V did not inhibit its regular discharges, nor the contractions of the velum, but activity did cease in the branchial muscles. Unilateral section of the brain stem stopped branchial contractions on the operated side, but velar movements continued. Homma suggested that pacemaker activity is generated near the V motor nucleus and that this is transmitted in part to the velar motoneurones of the opposite side and in part ipsilaterally to the motor nuclei of IX and X. However, the origin of this pacemaker activity has not been determined and the nucleus itself remains inactive when it is completely isolated from the rest of the brain. During metamorphosis, when the velum loses its respiratory functions, it seems that there must be a dramatic change in the site of the respiratory pacemakers. With the transition from the unidirectional respiratory current of the ammocoete to the tidal pumping system of the adult lamprey, respiratory functions are apparently transferred to the X motor nucleus. Thus, in isolated brain-gill preparations of adult lampreys, Rovainen observed powerful periodic bursts of EPSP's in the X motor nucleus, but similar discharges were not seen in the motor nucleus of V. These potentials occurred immediately before each branchial contraction, but their origin within the brain has not been established.

Velar motoneurones are the smallest motoneurones recorded in lampreys (5–10 μm) with the lowest conduction velocities (0.2 m s^{-1}), producing EPSP's with one-to-one contractions in the longitudinal or oblique velar muscles. Motoneurones to the branchial basket of the ammocoete are also small cells (less than 10 μm in diameter), but with somewhat higher conduction velocities than the velar motoneurones (0.6 m s^{-1}). Most of the identified branchial motoneurones of the adult lamprey were located in the rostral part of the X motor nucleus. These were about the same size as the myotomal and fin motoneurones with similar input resistances and time constants. However, the much lower conduction velocities of these branchial motoneurones suggests that their axons must have much smaller diameters.

8.2.2 The cerebellum

The vertebrate cerebellum is a correlation centre developed from the dorsal sensory columns of the medulla and usually consisting of a median corpus cerebelli and laterally evaginated auricles. On one interpretation, it has evolved from a fusion or bridge between the acoustico-lateralis areas on either side, but an alternative view regards the corpus cerebelli as derived from the general somatic sensory area of the trigeminal.

The cerebellum of lampreys is a simple transverse plate, continuous laterally with the acoustico-lateralis area of the medulla, and at least one of the nuclei of the lateral line nerves extends into the grey matter of the cerebellum. The main inputs consist of primary fibres from the lateral line, and nerves V and VIII, as well as some secondary fibres from the last two systems (Fig. 8.7a). Other afferents come from the lateral columns of the spinal cord (spino-cerebellar tracts) passing through the medulla, from the optic tectum of the midbrain (tecto-cerebellar) and from the hypothalamus (lobo-cerebellar). The output from the cerebellum arises from a larger cell type which has been regarded as a precursor of the Purkinje cells of higher vertebrates. Efferent fibres pass to the optic tectum (cerebello-tectal), to the tegmental motor nuclei of the midbrain, to the motor nuclei of the medulla and the nucleus of the oculomotor (III). Others join cerebello-spinal tracts descending through the medulla or join the medial longitudinal bundle of the reticular system (Fig. 8.7b).

In vertebrates reflex movements co-ordinate the movements of the eyes in association with the labyrinth, allowing the animal to stabilize an object in its field of vision in spite of movements of the head. These vestibulo-ocular reflexes have already been developed by the lampreys, producing opposing eye movements and turning of the body itself, when the animals are rotated. Electrophysiological studies have shown that these reflex movements can be produced, both by direct mechanical stimulation of the labyrinth or by electrical stimulation of either the anterior or posterior branches of the vestibular nerve (Rovainen, 1976). For example, mechanical stimulation of

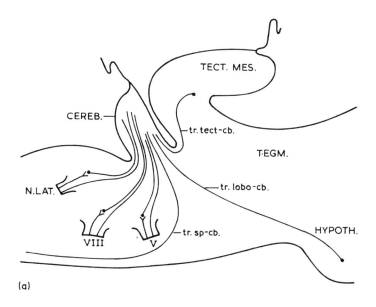

(a)

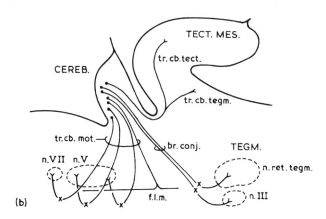

(b)

Fig. 8.7 Afferent (a) and efferent tracts (b) of the lamprey cerebellum. From Nieuwenhuys, 1967a.
Cereb., cerebellum; com. cb., cerebellar commissure; f.l.m., fasciculus longitudinalis medialis; N.LAT., anterior lateral line nerve; n.ret.tegm., tegmental
reticular nucleus; tr.sp.cb., spino-cerebellar tract; tr. tect. cb., tecto-cerebellar
tract; tr. lobo cb., lobo-cerebellar tract; tr.cb.tect., cerebello-tectal tract; tr.
cb.tegm. cerebello-tegmental tract; tr. cb. mot., cerebello-motor tract; TEGM.,
tegmentum; TECT.MES., mesencephalic tegmentum.

the anterior part of the labyrinth caused a rotation of the ipsilateral eye in a dorso-caudal direction, accompanied by a ventro-rostral rotation of the opposite eyeball. Mechanical stimulation of the posterior labyrinth caused a similar dorsal and ventral rotation of the ipsilateral and contralateral eyes, but in this case, the former moved rostrally and the latter caudally. Unlike other vertebrates, fast recovery movements of nystagmus were not seen. Although the pathways for these movements are not known, Rovainen believes that they represent direct and specific connections between the individual ampullae and the various oculomotor muscles. The interneurones are thought to be certain large cells in the vestibular nucleus which are intimately related to the axons of the vestibular nerve and which have projections towards the oculomotor and other more posterior motor nuclei (cf. Fig. 8.7b).

The existence of a cerebellar rudiment in myxinoids has been the subject of widely divergent views. It is possible that this region may be represented in *Myxine* by the posterior tectal commissure and, in *Eptatretus*, a small acoustico-lateralis commissure has been described, containing cell strands continuous with the acoustico-lateralis area, but not reaching the median plane. Little correlation exists throughout the vertebrate series between the degree of cerebellar differentiation and phylogenetic status. For this reason, the poorly developed cerebellum of the lamprey and the virtual absence of this structure in hagfishes, may have little or no evolutionary significance. As Kuhlenbeck (1975) has remarked, it is perhaps the 'least relevant region of the vertebrate brain, representing a suprasegmental structure, controlling or smoothing movements, but not initiating them and connected in parallel with the main input and output channels of the neuraxis'.

8.2.3 The mesencephalon

In the lower vertebrates this region has been regarded as the dominant directing centre of the brain; the supra-segmental tectal cortex being a centre where sensory inputs from wide areas including the optic system are processed, while the bulbo-spinal and reticulospinal tracts provide the efferent channels through which this processed information can be transmitted through the spinal cord. The relative importance of the visual system is shown by the fact that the tectum tends to be largest in those fishes in which visual cues are most important in courtship or hunting and smaller in nocturnal or abyssal forms or in genera that make more use of other senses in feeding or territorial behaviour (Pearson and Pearson, 1976). Its significance is also illustrated by the four-fold growth of the optic tectum during the metamorphosis of the lamprey accompanying the functional development of the functional eyes (Fig. 8.8).

In its possession of a choroid plexus, the lamprey mesencephalon is unique amongst vertebrates and it is interesting to recall that a similar condition is thought to have existed in the Heterostraci (Halstead, 1973a, b). The posterior

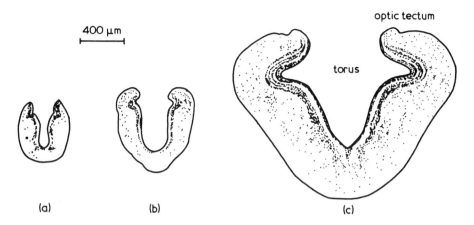

optic tectum

torus

400 μm

(a)   (b)   (c)

Fig. 8.8 Successive stages in the growth and differentiation of the optic tectum of the sea lamprey, *Petromyzon marinus* during metamorphosis. Redrawn from Kennedy and Rubinson, 1977.
(a) 70 mm ammocoete. (b) 105 mm ammocoete. (c) Adult lamprey after the completion of transformation. The lamination of the tectum which begins to develop in the larger ammocoete, becomes much more distinct after metamorphosis.

boundary of the mid-brain is marked by the posterior commissure and in the pretectal region there are paired differentiations of the ependymal roof – the sub-commissural organ. Laterally and caudally, the tectum takes the form of paired structures uniting behind the choroid plexus and projecting backwards over the cerebellum. In hagfishes, the mesencephalon is compressed into a wedge-shaped structure, partially divided into two halves by a dorsal groove and with its lateral sides covered by the lateral horns of the medulla.

In its general architecture, the mid-brain still retains the basic topography of the rhombencephalon, divisible into a ventral tegmental motor area, regarded as an extension of the rhombencephalic basal plate, and a dorsal sensory correlation area, representing a continuation of the alar plate. In the lamprey, the motor tegmentum of the mid-brain contains the nucleus of III and the cranial extension of the reticular system with its mesencephalic Muller cells. That part of the mid-brain dorsal to the motor tegmentum contains the two large somatic sensory correlation centres – the semicircular torus and the optic tectum. The tectal cortex consists of 8–10 layers of neurones with their dendrites directed towards the surface and overlying the fibrous layer. Its chief input are the optic fibres, but it also receives general somatic afferents, passing through the bulbar lemniscus and the torus from the spinal cord and other parts of the mid-brain. The main efferent pathways lead towards the medulla, the habenulae, thalamus and hypothalamus. The torus, situated between optic tectum and motor tegmentum, receives its main afferents from the acoustico-lateralis area.

Because of the rudimentary condition of the eyes and optic nerves, the tectum of myxinoids cannot function as an optic correlation centre. This region in *Myxine* shows ill-defined cell groups not clearly separated from the semicircular torus. The input to these cells is assumed to come from ascending fibres in the lemniscus and descending fibres from the fore-brain. Rostrally, the cells of the tectum cannot be separated from those of the posterior thalamus. Among the cell masses in the basal plate, the nuclei of the cranial nerves supplying the non-existent eye muscles are lacking and as in lampreys, the mesencephalic nucleus of the gnathostome trigeminal is also absent. This can be related to an absence of jaw muscles, to which in higher vertebrates these nerves supply proprioceptive fibres. The main efferent tract is the medial longitudinal bundle arising from the cells of the mesencephalic reticular nuclei. Further efferent fibres are carried in tecto-bulbar and tectospinal tracts.

In the lamprey, Schwab (1973) considers the caudal thalamus, pretectum and tectum to be a common functional system, with similar cytoarchitectural differentiation and relationships and above all with a common dominant efferent pathway – the optic tract. This view is supported by the simultaneous development of these areas at the time when the paired eyes are developing during metamorphosis. Thus, in young ammocoetes the two halves of the tectum are still entirely separate and even in quite large animals their fusion is still incomplete. Also, in young ammocoetes the caudal thalamus, pre-tectum and tectum are continuous and without cellular differentiation. It is only when the optic tract is fully developed that these areas assume their definitive characteristics. Further support for these views has come from observations on the degeneration of the axons of retinal ganglion cells following section of the optic nerve and unilateral removal of the eye (Fig. 8.9) (Northcutt and Przbylski, 1973; Kennedy and Rubinson, 1977). The projections of these retinal axons have been identified contralaterally, in the posterior third of the dorsal thalamus, the pre-tectum and tectum. Especially in the adult, these axons in the dorsal thalamus have been observed within the area of the lateral geniculate nucleus. Similar results have been obtained by the use of the horseradish peroxidase technique, which has resulted in labelling of optic fibres in the tectum and geniculate nucleus as well as retrograde transport in cells of the dorsal and central tegmentum (Kosareva *et al.*, 1977). In the optic tectum itself, the pattern of the optic nerve terminals is said to be typical of other non-mammalian vertebrates, in so far as they are distributed contra-laterally and superficially, lying outside a central dense fibre zone. The tectum shows a distinct lamination of the medial and denser cell region (Fig. 8.8), which becomes much more distinct at metamorphosis, when there is a massive enlargement of the whole area of the brain, due partly to cellular proliferation, and partly to the migrations of cells from the periventricular zone. Degenerating axons in the adult lamprey were located in the eighth and

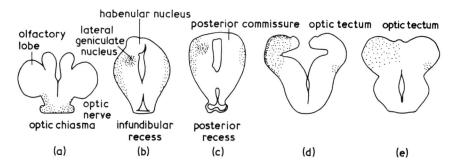

Fig. 8.9 Projections of retinal ganglion cells in the thalamus and optic tectum of the sea lamprey, *P. marinus* after section of the optic nerve and removal of the eye on one side, as seen in transverse sections of the brain at various levels. The site of the degenerating axons is indicated by the shaded areas. Redrawn from Kennedy and Rubinson, 1977.

(a) Diencephalon at the level of the optic chiasma. (b) Diencephalon at the mid-thalamic level. (c) At the level of the posterior commissure. (d) At mid-tectal level. (e) Caudal tectal level in the region of the ventral commissure.

more especially the ninth and most superficial layer, containing the optic tract fibres. At metamorphosis, a small and additional ipsilateral projection of optic fibres appears in an area at the ventrolateral margins of the pretectum and tectum, which in its location is said to have no counterpart in other vertebrates (Fig. 8.9).

The visual system of the lamprey has been investigated physiologically by stimulating the retina with light flashes of varying frequency or by direct electrical stimulation of the optic nerve (see Healey, 1972). Light flashes have resulted in responses, not only in the optic tectum, but more widely in the medulla and spinal cord (Veselkin, 1963, 1966). The fast oscillations recorded in the medulla still occur after removal of the tectum, but are abolished by incisions made between mid-brain and medulla. For these reasons, responses in the medulla and spinal cord are thought to be mediated through the tegmental motor nucleus of the mid-brain and the system of Muller neurones, thus implying the existence of visual afferents that by-pass the tectum (Fig. 8.10). This view is strengthened by recordings from electrodes deep in the mesencephalon, in the area of the tegmental nucleus, which showed potentials similar to those recorded from the tectum and with a similar latency. Medullary potentials have a latent period some $10-20\,\mu$s longer than those of the tectum and one or two of the later waves may be absent, suggesting the presence of an extra relay.

Responses in the tectum to light flashes show a positive wave of up to $200\,\mu$V with a duration of $60-80\,\mu$s and a latent period of $60-80\,\mu$s. This is followed immediately by a short negative wave and a series of variable oscillations. A number of observations point to a different origin of these components. With increased frequency of flash stimulation, the amplitude of the first wave in the tectal response decreases, whereas the amplitude of succeeding waves tended to increase but with decreased latency. More frequent stimulation thus selectively shortens the first wave. Anaesthetization with nembutal abolished all but the first wave and similarly, after interrupting the water supply irrigating the preparation, all components except the persistent first wave showed a decrease in amplitude and increased latency. Correlation with simultaneous recordings of activity in the optic nerve during light stimulation suggested that the primary tectal wave was due to the discharge of 'on units' in the retina and that the subsequent components represented the discharge of 'off units'. Direct electrical stimulation of the optic nerve results in a fast component interpreted as the presynaptic potentials of retinal ganglion cells, followed by a slow negative or positive-negative wave, thought to represent dendritic potentials within the radially arranged neurones of the tectal cortex (Karamyan *et al.*, 1966).

These experimental observations on the retino-tectal visual system of the lampreys illustrate the primitively diffuse organization of its nervous system, in which responses are spread over wide areas of brain and spinal cord. Although earlier work had suggested that these responses did not extend to the telencephalon, more recent studies have indicated that a retino-thalamic-telencephalic visual system may exist in lampreys and other lower vertebrates (Karamyan *et al.*, 1975), in which the responses to photic stimulation of the retina have much longer latent periods than those of the tectum. No details are known of these pathways, but it has been suggested that, as in the elasmobranchs, this system may be independent of the tectum (Fig. 8.10).

The relative importance of the cyclostome mesencephalon has been emphasized by Karamyan (1975) in a survey of the reflex mechanisms of the vertebrate brain. This author distinguishes five successive phylogenetic stages in the evolution of the integrative activity of the CNS, beginning with the

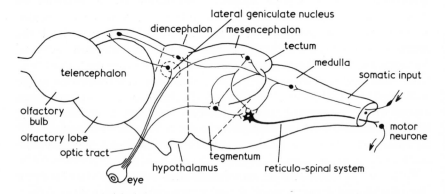

Fig. 8.10 The visual system of the lamprey and the integrative role of the mesencephalon. Based on the interpretations of Karamyan, 1975 and Karamyan *et al.*, 1975.

purely spinal level of the cephalochordates and reaching a climax in the neocerebellar-neothalamic-neocortical stage attained by the mammals. In this scheme, the lamprey represents a second stage, in which the dominant role in the establishment of reflex linkages is provided by the mesencephalic-bulbar system, where distance receptor afferents are switched to motor neurones through a single system of links – the reticulo-spinal system. This may be contrasted with conditions in fishes, where the well-developed cerebellum shares with the mid-brain the main burden of integration.

8.2.4 The diencephalon

Comprising those parts of the brain surrounding the third ventricle, the diencephalon may be divided into a dorsal epithalamus with a thin ependymal roof, the dorsal and ventral thalami forming the walls, and a ventral hypothalamus. Apart from the epiphysis which is evaginated from the roof of the lamprey diencephalon, the most conspicuous features of the epithalamus are the habenulae, whose relatively large size reflect the dominance of the olfactory system. The main input to the habenulae are olfactory fibres of second and third order, reaching it in widely dispersed olfacto-habenular tracts (Fig. 8.13). The main efferent channel is the conspicuous fasciculus retroflexus or Meynert's bundle, which appears to end around the interpeduncular nucleus, although some fibres may pass to the ventral thalamus and the posterior tubercular nucleus.

Developing from the epiphysis in the lamprey are the light-sensitive pineal and parapineal organs (Fig. 8.11). An epiphysis, absent altogether in *Myxine*, is present in the embryo of *Eptatretus*, whereas in the adult it is represented only by a rudiment, often containing a small ventricular recess. In lampreys of the genera *Lampetra* and *Petromyzon*, the smaller parapineal lies below the pineal, whereas in the adult *Geotria* it is located somewhat forwards and slightly to the left. However, this appears to be only a secondary condition and in the ammocoete of *Geotria*, the parapineal has a position very similar to that of the lampreys of the Northern Hemisphere (Eddy and Strahan, 1970). Above the pineal complex is an area free of myotomal muscles and covered by a translucent patch of unpigmented skin.

The pineal consists of a hollow vesicle connected to the pineal stalk. The vesicle has a dorsal pellucida, (which, in *Geotria*, has a lens-like shape) and ventrally a retina, consisting of an inner pigmented epithelial layer containing sensory cells and an outer layer of ganglion cells and nerve fibres. The comparatively rare photoreceptors are said to be of the cone type, with a few discs like those that support the photopigments of other retinal receptors (Meiniel, 1971). The ganglion cells form synaptic connections and have been implicated in the transmission of photic stimuli. Thus, although the structure of the pineal suggests that it retains some photoreceptive capabilities, it gives the impression of being an organ that is undergoing functional regression. In

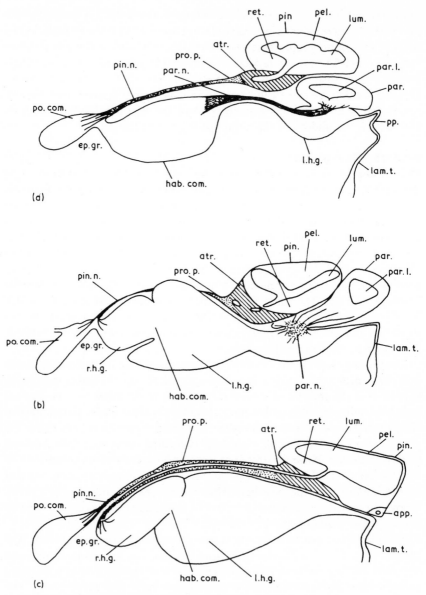

Fig. 8.11 Diagrammatic sagittal sections through the epiphyseal complexes of *Lampetra* (a), *Geotria* (b) and *Mordacia* (c). From Eddy and Strahan, 1970.

app. pineal appendix. atr. pineal atrium; ep. gr. ependymal groove; hab. com. habenular commissure; lam. t. lamina terminalis; r.h.g. right habenular ganglion; l.h.g. left habenular ganglion; lum. lumen of pineal; p.p. paraphysis; par. parapineal; par.l. lumen of parapineal; pin.n. pineal nerve; par.n. parapineal nerve; pel. pineal pellucida; pin. pineal; po. com. posterior commissure; pro. p. proximal part of pineal stalk; ret. retina.

the smaller parapineal, pellucida and retina are much less distinctly differentiated and the lower wall may contain only a few poorly developed sensory elements. In *Lampetra*, the pineal stalk contains about 600 nerve fibres forming the pineal nerve and connected directly to the right habenular ganglion and posterior commissure and in *Geotria* at least, some branches are said to join the bundle of Meynert. The parapineal is connected to the left habenular ganglion through an adjoining ganglionic mass, some of whose neural elements may have been derived by rostral migration from the habenular ganglion itself (Meiniel and Collins, 1971). In the Southern Hemisphere genus *Mordacia* the parapineal is absent, but attached to the front of the pineal vesicle is a small structure – the pineal appendix. This contains a few ganglion cells and nerve fibres and has sometimes been regarded as a vestigial parapineal, although this possibility has been questioned by Eddy and Strahan, mainly on the grounds that the appendix has no independent connection with the left habenular ganglion. The 'paired' nature of the relationship between the two elements of the pineal complex and the right and left habenular ganglia has naturally given rise to the belief that the parapineal and pineal have been derived from an ancestral form in which these structures were paired and probably symmetrical. On the other hand, it has been pointed out that if such a condition ever existed in the vertebrate lineage, it must have predated the known fossil agnathans, none of which shows paired epiphyseal foramina.

The condition of these structures in the lampreys of the Northern and Southern Hemispheres raises interesting problems of phylogeny. From the point of view of the histological differentiation of the pineal and parapineal and their photosensory functions, Eddy and Strahan suggest that *Geotria* represents the most developed form, *Mordacia* the least and that *Lampetra* (and presumably other Northern genera) would be intermediate in these respects. The complete (or almost complete loss) of the parapineal in *Mordacia* can hardly be other than a secondary regression, presumably indicating a long separation between this genus and the rest of the lamprey stock, but relative to *Lampetra* should we regard the more developed pineal of *Geotria* as a secondary improvement or as a more primitive state? However, the first of these alternatives would seem to imply a reversal of a previous evolutionary trend, involving the renewed elaboration of an organ that had already undergone considerable evolutionary regression.

Following illumination of the isolated pineal of *L. fluviatilis*, slow electrical discharges have been observed, as well as the impulse discharges of ganglion cells and nerve fibres (Morita and Dodt, 1973). Weak light stimuli cause a slowly rising positive response and a smaller deflection after the light is switched off. Stronger stimulation produces a faster positive deflection preceding the slow wave. The dark adapted threshold intensity was about 10^{-3} lm m^{-2} and above this threshold spontaneously active units were inhibited by light pulses of all wavelengths. During exposure to constant light of greater intensity, these units respond with off responses. The pineal is therefore described as a photoreceptor adapted to functioning at low light intensities and under these conditions, light signals of all wavelengths result in the inhibition of nervous discharges. Following pineal illumination, a slow negative wave appears in the optic tectum with a long latent period of about 140–150 ms, but nothing is known of the precise pathways that may be involved (Karamyan *et al.*, 1966).

In addition to its functions as a photoreceptor, it is now clear that the pineal is also a secretory organ, producing substances, which among other functions less well understood, are involved in the contraction of the melanophores and the diurnal changes in the skin colour of the ammocoete (Section 4.2.1). These rhythms are abolished by pinealectomy, but the normal skin pallor in the dark can be elicited by administering pineal extracts or melatonin (Eddy and Strahan, 1968; Eddy, 1972; Joss, 1973). A pineal enzyme, hydroxyindole-O-methyltransferase (HIOMT) converts serotonin to melatonin and the activity of this enzyme has been shown to reach a peak within 4 hours after the onset of the dark phase in *Geotria* ammocoetes (Joss, 1977). This short burst of melatonin produced melanophore contraction, but skin pallor did not persist over the whole of the dark phase and in fact skin darkening began before the onset of the next light period. Since the melatonin surge was so short-lived (Fig. 8.12), it is suggested that its wider function may be the timing of the circadian activity rhythms; the pineal acting as a biological clock or

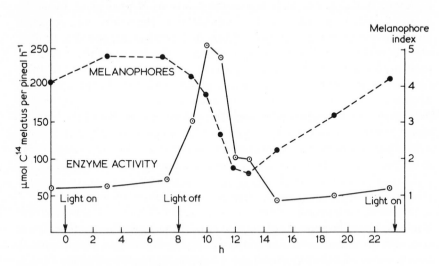

Fig. 8.12 Surge of hydroxyindole-O-methyl transferase activity in the pineal of the ammocoete of *Geotria* during a 24 hour cycle, correlated with changes in the contraction of the melanophores. Redrawn from the data of Joss, 1977.

neuroendocrine transducer converting diurnal changes in light intensity (or even perhaps seasonal changes in day length) into hormonal signals with far-reaching effects on the animal's metabolism. This might help to explain reports of the inhibition of metamorphosis in pinealectomised ammocoetes, since this is an event that is highly seasonal in its incidence (Eddy, 1969). Around the time of hatching and before the paired eyes are functional, the pineal of *Xenopus* tadpoles has been implicated in the reflex initiation of swimming activity; a response that can be produced by shading the pineal (Roberts, 1978). Similar responses could be invaluable in the 'blind' ammocoete stage and might have been developed by the benthic ancestral agnathans as a protection against predators.

The asymmetry of the habenular ganglia, in which the larger right member lies in front of the left, has been related to the connection of the former to the more developed pineal and the two ganglia are said to be more equal in size in earlier ontogenetic stages. On the other hand, the same kind of asymmetry also occurs in the myxinoids, where it has been attributed to pressure exerted by the growth of the olfactory lobes. In this connection it may be recalled that in their interpretation of the heterostracan brain, Whiting and Tarlo (1965) have shown a symmetrical epithalamus, with equally sized habenulae.

Behind this region, the medial part of the dorsal diencephalon of the lamprey is assumed to have important optic functions, including the auxilliary optic tract to the hippocampal region (Schwab, 1973). In this area, the posterior commissure connects the brain walls of the two sides, containing fibres ending in the pretectal region or running more ventrally towards the hypothalamus or the mesencephalic tegmentum. This dorsal thalamus also contains a dorso-medial cell group of bipolar neurones—the lateral geniculate nucleus, which receives the central projections of some axons of retinal ganglion cells (Kennedy and Rubinson, 1977).

With its reduced eyes, the myxinoid diencephalon is presumably dominated by olfactory afferents. Within the thalamus, the dorsal region is an important afferent area, receiving ascending fibres from the bulbar lemniscus and the tectum. In ventral areas there are important efferent centres with connections to the tegmentum via the post-optic commissure and with bulbar regions. In the hypothalamus, the post-optic nucleus receives important olfactory tracts and has connections with the tegmentum and medulla.

8.2.5 The telencephalon

The cyclostome telencephalon may be regarded as primarily, but not exclusively a rhinencephalon or nose brain, dominated by olfactory inputs through first order olfactory fibres entering the olfactory bulbs and by second or third order fibres to the olfactory lobes (Figs. 8.13, 8.14). As Andres (1975) has emphasized, there are striking parallels in the synaptic connections and neuronal circuitry of the olfactory bulb throughout the vertebrate series from

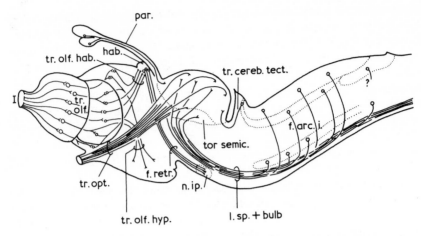

Fig. 8.13 Primary, secondary and some higher order olfactory connections and the principal afferents to the tectum mesencephali of the lamprey. After Nieuwenhuys, 1977.
f. arc. i. internal arcuate fibres; f. retr. fasciculus retroflexus; hab. habenula; l. sp. + bulb. spino-bulbar lemniscus; n. ip. interpeduncular nucleus; par. pineal nerve; tor. semic. semicircular torus; tr. cereb. tect. cerebello-tectal tract; tr. olf. olfactory tract; tr. olf. hab. olfacto-habenular tract; tr. olf. hyp. olfacto-hypophysial tract; tr. opt. optic tract; I. olfactory nerve.

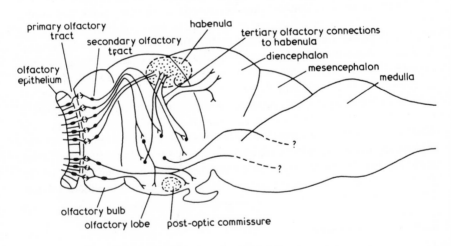

Fig. 8.14 Olfactory tracts of the hagfish. Based on Jensen, 1930.

the cyclostomes to the mammals, although in the higher vertebrates there is an increasing trend towards greater complexity of its layered structure and the differentiation of its cellular elements.

Primary fibres from the olfactory epithelium enter the olfactory bulb,

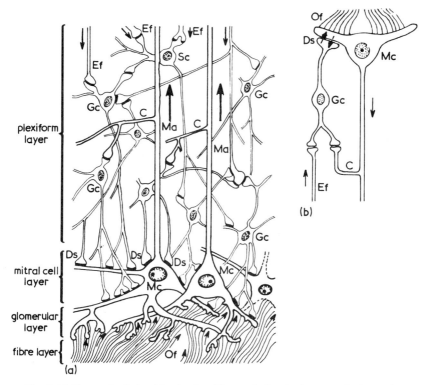

Fig. 8.15 Neuronal relationships in the olfactory bulb of a lamprey. Redrawn from
Andres, 1970, 1976.
(a) Section through the olfactory bulb. (b) Synaptic connections within a
glomerulus.
c. collateral of mitral cell axon; Ds. double two-way synapse; Ef. efferent fibre
of uncertain origin; Gc. granular cell; Ma. mitral cell axon; Mc. mitral cell; Of.
fibres of olfactory nerve; Sc. stellate cell.

forming a superficial stratum nervosum (Fig. 8.15). This olfactory information is conveyed to the dendrites of the mitral cells in what is termed the
glomerular layer. Both the glomeruli and the mitral cells are said to be better
differentiated in hagfishes than in lampreys. The mitral cell axons constitute
the main, if not the sole, olfactory output from the olfactory bulb, forming the
secondary olfactory tract to the olfactory lobes. Below the mitral cell layer is a
plexiform zone containing granular and stellate cells with short axons.
Processes from the granular cells establish connections with collaterals from
the mitral cell axons as well as two way or bipolar synapses with their
perikarya and main dendritic branches (Fig. 8.15). Although the dominant
input to the olfactory bulb is through the primary olfactory fibres, this
information can be modulated by wider inputs reaching the olfactory bulb via
the synaptic connections described by Andres between the efferent endings of

uncertain origin and the granular and stellate cells of the plexiform layer. These efferent channels are said to become progressively more numerous in the higher vertebrates. In these ways, the complex neuronal circuits of the olfactory bulb make it possible to visualize, not only the amplification of weak olfactory signals through the reciprocal connections of the mitral and granular cells, but also the possibility for the damping down of strong stimuli through the activity of efferent impulses from the higher centres (Fig. 8.15).

Although there have been widely varying interpretations of the cellular areas of the lamprey hemispheres, Nieuwenhuys (1967b) identified only a primordium hippocampi, pallium, corpus striatum and pre-optic nucleus (Fig. 8.16). Because it receives numerous secondary olfactory fibres, the evaginated pallial region is often referred to as the olfactory lobe. Some of these fibres decussate in the olfactory commissure and pass to the primordium hippocampi; others pass more deeply to the striatal and pre-optic regions. In addition to these secondary olfactory tracts ending in the telencephalon, others including mitral cell axons, are believed to reach the thalami and hypothalamus, or even as far as the tegmentum of the mid-brain.

In the solid hemispheres of the hagfish, Nieuwenhuys distinguished a basal area or striatal primordium, a hippocampal primordium and a pallial region (Fig. 8.17); the latter showing a very distinct five-layered structure in which the second layer was considered to be a true pallial cortex. A similar point of view

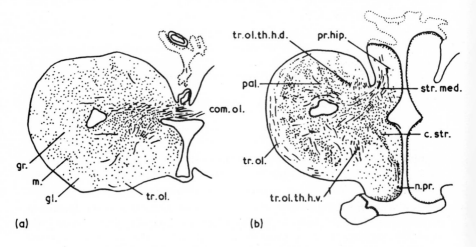

Fig. 8.16 Transverse sections through (a) the olfactory bulb of the lamprey and (b) through the olfactory lobes (cerebral hemisphere). From Nieuwenhuys. 1967b. com. ol. olfactory commissure; c. str. corpus striatum; gl. glomerular layer; gr. granular layer; m. mitral cell layer; n.pr. preoptic nucleus pal. pallium; pr. hip. primordium hippocampi; str. med. stria medullaris; tr. ol. olfactory tract; tr.ol.th.h.d. olfacto-thalamic and olfacto-hypothalamic tracts (dorsal); tr.ol.th.h.v. olfacto-thalamic and olfacto-hypothalamic tracts (ventral).

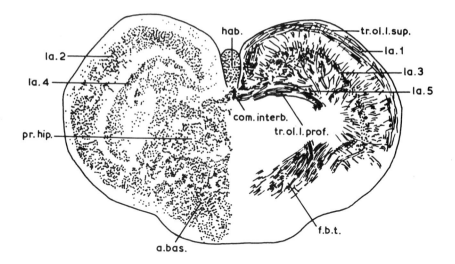

Fig. 8.17 Transverse sections through the telencephalon of *Myxine*, with the cell patterns on the left and the fibre tracts on the right. From Nieuwenhuys, 1967b.
a. bas. basal area; com.interb. interbulbar commissure; F.b.t. basal bundle; hab. habenula; la. 1–la. 5 layers of olfactory bulb; pr. hip. primordium hippocampi; tr.ol.l. prof. deep lateral olfactory tract; tr.ol.l.sup. superficial lateral olfactory tract.

was adopted by Kuhlenbeck (1975) and Schober (1966), who noted a tendency in the lamprey hemisphere for the nerve cells to migrate towards the surface as small cell clusters. This trend towards surface stratification outside the white matter of the evaginated hemisphere, was felt to justify the use of the term 'precortex'. As in the lamprey, all regions of the myxinoid hemisphere receive secondary olfactory fibres, carried in large lateral and smaller medial and ventral tracts. The lateral tracts have superficial and deep divisions corresponding to the first and fifth layers of the pallium and, from these, connections are established with cells in the second and fourth layers. Others pass to the opposite side in a commissure located immediately behind the habenulae. Fibres from the ventral tracts descend towards the pre-optic region, but a majority run in a basal forebrain bundle towards the hypothalamus. As in lampreys, there are also said to be olfactory connections with the tegmentum of the mid-brain (Fig. 8.14). The absence of lateral ventricles in the hagfish forebrain is a secondary condition, arising during embryonic development by the invagination of the hippocampal primordia and their fusion with the rudiments of the olfactory bulbs.

The hippocampal primordium of the lamprey is the dorso-medial and unevaginated part of the telencephalon. This area is said to retain the more primitive architecture of the CNS with nerve cells confined to a periventricular layer, which in this region lacks a typical ependyma. Lateral to this layer is a fibre zone. The main afferent channels of this area are in the stria medullaris, crossing from the dorso-lateral telencephalon (Fig. 8.16). In addition, there are connections with the hypothalamus and possibly with the mesencephalic tegmentum, while at the caudal end there are further links with the caudal thalamus and the habenulae. Schwab (1973) believes that this region has primary olfactory functions and forms an integral part of an olfactory system, embracing also the olfactory bulbs and habenulae: A quite different view has been adopted by Bone (1963) in relation to the hippocampal primordium of *Myxine*. This he considered to be relatively free of the olfactory dominance that characterizes other regions of the telencephalon, although it was thought to receive a few third order olfactory afferents from the olfactory lobes. For this reason, Bone suggests that it has retained a relatively primitive and undifferentiated condition, which in the phylogeny of the vertebrate brain would have permitted its further development as an associative cortex.

In both groups of cyclostomes, as in fishes, the immediate electrical response of the olfactory epithelium to irrigation by water containing odoriferous stimulants is a monophasic action potential, consisting of a fast 'on response' followed by a slow exponential decline and a return to baseline levels when the stimulus is removed. In lampreys, these responses have been observed to amines or crushed fish tissues (Kleerekoper, 1972) and in *Myxine* to various amino-acids including L-glutamine and L-alanine (Doving and Holmberg, 1974).

Electrical activity in various parts of the olfactory system of the lamprey has been monitored after stimulation of the olfactory nerve fibres (Bruckmoser, 1971; Bruckmoser and Dobrylko, 1972). The potentials observed are clearly very similar to those recorded in other vertebrates (Fig. 8.18), emphasizing the basic conservatism in the structural organization of the olfactory system. Recordings in the olfactory bulb showed four, predominantly negative components. The first of these, smaller than the others, was at its largest on the surface of the bulb and in the neighbourhood of the fibres of the olfactory nerve and has been interpreted as the presynaptic potential of the olfactory fibres. Estimates of the velocity of conduction in these fibres, based on their average length and on the observed latency of about 10 ms gave a figure of $0.1–0.15$ m s^{-1} which is similar to values reported for other vertebrate olfactory nerve fibres. The second component is considered to be the potential of the secondary neurones in the olfactory pathway, i.e. the mitral cells, and at the depth of the bulb where these cell bodies occur, it is replaced by the earliest signs of spike activity. The third component is believed to represent the activation of periglomerular granular cells through two-way synapses connecting these cells with the mitral cells and also via recurrent mitral cell collaterals (Fig. 8.15). The final fourth and slow potential, which is thought to originate in the nucleus of the olfactory bulb, has a latency of about 60 ms. It

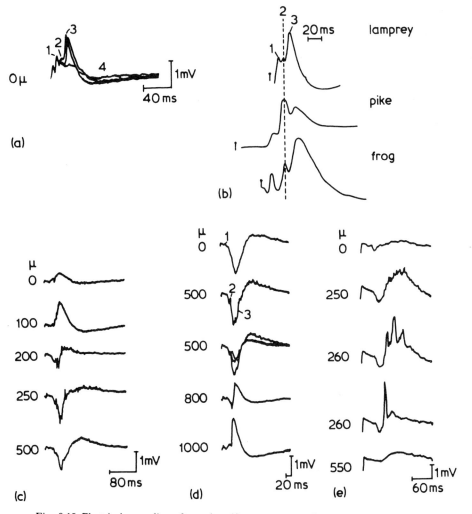

Fig. 8.18 Electrical recordings from the olfactory system of the lamprey following stimulation of the olfactory nerves. From Bruckmozer, 1971.

(a) Ipsilateral responses from the surface of the olfactory bulb. 1–4 are the four components. With repeated stimulation all except the first component are fatigued.

(b) Comparisons of olfactory bulb potentials of lamprey, pike and frog. The curves have been adjusted so that the second component coincides.

(c) A recording from different depths in the olfactory bulb showing the reversal of the second and third components at the depth of the dorsal glomerular layer.

(d) Reversal of second and third components in the ventral glomerular layer. At 500 μm the potentials are superimposed at a frequency of stimulation of 0.5 s^{-1} and components 2–4 are fatigued.

(e) Recording through the olfactory lobe as far as its ventral surface (500 μm). At 260 μm sharply localized rhythmic activity is seen.

reaches its greatest amplitude in the centre of the olfactory bulb and shows indications of spike activity. A characteristic feature of the second and third components is the reversal in their polarity that occurs when the electrode reaches the level of the glomerular layer and it is significant that the sequence of these changes in polarity is reversed, according to whether the electrode is penetrating this layer from above or below.

Recordings from the olfactory lobe (hemisphere) show a negative wave at all depths with a latency of 60–80 ms and a maximum amplitude at the centre of the lobe, decreasing towards both dorsal and ventral surfaces. Recording from various areas gave no physiological evidence of a topographical differentiation of the olfactory lobe. After olfactory nerve stimulation, responses have also been observed in the hippocampal region. These consisted of a first rapid and second slow negative component with a latency of 70–90 ms and attributed to the activation of hippocampal elements by mitral cell axons.

Opposing the view that the telencephalon of cyclostomes is solely concerned with olfactory afferents and that the neocortex of higher vertebrates has evolved from the 'olfactory brain' of primitive forms, Karamyan *et al.* (1975) have adduced experimental evidence that both visual and somatic afferent systems are represented in the lamprey telencephalon and that stimulation of the spinal cord evokes excitatory potentials in this region of the brain as well as in the diencephalon and brain stem (Fig. 8.19). Similarly, stimulation of the optic nerve or illumination of the retina also evokes telencephalic responses which these authors believe to be the result of the presence of visual afferents in the hippocampal primordium. A possible anatomical basis for this contention may be the small decussation of optic fibres described by Schwab (1973) immediately dorsal to the optic chiasma and which he was able to trace to their endings in the hippocampal region.

8.3 General characteristics of the cyclostome central nervous system

Anatomists and physiologists alike have been impressed by the apparently diffuse nature of the connections within the cyclostome CNS. In his morphological analysis of the myxinoid brain, Jansen (1930) remarked on the wealth of diffuse connections that exist in this structure, in addition to better organized and more distinct tracts. This condition, which he compared to the nerve nets of invertebrates, was attributed to the existence of long, branching dendritic processes and to the large numbers of collaterals that are given off along the course of the axons. A similar point was also made by Bruckmoser (1971), who stressed the comparatively undifferentiated and diffuse character of the lamprey olfactory system, as demonstrated by the uniform activation that occurs throughout the telencephalon after olfactory nerve stimulation.

The obvious similarities in the general organization of the CNS of

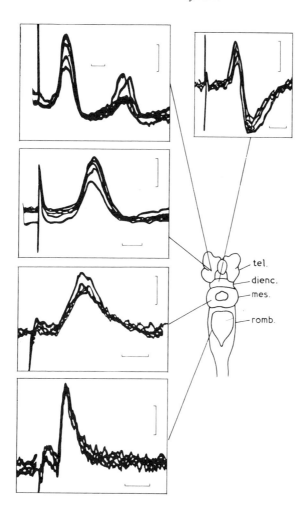

Fig. 8.19 Evoked potentials in the lamprey brain following stimulation of the spinal cord.
From Karamyan, 1972.
   The diagram also shows the position of the recording electrodes in the brain
Tel. telencephalon; dienc. diencephalon; mes. mesencephalon; romb. rhomben-
cephalon. Scales 100 mV and 40 ms.

myxinoids and lampreys, as well as the unique shape of the adult spinal cord,
are sufficient reminders of a common, though remote ancestry. The strange
external appearance of the hagfish brain, the reduction of the ventricular
system and most of the other divergences from the lamprey pattern can be
accounted for by morphogenetic changes in early embryonic development, by

reduction in the eye, ear or lateral line organs or by the excessive development of the olfactory and tactile senses. Thus, the rudimentary lateral line and degenerate eyes are reflected in the absence of the cerebellum and the small and poorly differentiated mid-brain region, while the dominance of the olfactory sense is parallelled by the greater histological differentiation of the myxinoid olfactory centres.

The reticular system has been regarded by Kuhlenbeck (1975) as part of a final common pathway through which sensory inputs from a wide variety of sources are able to converge on the motor units. Viewed in this light it may not be too fanciful to imagine that the apparent lack of differentiation in the Muller and Mauthner neurones of the hagfish, may be part of the general regression that has overtaken so many of its sensory systems. In this connection it may be significant that in lampreys, the olfactory system so dominant in myxinoids, is the only major sensory system for which Rovainen (1967) did not detect effects in reticular neurones after cranial nerve stimulation. On the other hand, EPSP's have been reported by Wickelgren (1977a,b) after electrical stimulation of the olfactory nerves, although it may be significant that these Muller cell responses showed much longer latencies than those that followed stimulation of other cranial nerves. Moreover, no responses followed the irrigation of the olfactory organ by water containing fish odours, whereas excitation did occur when appropriate natural stimuli were applied to optic, vestibular or touch receptors. The absence of Mauthner neurones in the myxinoids is, however, in line with their tendency to atrophy in bottom-dwelling fish species and those showing reduced swimming activity (Zottoli, 1978).

9

Osmotic and ionic regulation

The entirely marine hagfishes, living in water of full salinity and often at considerable depths, maintain their body fluids at concentrations close to, if not completely identical with those of their environment (Robertson, 1954). Thus, like the marine invertebrates and protochordates they are described as isosmotic and osmotic conformers. While they are in freshwater, lampreys on the other hand must be able to sustain their body fluids at concentrations far above those of the environment (hyperosmotic regulation), whereas in their marine phase they maintain internal concentrations little more than a third of those of the sea water (hyposmotic regulation).

9.1 Osmotic and ionic concentrations

9.1.1 Body fluids

In the concentration of their body fluids, the two groups of cyclostomes represent almost the extreme poles of the vertebrate series. The hagfishes have the highest blood concentrations of any vertebrate (1060 mOsm) except for the elasmobranchs, whereas the ammocoete of *L. planeri* has a concentration of only 205 mOsm, parallelled only by polypteroids and lungfishes (Lutz, 1975b). The high concentrations of hagfish blood are almost entirely due to inorganic ions, which account for about 98 % of the osmolar concentration (Robertson, 1976), the small deficit being made up mainly by urea and trimethylamine oxide. Yet, in spite of this osmotic equilibrium between the blood and the sea water there are some significant differences in the distribution of various ions, indicating some active regulation of the internal ionic composition.

In all species of lamprey that have been studied, the lowest blood concentrations are found in the larval stages and serum osmolarity begins to increase during and after metamorphosis coinciding with the change from the larval to adult kidney (Ooi and Youson, 1977). Comparisons of the freshwater serum concentrations in the three species, *L. planeri*, *L. fluviatilis* and *P. marinus* suggests that values tend to increase with body size, although in this respect, the dwarf freshwater race of the sea lamprey, *P. marinus* may be exceptional in showing somewhat higher osmolarities than the larger anadromous form of this species. However, the freshwater values for the latter represent a somewhat artificial situation and their lower serum osmolarity may reflect a failure of these animals to develop completely the mechanisms of hyperosmotic regulation at a period in the life cycle when these would normally be superseded by adaptations to life in a marine environment. In both forms of the sea lamprey, the blood concentrations are high when the animals first enter the rivers on their spawning migration, decreasing with sexual maturity and falling precipitously after the completion of spawning.

9.1.2 Ionic composition of the tissues

Indirect estimates have been made of the osmotic composition of hagfish muscle, based on the volume and composition of the extracellular fluid. These show the usual differences in the ionic composition of cell contents and extracellular fluids. Thus, sodium and chloride are in higher concentrations in the fluids bathing the surfaces of the cells, whereas concentrations of potassium and phosphate are higher within the cell than outside. A striking feature is the fact that inorganic and organic ions together account for rather less than half the intracellular osmotic activity; the remainder being due mainly to organic nitrogenous compounds, notably amino acids, trimethyl-amine oxide and betaine (Robertson, 1976). This is precisely the kind of situation that exists amongst marine invertebrates, such as the decapod crustaceans, where a high proportion of the osmotic activity of their muscle tissue is also attributable to these organic compounds.

9.2 Osmotic relations with the environment

9.2.1 Myxinoids

In their natural environment, hagfishes are likely to encounter only minimal changes in the concentration of the sea water and to this extent, experiments involving the dilution or concentration of the ambient water could be regarded as unphysiological. Nevertheless, these laboratory experiments have provided some clues to normal physiological processes or properties, including the permeability of the body surfaces to water or ions and the animal's potential for regulating its body volume. It should also be borne in mind that when the hagfish is feeding, the ingestion of fish tissues and body fluids might well create a temporary, although slight, osmotic load.

In spite of their tendency to suffer from abrupt changes in salinity, myxinoids have shown a surprising capacity to tolerate slow and gradual acclimation to dilution or concentration of the sea water. Both *Myxine glutinosa* and *Eptatretus (Paramyxine) atami* die quite quickly if they are transferred directly from normal sea water to dilutions of 20–25 ‰. On the other hand, *Myxine* may be acclimated successfully to slowly increasing or decreasing salinities and has been maintained for several weeks in a range of salinities from 600–1500 mOsm (Cholette *et al.*, 1970). In general, it appears that at least some members of the genus *Eptatretus* tend to be more tolerant of salinity changes than *Myxine*. The Pacific hagfish, *E. stouti* has tolerated direct transfer to 80 % or 120 % sea water for at least 7 days, after which they were returned to normal sea water. In 80 % sea water the body swelled rapidly and during the first day, their weight increased by about 10 % (McFarland and Munz, 1965). This was followed by a slow return to their original weight over a period of several days, but when replaced in normal sea water, the animals died. In 120 % sea water the animals initially lost about 25 % of their body weight, subsequently remaining at this level. When finally returned to full strength sea water, they recovered or even slightly overshot their initial weight (Fig. 9.1).

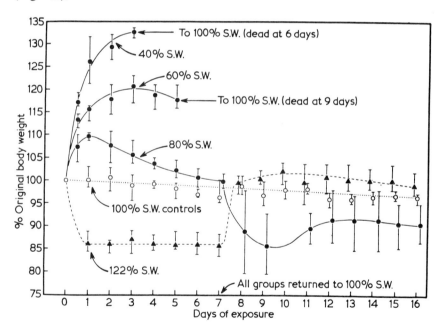

Fig. 9.1 Changes in the body weight of the hagfish, *Eptatretus stoutii* during exposure to concentrated or dilute sea water. From McFarland and Munz, 1965.
Open circles; controls maintained in 100 % sea water (950 mOsm).
Black circles; exposed to dilute sea water at concentrations as indicated.
Black triangles; concentrated sea water 122 %.
Except for the hagfish in 40 % and 60 % sea water, all were returned to normal sea water after 7 days.

All the experiments carried out on hagfishes in varying concentrations of the external medium agree that the animals remain virtually isosmotic throughout, behaving as almost perfect osmoconformers. The rapidity of the weight changes in dilute or concentrated sea water indicate a high degree of permeability to water of the body surfaces, including the gills. This has been confirmed by measurements of the water flux using tritiated water (Rudy and Wagner, 1970). In *Eptatretus*, an exchange rate of 2287 ml kg^{-1} h^{-1} has been observed. This is some 5–10 times higher than the rates for freshwater teleosts measured by the same techniques and 20–50 higher than those recorded in marine teleosts (Fig. 9.2). In these respects, a parallel has been drawn between the conditions in the myxinoids and in the decapod crustaceans. In the latter group, the freshwater forms such as the crayfish show much lower water permeability than the isosmotic marine crabs, whose high exchange rates are similar to those of the hagfish. High water permeability of this order could only be tolerated by animals which would not normally experience serious

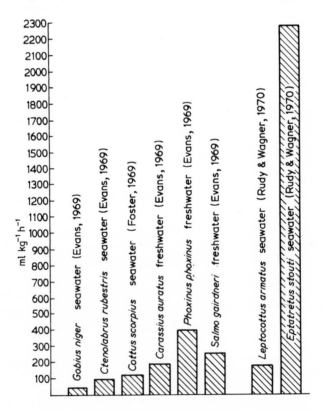

Fig. 9.2 Water influx in the hagfish, *Eptatretus stouti* compared with that of marine and freshwater teleosts. From Rudy and Wagner, 1970.

osmotic gradients and where there would therefore be little osmotic movement of water through the body surfaces, to and from the external medium.

Experiments involving the subjection of hagfishes to varying salinities have also demonstrated how osmotic stresses are transmitted to the tissues of an isosmotic animal through the medium of the extracellular fluids. These stresses are expressed by the transfer of water from cell to extracellular fluid with increasing concentration of the external medium and a movement in the reverse direction when the external medium is diluted. When *Myxine* was subjected to concentrations of sea water from 600–1500 mOsm, a linear relationship was observed between the volume of the extracellular fluid compartment of the parietal muscles (Cholette *et al.*, 1970) and the osmotic concentration of the external medium. Thus, when the sea water concentration was doubled from 700–1400 mOsm the extracellular space, expressed as a percentage of the body water, increased threefold and over the same range, the water content of the muscle increased from 66–78 g water 100 g tissue^{-1}. Assuming that tissue cells and extracellular fluid are in osmotic equilibrium, the observed gains or losses of intracellular water do not account completely for the changes that must have occurred to maintain the isosmotic condition. This indicates some capacity for regulating intracellular osmotic concentrations, in which the pool of free amino acids is believed to play a significant part.

9.2.2 Lampreys

Twice in their life the anadromous lampreys are exposed to drastic changes in their external environment. The first of these crises occurs after metamorphosis, when the newly transformed adults or macrophthalmia drift downstream towards the estuary or open sea, necessitating a switch from freshwater hyperosmotic regulation to marine hyposmotic regulation. The second episode occurs at the onset of the spawning migration when these regulatory mechanisms are once more reversed. Although we can safely assume that the anadromous habit and the mechanisms of marine regulation have been secondary introductions into the life cycles of lampreys, we cannot be sure whether existing freshwater species such as *Ichthyomyzon, Tetrapleurodon* or *Eudontomyzon* have retained the primitive condition or, like some of the non-parasitic lampreys, have descended from anadromous forms. In the latter case, it might be expected that these freshwater species would retain at least some limited capacity for regulation in saline media, but this has not yet been tested experimentally. In this connection, comparisons of the hyposmotic abilities of the closely related freshwater and anadromous forms of the sea lamprey, *P. marinus* are of particular interest in showing that a measurable change in osmotic performance may occur within a relatively short period of time.

At the larval stage, lampreys are quite incapable of hyposmotic regulation and, in ammocoetes of both anadromous and freshwater sea lampreys, serum osmolarity rises abruptly in water of more than $10\%_{oo}$ salinity (about 256 mOsm). In the anadromous form, the ammocoete values in freshwater (225 mOsm) or in $8\%_{oo}$ sea water (233 mOsm) are not significantly different, but in $10\%_{oo}$ sea water they reached a serum osmolarity of 299 mOsm after 8 days (Beamish *et al.*, 1978). During this gradual acclimation to increasing concentrations, body weight decreases with the osmotic loss of water to the external medium, blood volumes are drastically reduced and the concentration of the blood, although remaining hyperosmotic, rises with that of the water.

The transition from freshwater to marine regulation is dependent on morphological and physiological changes that occur during metamorphosis, particularly those affecting the gills, kidneys, gut and general body surfaces. Only when these changes have been completed are the animals able to move towards saline environments. At this time the macrophthalmia of *L. fluviatilis* or *P. marinus* can usually be transferred direct from freshwater to full strength sea water. However, even then the switch from freshwater to marine regulation is not irreversible and downstream migrant stages are able to survive a return from salt water to fresh.

In their studies on the landlocked sea lamprey, Mathers and Beamish (1974) have shown that marine regulation is first developed after metamorphosis, when the macrophthalmia, unlike the ammocoete, are able to regulate successfully in $10\%_{oo}$ sea water, maintaining levels of serum osmolarity similar to those of the adults in freshwater (Fig. 9.3). On the other hand, in water of this concentration the blood concentration of the ammocoete increased to about 10% above freshwater values. Young adult lampreys, that had just begun to feed, were able to regulate at least for short periods in higher salinities up to normal sea water, but only the larger individuals were able to cope for long periods with salinities above $10\%_{oo}$. At a salinity of $26\%_{oo}$ animals with lengths of 127–188 mm suffered a 50% mortality within a 10 day period and only individuals of 280 mm or more were able to survive in full-strength sea water. This increase in regulatory capacity with age and body size is no doubt partly attributable to a decrease in the relative body surface, effectively reducing the area exposed to ionic and osmotic gradients.

In parallel experiments on the anadromous sea lamprey (Beamish *et al.*, 1978), small feeding adults only recently transformed and with body lengths of 135–140 mm were subjected to a series of increasing salinities from freshwater up to normal sea water (Fig. 9.3). Until salinities exceeded $16\%_{oo}$ serum osmolarities showed no increases over the freshwater values and even beyond this point, increases in blood concentration were only slight. In $34\%_{oo}$ sea water, serum osmolarity averaged 263 mOsm, representing a concentration less than a third that of the external medium. In a smaller series of large feeding adults with body lengths of 260–310 mm, serum concentrations rose

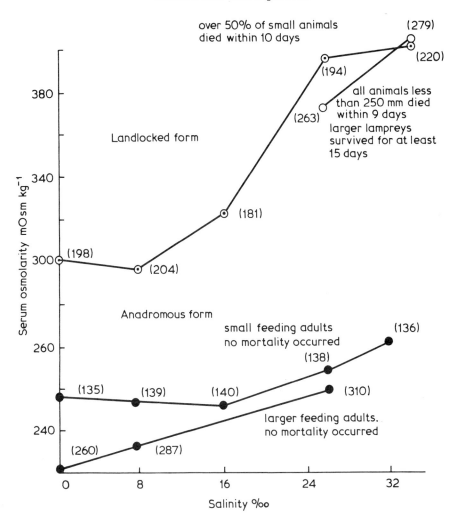

Fig. 9.3 Comparisons of the serum osmolarities of landlocked and anadromous sea
lampreys (*Petromyzon marinus*) exposed to various salinities. Redrawn from the
data of Mathers and Beamish, 1974 and Beamish *et al.*, 1978.
Figures in brackets indicate the mean lengths of the experimental animals.

from 200 mOsm in fresh water to about 210 mOsm at 8‰ and about 250
mOsm in 26‰. In upstream migrants nearing sexual maturity, the capacity
for marine regulation is clearly lost and serum osmolarities increased sharply
in 16‰ sea water. This rise continued in 26‰ sea water and was
accompanied by heavy mortality.

In these experiments, the serum osmolarities for the smaller sea lampreys
are lower than those suggested for large and mature specimens of the same

species, for which determinations have been made soon after the animals first entered the rivers (Fontaine, 1930; Sawyer, 1957; Pickering and Morris, 1970). When measured in freshwater, these have given values varying from 220–291 mOsm. For sea lampreys actually taken in sea water we have only two determinations; the first (when recalculated from the freezing point) gives a value of 315 mOsm (Burian, 1910) and the second suggests a concentration of about 346 mOsm (Robertson, 1974), although in the latter case there is an element of uncertainty since haemolysis had already occurred. While it is difficult to form a clear picture from this data of conditions during the marine phase, the information seems to suggest that the values obtained by Beamish *et al.*, for recently transformed animals under laboratory conditions may represent rather low values for the blood concentrations of larger animals in the sea and in this connection it may be noted that the 34‰ sea water used in these experiments gave an osmolar concentration of only 919 mOsm kg^{-1}.

Nevertheless, in spite of these uncertainties, comparisons with the land-locked sea lamprey, show quite clearly that this dwarf form has lost some of its capacity for hyposmotic regulation and at higher salinities its performance is markedly inferior to that of the ancestral anadromous form. Once the salinity of the external medium has risen beyond the isosmotic point, serum osmolarity for the anadromous lamprey was always well below the levels observed in the landlocked form (Fig. 9.3). It has been widely assumed that the origin of the freshwater race can be traced back to the recession of the Wisconsin ice sheet, which allowed the anadromous parent stock to invade Lake Ontario. If this assumption is valid, the quite pronounced changes in the osmoregulatory ability of the landlocked race, must have been accomplished within a period of separation of no more than about 8000 years. This interpretation would be in agreement with a suggestion that the differentiation of the freshwater forms has involved selection operating within the parent population in favour of smaller body size, shorter parasitic phase and lower potential fecundity, combined with some restriction in migratory habits and capacity for marine osmoregulation (Hardisty, 1963, 1964).

9.3 Physiological mechanisms

9.3.1 Cyclostome kidneys

Cyclostomes are characterized by their possession of a persistent pronephros, although in lampreys this is functional only in the late embryonic and pro-larval stages and, in the hagfish, the pronephric tubules do not retain a connection with the main kidney duct and cannot therefore contribute to the formation of urine. The lamprey pronephros consists of 3–5 coiled tubules, served by a single glomus, receiving blood from the dorsal aorta. Each tubule opens by a ciliated funnel into the pericardial coelom and is connected at its distal end with the nephric duct. After hatching, the functions of the pronephros are gradually taken over by the continuous development of mesonephric tubules, some 5–6 segments behind its posterior end and in the adult lamprey, although the funnels remain, the tubules have disappeared (Fig. 11.11).

In the hagfish, the pronephric tubules are initially segmental, but are then increased in number by branching until, in the adult, several hundred small tubules are present, opening by ciliated funnels into the pericardial coelom and communicating internally with the central mass of lympho-myeloid tissue. The single large glomus is thought to be concerned in the formation of the peritoneal fluid (Fänge, 1973) and there is some evidence that the central mass acts as a phagocytic filter between the peritoneal fluid and the blood.

During the larval life of the lamprey, new mesonephric tubules continue to be differentiated at the posterior end of the kidney as those at the cranial end regress, but during metamorphosis the ammocoete mesonephros breaks down completely and is wholly replaced by newly differentiated adult tubules (Ooi and Youson, 1977). The significance of this curious transformation remains obscure. Further growth appears to involve the lengthening of existing tubules rather than the addition of new nephrons. A unique feature of the lamprey kidney is the union of the separate glomeruli to form a continuous rope-like strand surrounded by the capsules into which open the ciliated funnels of a number of nephric tubules. In the adult, unlike the ammocoete, these capsules lack the usual double-layered structure, the visceral layer consisting only of podocytes covering the glomerular surface (Youson & McMillan, 1970a). Each nephron has a short narrow neck segment, followed by the long convoluted proximal tubule, lined by columnar cells bearing microvilli (Fig. 9.4). The short distal segment consists of cells rich in mitochondria and this section, together with the terminal region of the proximal tubule is arranged in ascending and descending loops which have been compared to the loop of Henle of the mammalian or bird kidney (Notochin, 1977). The final collecting ducts are formed by the distal segments of four adjacent nephrons (Fig. 9.4) (Youson and McMillan, 1970b; 1971a,b).

The hagfish mesonephros is represented by little more than the longitudinal archinephric ducts and the segmentally arranged glomeruli (Fig. 9.4). In the more caudal regions glomeruli may be absent, but conditions in the embryo suggest that this may be a secondary loss. Similarly, there may be a short cranial section where the ducts end blindly and are devoid of glomeruli, while isolated glomeruli unconnected with the ducts may be present in the area between pronephros and mesonephros. This variability, together with the branching and convolutions of the ducts in the embryo of *Eptatretus* have sometimes been adduced as evidence of phylogenetic regression in the myxinoid kidney, both as regards its extension and tubular development. However, in spite of its rudimentary tubule, the hagfish kidney appears to be able to carry out some of the functions associated with the highly differen-

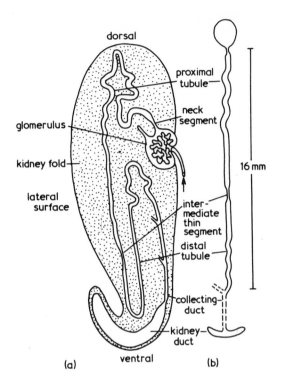

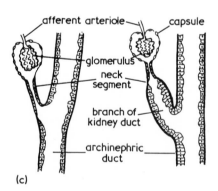

Fig. 9.4 Cyclostome kidney tubules.

(a) Arrangement of the lamprey tubule in relation to the kidney fold. Redrawn from Youson and McMillan, 1970.

(b) Scale drawing of a single tubule. Based on Goncharevskaya, 1975.

(c) Variant forms of the kidney tubule of the hagfish, *E. stouti*. Redrawn from Jollie, 1962.

tiated tubule of lampreys or higher vertebrates and there are indications that the archinephric duct is far from being simply a passive route for the passage of the urine. Throughout its length the epithelium has a brush border of microvilli with numerous apical pits, tubules and vesicles, suggestive of the structure of the proximal tubule of higher vertebrates (Heath-Eves and McMillan, 1974; Kuhn *et al.*, 1975). Moreover, these structural indications of endocytosis have been confirmed by the experimental uptake of exogenous macromolecules (Ericsson and Seljelid, 1968).

9.3.2 Marine regulation

(a) *Lampreys*

In their marine osmoregulatory mechanisms, lampreys have developed methods similar to those evolved by marine and euryhaline teleosts and these are capable of maintaining rather similar levels of hyposmolarity. This involves the replacement of the considerable volumes of water lost osmotically through the body surfaces by swallowing and absorbing sea water (Fig. 9.5). The rate at which water is swallowed is known only for river lampreys *L. fluviatilis* held in 50 % sea water. Under these conditions, rates varies from 5–22 % body weight day^{-1} of which about 75 % is absorbed mainly in the anterior intestine (Pickering and Morris, 1973). The mechanism of water absorption depends on the active uptake of sodium and chloride ions in the gut epithelium; the water then following passively along the osmotic gradient across the gut wall. Divalent ions – magnesium, calcium and sulphate – remain in the gut lumen and are apparently eliminated by this route, although it is possible that there may also be some active excretion of these ions into the luminal contents of the gut (Morris, 1960, 1972). Apart from the replace-

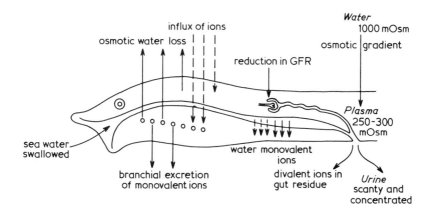

Fig. 9.5 Marine regulatory mechanisms of the lamprey.

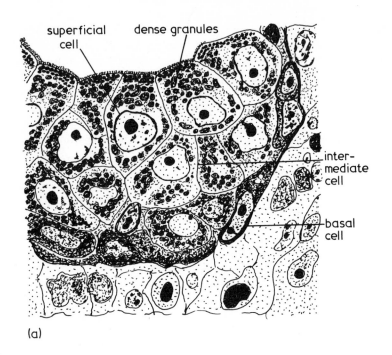

(a)

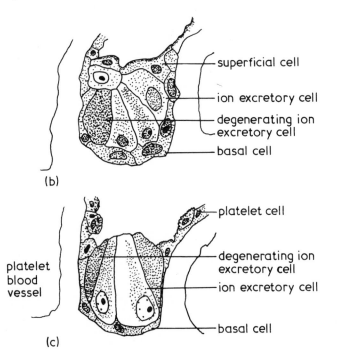

(b)

(c)

ment of water losses, the lamprey in sea water will also be permeable to ions diffusing through the body surfaces. As in teleosts, the elimination of monovalent ions is achieved extra-renally, through the activity of specialized chloride cells (Morris, 1957), located in the interplatelet areas of the gills (Fig. 9.6). These are characterized by their abundant mitochondria and prominent smooth endoplasmic reticulum (Pickering and Morris, 1976). In freshwater, the majority of these ion-transporting cells are covered by a layer of superficial cells, whereas in sea water their tips are exposed. In areas where teleost chloride cells occur, a sodium-potassium ATPase is thought to be involved in the extrusion of sodium ions and the inward movement of potassium. In lamprey gills, this enzyme has been detected in the gills of feeding adults of *P. marinus* and its activity apparently increased at higher salinities (Beamish *et al.*, 1978). On the other hand, in freshwater migrant stages this Na-K ATPase activity was observed only rarely.

In 50 % sea water the river lamprey (*L. fluviatilis*) swallows water and excretes chloride ions at a rate similar to that of the eel in normal sea water, but for the teleost, the osmotic gradient would be three times greater than for the lamprey (Morris, 1960). This indicates a much higher permeability to water and ions in the lamprey, which to this extent is less efficiently adapted to marine life. In half strength sea water only small volumes of urine are produced ($0-6.2$ ml kg^{-1} day^{-1}) and this is hyposmotic to the blood. Judging from the high content of divalent ions, it is believed that these may be actively secreted by the kidney tubule, whereas the monovalent ions are excreted mainly by the gills.

(b) *Myxinoids*

In its natural habitat, the isosmotic condition of the hagfish would involve little or no osmotic influx or efflux of water. Urine volumes are therefore small and of a similar order to those of the lamprey, *L. fluviatilis* in 50 % sea water. Thus, in *Myxine*, rates of urine flow of $1-11$ ml kg^{-1} day^{-1} have been recorded (Rall & Burger, 1967) and $4-10$ ml kg^{-1} day^{-1} in *Eptatretus* (Munz and McFarland, 1964; Riegel, 1978). The rate of urine flow is thought to be dependent on the glomerular filtration rate, with little or no reabsorption of water in the rudimentary tubule or archinephric duct (Munz and McFarland, 1964; Morris, 1965; Eisenbach *et al.*, 1971). Measurements of the filtration

Fig. 9.6 Presumptive ion-transport cells in the gills of the lamprey.

       Presumptive chloride uptake cells in the interplatelet area of the gills of an ammocoete of the brook lamprey, *Lampetra planeri* × 2500 (a). Based on an electron micrograph in Morris and Pickering, 1975.

       Interplatelet areas from a macrophthalmia stage of *Lampetra fluviatilis* adapted to freshwater (b) and to sea water (c) × 750. From Pickering and Morris, 1976.

rate in single nephrons of the kidney of *Myxine* show that at the normal glomerular arteriolar pressure of 5.4 cm H_2O, the mean filtration rate is 23×10^{-6} ml min^{-1} (Stollte and Eisenbach, 1973) and very similar values have been obtained for the kidney of the Pacific hagfish, *E. stouti* (Riegel, 1978). If these filtration rates are applied to the whole hagfish kidney with some 30 glomeruli connected to each duct, they would imply rates of urine flow considerably in excess of those observed experimentally. Since no evidence of reabsorption has been forthcoming, it has been suggested that variable glomerular recruitment may play a part in the regulation of urine flow. The fact that small increases in glomerular hydrostatic pressure result in substantial increases in urine flow would explain the ability of hagfishes to regulate their body weight when the external medium is diluted (Section 9.2.1). On the other hand, the low rates of urine production in normal sea water, together with the apparent inability of the nephron to transport water from the glomerular filtrate, are consistent with the inability of the hagfish to regulate its body weight in concentrated sea water. It is uncertain whether these animals normally swallow sea water to replace the small water losses via the urine (Morris, 1965) and there is no evidence of water absorption in the gut, whose contents appear to be isosmotic with the blood.

The significant differences in ionic composition of blood and sea water imply the existence of ionic regulatory mechanisms, still incompletely understood. In the blood, sodium and bicarbonate concentrations are higher than in the water, while the levels of magnesium, calcium and sulphate are higher in sea water (Fig. 9.7). The role of the kidney in ionic regulation is shown by differences in the distribution of ions in plasma and urine. Thus, potassium, magnesium and sulphate are in higher concentrations in the urine, but, sodium levels are lower. In spite of this difference in sodium concentrations, it had generally been thought that the myxinoid kidney lacks the ability to recover this ion from the glomerular filtrate (Robertson, 1974), although sodium concentrations in the ultrafiltrate are reported to decrease along the length of the kidney duct (McInerney, 1974). At least some of the divalent ions are believed to be actively secreted from the nephron and potassium may be transported in the narrow neck segment (Eisenbach *et al.*, 1971). Calcium and magnesium are in high concentration in the bile relative to the blood and unless they are subsequently reabsorbed in the gut, this could be a route for their excretion. Analyses of slime gland secretions have also raised the possibility that these organs may play a part in excreting calcium, magnesium and potassium.

9.3.3 Osmotic regulation of lampreys in freshwater
The basic physiological problems confronting the lamprey in freshwater are the reverse of those that it encounters in the sea. The maintenance of an osmotic gradient of 200–300 mOsm between the body fluids and the

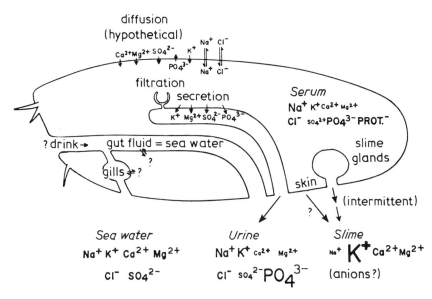

Fig. 9.7 Proposed mechanisms of ionic regulation in the hagfish, *Eptatretus stouti*. The size of the symbols for the various ions indicate their concentration in serum, urine or slime relative to that of the sea water. In the case of phosphate or protein the reference level is the serum. Relative degrees of permeability are suggested by the lengths of the arrows. From McFarland and Munz, 1965.

environment must involve a very large influx of water through the relatively permeable body surfaces, whereas the ionic gradients in the same direction presuppose a continual loss of ions by diffusion; losses that can easily be replaced the animals are still feeding, but which must be counterbalanced by the active uptake of ions from the low concentrations in freshwater during the long periods of starvation that occur in the life cycle (Fig. 9.8).

In the early stages of their spawning migration, anadromous lampreys show a progressive breakdown of marine regulatory mechanisms and their replacement by freshwater adaptations. Some of these changes have been analysed in detail in the case of the river lamprey, *L. fluviatilis* which, when it first enters the rivers in the autumn may retain for a time some limited ability to regulate in saline media (Morris, 1956, 1958). This period of transition is marked by an increase in the permeability of the body surfaces to water and probably to ions. With the cessation of feeding there is a progressive atrophy of the intestine as the gonads mature, involving a dramatic reduction in its diameter and degenerative changes in its epithelial and muscular tissues. This is accompanied by a loss of the sea water swallowing habit and the ability actively to transport sodium ions across the intestinal wall, on which the absorption of the swallowed water depends (Pickering and Morris, 1973). It is

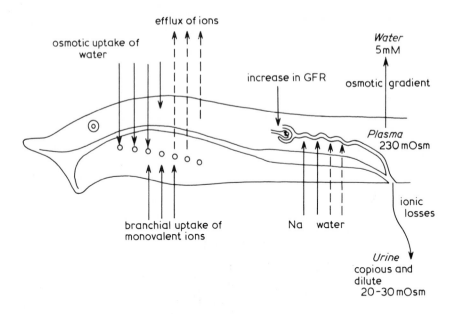

Fig. 9.8 Freshwater regulatory mechanisms of the lamprey.

interesting to find that this degeneration of the gut and the loss of water swallowing and reabsorption can be inhibited by removal of the gonads. On the other hand, gonadodectomy does not prevent the increase in the permeability of the body surfaces during the period of the upstream migration (Pickering and Dockray, 1972). A further characteristic of this transitional stage in the development of freshwater adaptation is the replacement of the ion excretory cells of the early upstream migrants by a different cell type which appears to be identical with the presumptive ion uptake cells described by Morris and Pickering (1975) in the ammocoete gills (Fig. 9.6). These cells differ from the ion transport cells of the marine phase in the absence of the extensive tubular, smooth endoplasmic reticulum and in the presence of characteristic cytoplasmic granules. These cells are believed to be responsible for the active uptake of monovalent ions in the gills that is such an important element in the physiological adaptation to freshwater life (Wikgren, 1953; Hardisty, 1956).

The osmotic flux of water through the body surfaces is balanced by the production of large volumes of dilute urine. According to the various techniques used to measure urine flow, estimates of the rate have varied widely from 48–423 ml kg^{-1} day^{-1}, but in the majority of cases the values represent volumes of 15–40% of the body weight day^{-1} and are considerably higher than those of freshwater teleosts (Robertson, 1974). In freshwater, a mean single nephron filtration rate of 0.007 ml min^{-1} has been recorded for the kidney of *L. fluviatilis* (Moriarty *et al.*, 1978). This is some 350 times higher that the comparable rates for the hagfish glomerulus (Section 9.3.2b). Regulation of urine flow in response to changes in the osmotic gradient are achieved through alterations in the glomerular filtration rate, brought about by changes in the hydrostatic pressure in the glomerular arterioles. Unlike the kidneys of teleosts or amphibians, urine flow is not adjusted by varying the rate of reabsorption in the tubule. This is said to remain constant for a given temperature and accounts for 36–60 % of the volume of the glomerular filtrate (Bentley and Follett, 1963; Morris, 1972). This difference in kidney function has been related to the absence in the cyclostomes of a renal portal system; an arrangement which makes the tubular circulation independent of the glomerulus and which can be regarded as an adaptation to water conservation. As in other vertebrates, tubular water absorption is a passive process, dependent on the active transport of sodium across the tubule. In the adult lamprey, the main site of water transport is thought to be the distal and collecting tubules, although it is possible that the kidney duct may also be implicated (Youson, 1975b,c). In the ventral part of the kidney, a parallel arrangment of these ducts with respect to the vascular sinusoids recalls the arrangement in the medulla of the mammalian kidney and suggests that a countercurrent mechanism may be present (Goncharevskaya, 1975). It may be significant that these structural arrangements are said to be absent in the kidney of the freshwater ammocoete stage.

An important aspect of kidney function in freshwater is its ability to conserve salts. This would be vitally important in the 6–8 months of fasting that occurs during the spawning migration of anadromous lampreys and in the long interval between metamorphosis and spawning in non-parasitic species. In the freshwater phase, the total concentration of the urine is from 7–15 % that of the blood so that, in spite of the reabsorption of ions from the glomerular filtrate, the loss of ions, particularly sodium, potassium and chloride, represent a significant rate of ionic depletion for an animal that is no longer feeding. In addition, it is likely that extra-renal losses through skin and gills are even more serious than urine losses, emphasizing the crucial importance of the highly developed ion uptake mechanisms in the gill epithelium.

As regards freshwater adaptation, the non-parasitic lampreys are a rather special case. Their confinement to fresh water is a necessary consequence of the abandonment of an adult feeding stage. In these species, in which the growth of the gonads begins in the later stages of metamorphosis, the intestine is never functional and like that of the upstream migrant parasitic lamprey undergoes almost complete atrophy (Fig. 2.3). Meanwhile, the adult foregut, which develops at metamorphosis, generally remains a solid cell cord without a lumen, so that both feeding and water swallowing are totally excluded, in all but perhaps exceptional or aberrant individuals; developments that would exclude any possibility of regulation in hyperosmotic media.

9.4 Osmotic conditions during the embryonic development of the lamprey

At the time of ovulation, the osmolar concentration within the egg of the lamprey is similar to that of the body fluids of the female. When deposited in freshwater therefore, the egg faces somewhat similar osmotic problems to those that confront the parent organism. Initially, the large osmotic gradient (about 210 mOsm) between egg and water leads to a osmotic influx of water, resulting in the accumulation of fluid in the perivitelline space between the egg surface and the outer egg membranes (Hardisty, 1957). Within the first few hours there is also a rapid uptake of water by the egg itself, but subsequently, the permeability of the vitelline membrane decreases sharply, remaining at about the same level throughout the first few days of embryonic development. The rapid initial influx of water into the perivitelline space and, to a lesser extent, the egg itself is reflected by a sharp drop in the osmolar and chloride concentrations. Since dilution of the egg contents does not entirely account for the observed decreases in concentration, some ions are presumably lost by diffusion.

The high permeability of the egg both to water and to salts is clearly only tolerable if a relatively short period elapses before the embryo is able to develop compensatory mechanisms, enabling it to take up ions from the water and eliminate water entering the egg osmotically. These characteristics would appear to be incompatible with marine development, where the osmotic and ionic gradients would be more severe. In fact in its osmotic relations and permeability, the egg of the lamprey bears a closer resemblance to that of the amphibians than marine teleosts and this would be consistent with the view that this group of cyclostomes has a freshwater ancestry.

9.5 Hormonal factors

Vertebrate hormones principally involved in water and electrolyte balance are prolactin from the adenohypophysis, the neurohypophysial octapeptides and the steroids of the adrenal cortex. Among the target organs for these hormones are the body surfaces (including the gills), the bladder and the kidney tubules, which may be affected either by changes in permeability or in their ion transport mechanisms.

Such fragmentary indications as we have of the possible involvement in ionic and osmotic regulatory processes of cyclostome hormones have come mainly from observations on the effects of cyclostome pituitary extracts on the target organs of other vertebrate groups or the effects of exogenous (usually mammalian) hormones on certain aspects of cyclostome physiology. In either case, there is often a considerable element of doubt as to the significance that can be attached to what have often been essentially pharmacological procedures and in many instances the dosages employed have probably been quite unphysiological.

Prolactin, which plays an important role in the water and electrolyte metabolism of some teleost species has not been identified in cyclostome pituitary extracts (Nicoll and Bern, 1968), although injections of mammalian prolactin have been reported to alter the ionic composition of hagfish serum and tissues (Chester Jones *et al.*, 1962). The apparent absence of this hormone in the cyclostomes is all the more surprising in view of the very close parallels between the osmoregulatory mechanisms of teleosts and lampreys.

Among the range of vertebrate neurohypophysial peptides, only arginine vasotocin has so far been identified in lamprey pituitary extracts (Rurak and Perks, 1976, 1977). Injections of this hormone in *L. fluviatilis* were found to have no effects on the rate of urine flow, neither did they produce an increase in body water as in the amphibians (Bentley and Follett, 1963). On the other hand, injections of the mammalian preparation, pituitrin provided some evidence for increased sodium and potassium losses in the urine and similar effects have been reported following the administration of arginine vasotocin. Extracts of lamprey pituitaries has caused water retention in frogs, accelerated the transport of water across the toad bladder and exerted an antidiuretic effect on the rat (Lanzing, 1954). Arginine vasotocin has also been identified in the pituitary of the hagfish (Rurak and Perks, 1974) and pituitary extracts from *Myxine* had similar effects to those of the lamprey, when tested on the water and sodium balance of frogs (Follett and Heller, 1964). More direct methods, involving the administration of homologous pituitary extracts to the hagfish previously acclimated to dilute or concentrated sea water, have given some indication of changes in blood concentrations.

In view of uncertainty surrounding the steroidogenic capabilities of the lamprey interrenal (Chapter 11) and the, as yet, poorly understood adrenocortical homologue of the myxinoids, a possible role for corticosteroids in the salt and water balance of the cyclostomes must be very problematical. Of the range of vertebrate corticosteroids, the potent mineralocorticoid aldosterone has not been identified in cyclostomes. Nevertheless, injections of this steroid have produced slight reductions of urine flow and decreased losses of sodium in the lamprey, *L. fluviatilis*, perhaps through an effect on branchial sodium transport (Bentley and Follett, 1963). In *Myxine*, only large doses of aldosterone have been found to be effective in reducing the sodium and potassium content of the muscles (Chester Jones *et al.*, 1962). Neither in lampreys nor in hagfishes have cortisone or cortisol produced anything other than very slight effects on salt concentrations or kidney function.

9.6 Cyclostomes and the environment of the early vertebrates

9.6.1 Palaeontological and geological evidence
The oldest vertebrate fossils are the exoskeletal fragments of *Anatolepis* from marine Upper Cambrian deposits of North America (Repetski, 1978). The

much more numerous remains from the Harding sandstones of North America represent later, middle Ordovician deposits, believed by White (1958) to be littoral deposits. This question has since been re-examined in detail by Spjeldnaes (1968) from the point of view of palaeoecology and salinity regimes. From the boron content of the clay minerals, he considers that these fossil heterostracans had been deposited under marine conditions, but in areas such as tidal flats, where shallow pools subjected to alternate evaporation or to dilution by rain, would have produced short-term changes in salinity; a type of habitat that might be expected to produce the necessary selection pressures favouring the development of regulatory mechanisms.

To what extent geological findings are relevant to the general problem of the environment of the vertebrate ancestors depends to a certain extent on the view that is taken of the relationships of the agnathans and the gnathostomes. If it is assumed that the agnathans were directly ancestral to the gnathostomes, or that the heterostracans in particular are close to the line from which the higher vertebrates originated, then the marine habit of the Cambrian or Ordovician heterostracans would be decisive. On the other hand, should we prefer to regard the agnathans and the gnathostomes as sister groups, both descended from a common ancestor, the geological evidence from these early heterostracans would scarcely be relevant and the absence of contemporary jawed vertebrates would be attributed to the imperfections of the fossil record and perhaps to their lack of calcification.

In the Silurian and Devonian, the anaspids and cephalaspids seem to have been predominantly freshwater forms and some of the heterostracans may even have developed euryhaline mechanisms (Moy-Thomas and Miles, 1971). The earliest gnathostomes to appear in the fossil record – the acanthodians – were primarily a marine group in the Silurian, but had become an important freshwater group in the Devonian. These early penetrations of agnathans and gnathostomes into estuarine or freshwater habitats may have been connected with changes in their feeding or breeding habits. Just as today, estuarine regions are often a nursery ground for larval and juvenile stages of many marine fish species, so these habitats may have provided a safer environment for eggs and young and one where competition may have been less intense. If, as has been suggested, some at least of these agnathans were microphagous feeders, using the detritus deposited in slow-moving currents of rivers and estuaries, the adoption of these fresh or brackish habitats could hardly have occurred before the upper Silurian, when plant colonisation of these areas first occurred. At the same time we should not overlook the possible influence of global changes in palaeogeography on the environment of early vertebrate groups. For example, in a recent re-examination of the changes in the marine inundation of the Northern continents throughout the Phanerozoic (Hallam, 1977), it has been claimed that the maximum extent of inundation was reached in the Ordovician and Silurian, when between 50–60 % of North America and the USSR were covered by the sea. This was followed by a period of dramatic regression beginning in North America during the Silurian and continuing through the Devonian and Carboniferous, to fall to values of 20–30 % during the Permian. Thus, the periods when the land areas were being elevated appear to coincide with the era when the early vertebrates, both agnathans and gnathostomes were beginning to leave marine environments and colonise freshwater habitats.

9.6.2 Physiological considerations

The general picture that emerges from our information on the osmotic and ionic physiology of the myxinoids is that of a primarily marine group and as Robertson (1974) has pointed out, there are no precedents within the vertebrates for a reversion from osmotic independence to osmotic conformity. Many of the physiological characteristics of the hagfish are shared by marine invertebrates and may also have been common to the Cambrian or Ordovician agnathans. In this respect, the idea that the myxinoids stem from an early heterostracan stock would not be inconsistent with the physiological and palaeontological evidence. In their high concentrations of sodium and chloride ions and their isosmotic state, they have made the greatest possible economy in the energy costs of ionic and osmotic regulation, but at the price of evolutionary lability and the capacity for environmental adaptation. On the other hand, the plasticity of the petromyzonids manifested by the variety of their life cycles and migratory habits, may have been acquired at quite an early stage in their evolution, by adaptations to low or fluctuating salinities in brackish estuarine waters. This would not entirely conflict with the evidence from the Carboniferous *Mayomyzon*, whose fossil remains are said to be associated with a primarily marine fauna, but which might well have been transported from an estuarine or deltaic habitat. The very small body size of these fossil lampreys would inevitably have aggravated their osmotic stresses in water of high salinity and from what we know of the characteristics of present day lampreys, it would be difficult to imagine that they would have been capable of extensive migrations between fresh and salt water. Neither is it clear what selective advantages might have stemmed from such migrations. In present day forms, the seaward migration is able to take advantage of the greater availability and diversity of marine host fishes. For reasons discussed earlier (Section 3.1.3.), it is doubtful whether *Mayomyzon* could have been a parasitic feeder and the imperfectly developed sucker and tongue mechanism might have been related to a browsing feeding habit. The gradual development of predatory feeding may have accompanied a general increase in body size and the emergence of suitable host fishes, with an integument sufficiently delicate for the lamprey to penetrate. Parasitic feeding would in turn have been conducive to the development of hyposmotic regulation by facilitating the swallowing mechanisms. In addition, the ingestion of fish blood and tissues with similar concentrations to those of the lamprey itself, would make some contribution to its ionic and osmotic problems. This is all the more significant when we consider the fact that parasitic lampreys may spend most of their feeding life attached to host fishes (Parker and Lennon, 1956; Hardisty and Potter, 1971b).

The idea that the osmotic levels and ionic composition of vertebrates are a reflection of their early environment and marine history is mainly attributable to Macallum (1910). Impressed by the general similarities in the ionic composition of blood and sea water, this author interpreted this resemblance as an historic survival of an early period in vertebrate history when the body fluids would have been similar to the composition of the primeval seas. Such differences as now exist between blood and sea water were to be explained by the changes that were assumed to have occurred over geological time in the composition and salinity of the seas. Although there is now considerable doubt whether changes of this magnitude have occurred in sea water over the period covered by the evolution of the vertebrates, the idea that vertebrate body fluids have a 'historical component' has persisted and forms the basis of arguments advanced by Robertson (1957, 1959) and Lutz (1975a).

Basing his case on the total ionic content of the plasma (in practice the sum of sodium and chloride ions), Lutz suggests that these reflect an original adaptation by the tissues to particular ionic concentrations and the deve op-ment of a physiological dependence that would have limited the extent to which subsequent modifications could be tolerated when a particular vertebrate group moved from one type of environment to another. With the exception of the myxinoids, the range of vertebrate salt concentrations (200–500 mM l^{-1}) is very much smaller than that of the invertebrates 22–500 mM l^{-1}) suggesting that a commitment to a lower range of concentration may have been made at an early period in the history of the vertebrates, at the time of their initial radiation into brackish and freshwater environments. Groups that re-invaded the seas at a later time than the elasmobranchs and which therefore had a longer experience of freshwater life, might have become committed to lower salt concentrations, while the lowest values of all are to be found in the lungfishes and polypteroids, which have had a long and continuous history of life in freshwater (Fig. 9.9). Where vertebrate groups have subsequently developed a marine phase, the difference between their marine and freshwater blood concentrations are regarded as an indication of the antiquity of the marine habit; the most recent being the sturgeons where this difference is only (13 mM l^{-1}).

The lampreys do not fit easily into this hypothetical framework. On the limited evidence, their sea water values fall between those of the sturgeons and the teleosts, but the freshwater values are almost as low as those of the polypteroids and Dipnoi. Especially in view of other physiological evidence, the freshwater values of the lampreys are probably a safer guide to their

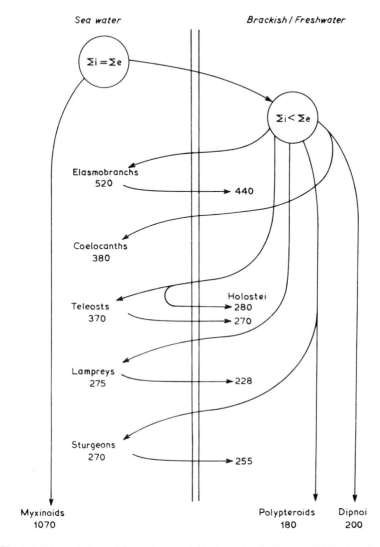

Fig. 9.9 Interpretations of the environmental history of cyclostomes and fishes based on ionic concentrations. Modified from the diagram by Lutz, 1975.
   The figures for the lamprey are based on the data of Beamish *et al.*, 1978 for young feeding stages of the anadromous sea lamprey.

evolutionary history than those of the marine phase. In very broad outlines, this picture, particularly as it affects the lampreys and freshwater teleosts, is in harmony with their respective freshwater and marine origins and with the assumption that the anadromous migrations of lampreys are, in geological terms, a comparatively recent development.

Reviewing the distribution within the vertebrates of osmoregulatory mechanisms, Sawyer (1973) has pointed out that the freshwater hyperosmotic mechanisms are much more uniform than those concerned with marine regulation. For example, the ability to take up sodium from the environment and to excrete an almost salt-free urine by re-absorption in the kidney tubule, is found in elasmobranchs, lampreys, actinopterygians, lungfishes and amphibians, although in elasmobranchs and lungfishes the recovery of sodium is barely adequate to maintain sodium balance in low concentrations of this ion. As regards the mechanisms of marine regulation, only lampreys and actinopterygians drink sea water, while the use of urea to maintain high blood concentrations has been achieved by elasmobranchs, holocephalans, coelocanths and anurans. This greater degree of uniformity in freshwater physiology would be consistent with an initial radiation of the vertebrates in freshwater and with their common ancestry, whereas the divergent modes of marine regulation reflect an independent return to the sea at varying geological periods.

Impressed by the adaptation of the glomerular kidney to the expulsion of large volumes of urine, it had been argued that this structure might have been first developed by vertebrates in freshwater to cope with the large osmotic influx. This argument appeared to be supported by comparative studies which showed that the glomeruli tended to be reduced or absent in some marine fish. However, the presence of a glomerular kidney in the myxinoids was a serious problem for this interpretation. What is more, in terms of body weight, the filtration surface of the hagfish kidney is apparently similar to that of freshwater teleosts and shows a high rate of filtration. As Robertson has emphasized, such a high rate of ultrafiltration is unlikely to have been developed solely as a means of eliminating nitrogenous compounds, since in fish, as in lampreys, this is mainly a function of the gill surfaces.

Further questions are posed by the peculiar features of the hagfish kidney – the very short rudimentary tubule and the apparent absence of water re-absorption – contrasting sharply with the highly developed nephron of lampreys. Is this simple structure a result of evolutionary regression or does it rather represent the primitive condition of the kidney in the marine vertebrate ancestors? Here it may be noted that in the kidneys of *Eptatretus* the ureters are described as serpentine, tending to form branched ducts leading to the renal corpuscle. This has been regarded as a more primitive condition than that of *Myxine* and suggestive of a phylogenetic regression from kidneys that originally consisted of branched ducts (Fänge, 1963). However this may be, at least some of the functions normally associated with the more complex kidney tubule of the higher vertebrates are taken over by the archinephric duct of the hagfish, which is well able to cope with the small urine volumes that this animal normally produces.

The divergent kidney structure and osmotic physiology of lampreys and hagfishes presents a number of difficult evolutionary problems. In lampreys, the parallels between their freshwater physiology and those of higher vertebrates extend also to the cytology of the kidney tubule, its topography and vascular arrangements (Youson and McMillan, 1970a, b 1971a, b). Are all these structural and functional parallels to be attributed solely to evolutionary convergence or were they perhaps derived from ancestors common to both agnathans and gnathostomes? In view of the overwhelming evidence, there is no reason to doubt that the divergent patterns of marine osmoregulation have been acquired independently by different vertebrate groups and at different times. If, on the other hand, we accept that the freshwater regulatory mechanisms (and the differentiated nephron structure on which they partly depend) had already been developed in a common agnathan-gnathostome ancestor during the initial radiation of the vertebrates in freshwater, how are we to account for the rudimentary tubule of the myxinoids and for the persuasive arguments in favour of their marine origins? If the simplicity of the hagfish kidney has been due to regression this would presumably imply a change in the environment and in the physiological demands placed upon this organ which is quite inconsistent with the physiological evidence. A more economical hypothesis, would be that the myxinoids have always been marine and that they arose from a vertebrate line distinct from that which gave rise both to the lampreys and to the gnathostomes; a view which is in complete harmony with most of the morphological and non-morphological evidence discussed at greater length in the final chapters.

10
The pituitary

10.1 Embryonic development

Throughout the vertebrate series the pituitary shows a consistent pattern of development, involving two distinct components; a neurohypophysis, derived from the neural tissue of the floor of the diencephalon, and the adenohypophysis, an epithelial component originating from the epithelium of the embryonic mouth cavity. The cyclostomes diverge from the typical vertebrate pattern in that the adenohypophysial tissue develops from a naso-hypophysial anlagen, which in early ontogeny, lies immediately in front of the stomodeal depression (Fig. 3.1). In all probability, this mode of development was also shared by the fossil agnathans and was already established at the time of the agnathan–gnathostome dichotomy. Nevertheless, although by slightly different routes, both vertebrate groups have attained the same association of nervous and glandular tissue that has made possible the integration of endocrine functions and the adaptation of hormonally controlled physiological processes to changes in the external environment.

In the gnathostome embryo, the stomodaeum develops a dorsal evagination – Rathke's pouch, directed upwards towards the floor of the brain – from which the adenohypophysial tissue is differentiated. Meanwhile, the pouch has made contact with a ventral extension from the floor of the hypothalamus – the infundibulum – from which the neurohypophysis subsequently develops. Although in the course of ontogeny, the oral connection of Rathke's pouch is normally lost, in some fish species an open connection with the mouth cavity may persist in the adult as a hypophysial duct. Neither in myxinoids nor in lampreys is a Rathke's pouch of the gnathostome type developed. Here, the adenohypophysial tissue originates from the epithelium of the nasohypophysial canal or in lampreys, from a cellular cord, within which this canal is subsequently formed. A strict homology between the nasohypophysial tract of the cyclostomes and the Rathke's pouch of the higher vertebrates remains somewhat controversial (Wingstrand, 1966), but it would not be difficult to visualize the derivation of both cyclostome and gnathostome modes of pituitary development from a common ancestral pattern.

In lampreys, that part of the nasohypophysial tract that extends posteriorly from the olfactory organ remains, throughout the larval period, a solid epithelial cell cord (Leach, 1951). The differentiation of glandular tissue from this structure seems to be a continuous process, which continues or is renewed during metamorphosis at the time when this solid cord of cells is being converted into a hollow tube by the formation and fusion of intracellular vacuoles (Larsen and Rothwell, 1972). No information is available on the earliest development of the pituitary in myxinoid embryos, but in the comparatively late stage described by Fernholm (1969) the naso-hypophysial rudiment was an open canal, from which adenohypophysial tissue appeared to be differentiating. Even in the adult, diverticula of the canal may penetrate into the glandular tissue (Olsson, 1959).

10.2 The adenohypophysis

10.2.1 General morphology

Because of the absence of any division into distinct regions and the undifferentiated state of its cellular elements, the myxinoid adenohypophysis is unique amongst the vertebrates (Fernholm and Olsson, 1969; Fernholm, 1969, 1972a). Such simplicity is unlikely to be a primitive feature and is consistent with the regressive trends that have affected so many aspects of myxinoid organization. The glandular tissue is in the form of a flattened plate lying in front of, and below the infundibular process (Fig. 10.1), from which it is separated by a dense zone of connective tissue.

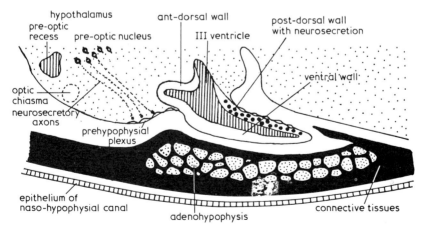

Fig. 10.1 Diagrammatic sagittal section through the pituitary of a hagfish.

In contrast to this simple organization, the adenohypophysis of the lamprey is (except in early larval stages) divided into three distinct regions, separated from one another by connective tissue septa (Fig. 10.2). The two more anterior lobes, the pro- and meso-adenohypophysis are separated from the floor of the brain by a layer of connective tissue, but this thins out over the posterior region – the meta-adenohypophysis – and in this region there is a capillary plexus (mantle plexus) at the interface between the adenohypophysis and the neural tissue of the infundibulum.

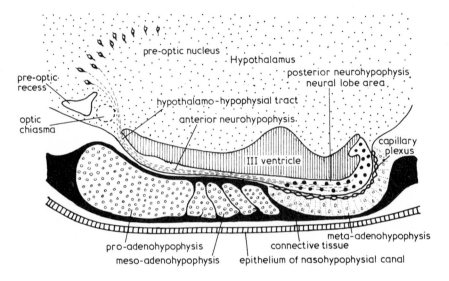

Fig. 10.2 Diagrammatic parasagittal section through the pituitary of a lamprey.

10.2.2 Cytological differentiation
In the description that follows, the cells of the adenohypophysis are referred to the three broad groups, whose staining reactions are described by optical microscopists as basophilic, acidophilic and chromophobic. However imprecise these terms may be in relation to cellular secretion, their use in descriptions of the cyclostome pituitary can hardly be avoided, since we know so little of the functions of the various cell types or the chemical nature of their secretions. By the term basophil is implied a mucoid cell, whose secretory granules give a positive reaction with the PAS technique for demonstrating the presence of material containing carbohydrates, in this case, muco- or glycoproteins. The same cells may also to varying extents, react to aldehyde-fuchsin (AF). The acidophils (carminophils) are serous cells, producing simple protein materials, whose granules might be expected to stain with orange G, azocarmine or

erythrosin. The term chromophobe is ambiguous in the sense that cells which
fail to respond to dyes, may nevertheless show secretory granules in electron
micrographs (Figs. 10.3 and 10.4). These may be described as 'active'

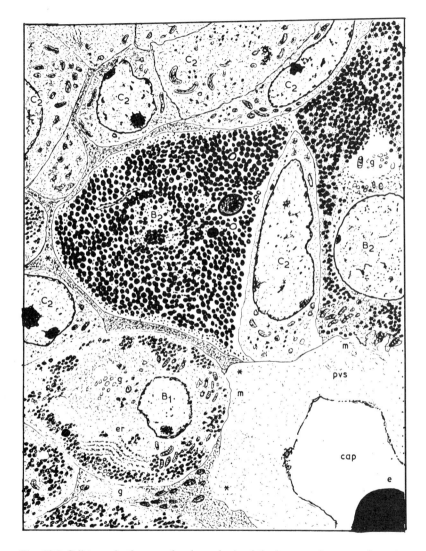

Fig. 10.3 Cell types in the pro-adenohypophysis of the lamprey, *Lampetra fluviatilis*,
during the period of the upstream migration × 5700. Original drawing from an
electron micrograph by Båge and Fernholm, 1975.
B_1 and B_2 basophils; C_2 granulated chromophobe. * thin strands of stellate
cells; m. mitochondria; cap. capillary; pvs. perivascular space; g. Golgi region;
e. erythrocyte; er, endoplasmic reticulum.

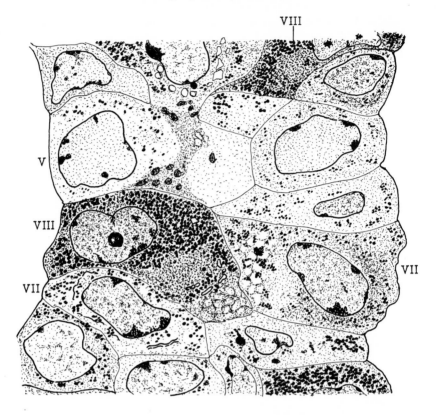

Fig. 10.4 Cell types in the meso-adenohypophysis of the lamprey, *Lampetra fluviatilis*, during the period of the upstream migration × 5000. Numbers given to the cell types are those used by Larsen and Rothwell, 1972.
VIII, meso-adenohypophysial basophil, V, and VII are chromophobes. Original drawing made from an electron micrograph in Larsen and Rothwell, 1972.

chromophobes or granular chromophobes to distinguish them from cells that show no indications of secretory activity.

(a) *Lampreys*
Both pro- and meso-adenohypophysis contain basophils, reacting to PAS, AF or alcian blue staining. Apart from the presence of chromophobes, many authors have also observed acidophils or carminophils, generally but not exclusively in the meso-adenohypophysis. This lobe also contains cells that react strongly with lead haematoxylin, used extensively to distinguish the corticotrophs of the teleost pro-adenohypophysis (rostral pars distalis). In the lamprey, a few cells in the pro-adenohypophysis may stain in the same way

and it seems likely from their size and shape that these may be identical with the lead haematoxylin cells of the meso-adenohypophysis.

Compared to this limited range of cells types that can be distinguished by optical microscopy, the application of the electron microscope has produced a great variety of so-called cell types, in striking contrast to the apparently limited hormonal repertoire of the cyclostomes. These have been distinguished on the basis of their general morphology, position within the gland, the development of their synthetic organelles and above all the size and character of their secretory granules. Unfortunately, these methods of analysis often make it difficult to decide whether particular cells are merely stages in the secretory cycle of the same cell, rather than distinct functional elements. With this in mind, it is safe to assume that the number of functional cell types will always be less than the number that have been distinguished on the basis of these criteria.

The majority of cytological studies on the lamprey pituitary have been confined to the upstream migrant stage of the parasitic species, when the animals are already approaching sexual maturity. For this reason the discussion that follows has been based mainly on the condition of the gland at this stage of the life cycle.

Pro-adenohypophysis This lobe consists of vertically arranged cell cords, separated by connective tissue (Fig. 10.3). In the sexually maturing stage, the dominant cell type is the basophil, whose granulation tends to become more intense as the animals approach the time of spawning. Although some authors have described several basophilic cell types, it now seems likely that these are only different stages in the development of a single cell. For example, in upstream migrants of *L. fluviatilis*, Båge and Fernholm (1975) described three basophils B^1-B^3 with different maximum and mean granule diameters (Fig. 10.5), but with almost identical minima. The synthetic organelles – Golgi and endoplasmic reticulum – were best developed in the B^1 cell with the smallest granules, but were reduced in the B^3 cells with the largest granules. Moreover, the B^1 cell predominated in the autumn and the B^3 cell in the winter and spring. Clearly these three cell types can be interpreted as successive stages in the development of a single cell, which accumulates its maximum store of secretory materials in the final stages of sexual maturation. This example should also serve as a warning against placing too much reliance on the size of secretory granules in the identification of cell types. In the same species, Larsen and Rothwell (1972) described two types of pro-adenohypophysial basophil (III, IV), which can almost certainly be equated with the basophils of Båge and Fernholm.

Two types of chromophobe have been identified in the pro-adenohypophysis. The first of these shows no signs of secretory activity and is a small cell with a thin rim of cytoplasm surrounding the nucleus. This can be

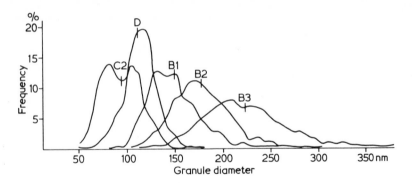

Fig. 10.5 Distribution of granule diameters in the basophils of the pro-adenohypophysis of *Lampetra fluviatilis*. From Båge and Fernholm, 1975.

B_1 predominates in autumn; B_2 in winter and B_3 in spring. C_2 is a chromophobe and D is a rare cell type which also occurs in the meso-adenohypophysis. Mean granule diameters are indicated by vertical lines.

confidently described as an undifferentiated stem cell. A second type of chromophobe described by Båge and Fernholm appears from its fine structure to be an active secretory cell, becoming prominent in late winter and spring and containing granules, whose diameters form a double peak at 80 and 105 nm. An additional, but rare, chromophobe was also seen. This cell (D) which has small granules, appears to reach its maximum development early in the migratory period and may be identical with the scarce lead haematoxylin cells of the pro-adenohypophysis and their more numerous counterparts in the meso-adenohypophysis. In its general morphology and the size of its granules it can probably be equated with the second type of chromophobe (type II) described in the same species by Larsen and Rothwell (1972).

Stellate cells, sometimes described as chromophobes, have been identified in the pro-adenohypophysis (Fig. 10.3). These are relatively small, non-granular elements which are apparently capable of amoeboid and phagocytic activity. This is shown by their elongated cytoplasmic extensions (containing bundles of microfilaments) that are to be seen amongst the secretory cells and along the outer surfaces of cell cords. According to Båge and Fernholm these cells are found only in the pro-adenohypophysis of *L. fluviatilis*, although in *P. marinus*, Leatherland (1975) has also found them closely associated with the mesoadenohypophysial basophils, and with fibrovascular septa, to which they are attached by desmosomes. Apart from their phagocytic functions it has been suggested that they may regulate the secretion of the basophils by opening and closing the gaps between the cell surface and the perivascular space.

Although ultrastructural studies have also been made on *L. (Entosphenus) tridentata* and *P. marinus*, it is difficult to compare these with the pituitary of *L. fluviatilis* in the absence of adequate light microscopic correlations. In the pro-adenohypophysis of *L. tridentata*, two types of granular cell were described (Tsuneki and Gorbman, 1975a); the first, (type I) thought to be in the storage phase and the second (type III) an active synthetic stage. In the first the granule diameters were 230–500 nm and 280–380 nm in the second. Similarly, two highly granular cells are described in the pro-adenohypophysis of the sexually mature *P. marinus* with granules 130–250/260 nm in diameter. In both species it seems likely that we are dealing with a single functional cell type.

Meso-adenohypophysis The cells of this region are more difficult to interpret than those of the pro-adenohypophysis and this is especially true of the several kinds of active chromophobes (Fig. 10.4). In *L. fluviatilis*, the meso-adenohypophysial basophil appears to have distinctive staining properties and its granules are somewhat smaller than those of the basophil of the pro-adenohypophysis (80–200 nm). These cells, unlike those of the pro-adenohypophysis, only develop their granulation after metamorphosis and in the spawning migrant are said to become depleted after breeding. In *L. tridentata*, the ultrastructural counterpart of the meso-adenohypophysial basophil of *L. fluviatilis* may be the type IV cell described by Tsuneki and Gorbman, which has a similar vacuolated cytoplasm and is packed with granules with diameters of 240–300 nm. A type V cell has been described by these authors with vacuolated cytoplasm, but without prominent granulation and may represent the same cell in a depleted state.

In *L. fluviatilis*, Larsen and Rothwell distinguished two types of active and granulated chromophobes in the meso-adenohypophysis (types, V, VII), of which the former is said to be the dominant type, with well-developed synthetic organelles and granules of 60–300 nm. This was thought to correspond to the lead haematoxylin carminophil described in this lobe by the light microscopists (van Oordt, 1968), but it now seems more likely that the latter is represented by the type VII cell. Like the type D cell of Båge and Fernholm, this cell is characterized by its small size, restricted cytoplasm and the small granule diameters (80–150 nm). In the corresponding stage in the life cycle of *P. marinus*, two types of active chromophobe are present in the mesoadenohypophysis; both small cells with small granules, either of which could correspond to the type VII cell of *L. fluviatilis*. In *L. tridentata*, no fewer than four types of granular chromophobes have been listed, of which the closest equivalent to the *fluviatilis* type VII appears to be type VI of Tsuneki and Gorbman, which has comparatively small granules of 110–180 nm.

Meta-adenohypophysis This is separated from the overlying pars nervosa by only a thin connective tissue layer (Fig. 10.6), containing capillaries, although

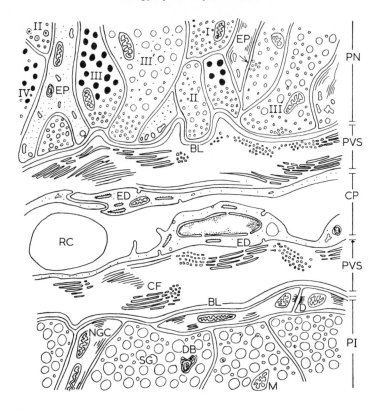

Fig. 10.6 Structure of the boundary region between the meta-adenohypophysis and the neural lobe region of the lamprey neurohypophysis. From Tsuneki and Gorbman, 1975b.

BL. basal lamina; CF. collagen fibrils; CP. capillary; D. desmosome; DB. dense body; ED. endothelium; EP. ependymal process; M. mitochondria; NGC. cytoplasmic process of non-granulated cell; PI. pars intermedia (meta-adenohypophysis); PN pars nervosa (neural lobe): PVS. perivascular space; RC. red blood cell; SG. secretory granules of meta-adenohypophysis. I–IV. axons.

the meta-adenohypophysis itself is avascular and is not penetrated by nerve fibres. In the ammocoete, this lobe of the pituitary is composed of chromophobic cells and the secretory granules (diameters 160–240 nm) do not appear until metamorphosis. In the adult lamprey, a variety of cell types can be observed, but there is general agreement that these represent stages in the secretory activity of a single cell type. Both the granular and agranular cells are elongated and arranged in a palisade-like fashion, with their cytoplasmic processes frequently extending vertically upwards to reach the fibrovascular space separating the meta-adenohypophysis from the pars nervosa above. In what is interpreted as their active secretory state, the secretory granules are concentrated in the apices of the cells, while the extensive lamellated rough endoplasmic reticulum is confined towards the base of the cell containing the nucleus. At spawning time the cells are said to be depleted of their secretory granules, which are presumably discharged into the capillary space between the neuro- and meta-adenohypophysis.

(b) *Myxinoids*

The myxinoid adenohypophysis consists of cell nests and follicles, some with a central lumen. Its outstanding features are the predominance of chromophobic cells and the scarcity of granular elements responding to the normal staining techniques of the light microscopists. The chromophobes, which may represent more than one cell type, are devoid of granules and have comparatively little cytoplasm. In the light microscope, a few basophils reacting to PAS and AF staining have been described in *Myxine*, and differences in their staining reactions have raised the possibility that two types of cell might be present. Other cells containing granules reacting to erythrosin would be classified as acidophils. In ultrastructural studies, only two granular cells can be distinguished; one with granules with a mean diameter of about 88 nm and the other with larger granules averaging 176 nm. A single basophil has been identified, which tends to be associated with follicles and cysts, whose luminal contents react to PAS (Fernholm, 1972a). These are now considered to be mucoid cells like those of the nasohypophysial canal.

Somewhat similar results have been reported in the pituitary of *E. burgeri*, where three types of granular cells have been recognized (Oota, 1974). The first has a distinct Golgi and well-developed rough endoplasmic reticulum with parallel lamellae. The electron-dense granules measure from 170–250 nm. A second type with variably shaped granules about 150 nm in diameter, shows cytoplasmic protrusions containing microfilaments extending among the glandular cells. This suggests that we are dealing with a stellate cell, like that reported in *Myxine*. A third and rare ciliated cell has well-developed synthetic organelles and granules of about 70 nm. These descriptions suggest that the first and third of these cells in *Eptatretus* may correspond to the two types of granular cell in *Myxine*. Based on average granule sizes, Tsuneki (1976) has also distinguished three granulated cell types in *E. burgeri*, although the extent of the overlap in granule diameters made it difficult to characterize them with precision. These types had granule diameters of 220–310, 170–220 and 100–170 nm respectively, presumably corresponding to the cell types described by Fernholm and Oota. In none was the rough endoplasmic reticulum well-developed and the degree of cytoplasmic activity, as judged by the state of the Golgi, mitochondria or the lysosomes, was very variable.

It now seems doubtful whether the myxinoid pituitary has any cellular elements comparable with the meta-adenohypophysis of other vertebrates. In larger (and older) specimens of *Myxine* some histological differentiation was noted at the caudal extremity of the adenohypophysis, where the connective tissue separating it from the neurohypophysis was thinner than elsewhere. In the gnathostomes, the aboral part of Rathke's pouch, first coming into contact with the infundibulum, differentiates as the pars intermedia and by analogy it was suggested that the region of the *Myxine* adenohypophysis in closest contact with the neural lobe might have a similar connotation. However, some doubt has been thrown on this view by the fact that similar regions of contact have not been seen in *Eptatretus*.

10.2.3 Functional interpretations

(a) *Lampreys*
In attempts to identify functional cell types in the adenohypophysis, a variety of experimental techniques have been attempted, including the use of hormonal inhibitors, changes in environmental conditions and the surgical removal of all or part of the gland (hypophysectomy). In addition, cytological comparisons at different points of the life cycle might be expected to yield some clues to the functional activities of particular cell types in relation to such events as metamorphosis or spawning.

Perhaps the most obvious example of a correlation between cellular activity and a particular phase of the life cycle, is the apparent increase in the secretory activity of the pro-adenohypophysial basophils during sexual maturation (Kamer and Schreurs, 1959; Szabó *et al.*, 1965; Lanzing, 1959; Rühle and Sterba, 1966). This leaves little room for doubt that these cells are involved in some aspects of reproductive physiology. The interpretation of the meso-adenohypophysial basophil is more difficult. These cells are said to become depleted of their secretory granules shortly before, or during the spawning season. Earlier suggestions that they might be thyrotrophs (Lanzing, 1959; Kamer and Schreurs, 1959; Rühle and Sterba, 1966) are difficult to sustain in view of the absence of any direct evidence for thyrotrophic activity in lampreys (Section 11.1.6). On the other hand, the effects of hypophysectomy on sexual maturation are indicative of some kind of gonadotrophic activity in the meso-adenohypophysis (Section 10.4.1). Assuming that this gonado-trophic activity is due to a glycoprotein, the meso-adenohypophysial basophil must be the most likely source. Furthermore, if as seems probable, the basophils of the pro- and meso-adenohypophysis are distinct functional cell types, we are forced to the conclusion that the lamprey pituitary produces two gonadotrophic hormones with different target organs and different physiological roles.

Of the various active chromophobes that have been described ultrastructurally, one of these almost certainly corresponds to the lead haematoxylin carminophil of the light microscopists. These have been found to be degranulated after injections of aldactone (an inhibitor of corticosteroid biosynthesis) (Molnár and Szabo, 1968), or when lampreys have been kept in chlorinated tap water. These responses to stress or corticosteroid inhibition suggest that we are dealing with a corticotroph and it may be significant that these cells are reported to increase in number at the time of spawning when some heightening of interrenal activity might be expected. On the other hand, to postulate the existence of a corticotroph is not without its difficulties in view of the absence of any evidence that lampreys are able to synthesize the usual range of vertebrate corticosteroids (Section 11.3.2). Although the type V cell of the meso-adenohypophysis has been suggested as a possible corticotroph (Larsen and Rothwell, 1972), the location and cytological characteristics of the type VII cell make this a more likely candidate. It appears that only a single functional cell type is present in the meta-adenohypophysis and that this is involved in the secretion of an MSH type hormone, but it is doubtful whether pigmentary control is its sole function. The suspicion that the meta-adenohypophysis may be involved in wider metabolic functions is strengthened by its comparatively large size, its close vascular relationships to the overlying neurohypophysis and the fact that its secretory activity becomes intensified at the time of metamorphosis.

In the sea lamprey *P. marinus*, Percy *et al.* (1975) have studied the fine structure of the pituitary from early larval stages through metamorphosis. In younger ammocoetes within two years of hatching, the adenohypophysis is a thin plate of chromophobic cells, without internal fibrovascular septa or regional divisions into distinct lobes. In electron micrographs, granular cells can be detected in the anterior third of the adenohypophysis, corresponding in position to the site of the future pro-adenohypophysis and underlying that part of the neurohypophysis where a median eminence-like structure has been described (Section 10.3). Differentiation of the gland into its three component lobes is delayed until the ammocoete reaches lengths of about 90 mm and at this stage, the active cells of the pro-adenohypophysis contain granules of a similar size to those of the younger ammocoetes (140–230 nm) or to the basophils of the upstream migrants. On the other hand, it is not until the ammocoete is approaching metamorphosis at lengths of 130–150 mm that the basophils can be recognized in the light microscope by their affinity for PAS and aniline blue staining.

Up to the onset of metamorphosis, the meso-adenohypophysis appears inactive and is almost completely devoid of granular cells. At this time there is a rapid growth in the adenohypophysis, which in the non-parasitic *L. planeri* is due mainly to a rapid enlargement of the meso-adenohypophysis. On the other hand, in the closely related and parasitic form *L. fluviatilis*, little or no pituitary growth occurs in the earlier stages of metamorphosis and it is not until the end of metamorphosis that significant increases have been noted in the volume of the adenohypophysis. The earlier and more marked increase in the volume of the meso-adenohypophysis in the non-parasitic form, raises the possibility that this may be related to the almost immediate onset of sexual maturation that occurs at this period in the brook lamprey species.

From the outset, transforming ammocoetes of *P. marinus* show an increase in both the numbers of pro-adenohypophysial basophils and in the amount of stainable secretory material that they contain. In the meso-adenohypophysis however, basophils do not appear until metamorphosis is well under way. In keeping with its earlier differentiation in the very young ammocoete, this development of synthetic activity in the pro-adenohypophysis might be due to inductive factors from the hypothalamus, reaching it through the so-called 'median eminence' region of the neurohypophysis (Section 10.3.3).

(b) *Myxinoids*

In *Myxine*, the response of the erythrosinophil with the smaller type of granules to inhibitors of corticosteroid biosynthesis has suggested that this cell may be a corticotroph (Fernholm and Olsson, 1969), or alternatively, that it produces a protein hormone related to the MSH, ACTH and LPH family, all of which share common aminoacid sequences and may have evolved from a single parent molecule (Fernholm, 1972a). This possibility is not entirely at odds with some experimental evidence for ACTH-like activity in the hagfish pituitary (Adam, 1963; Chester Jones, 1963). On the other hand, the suggestion that the second type of erythrosinophil might be involved in the secretion of a hormone related to growth hormone (STH) or prolactin (LTH) has little or no experimental support and prolactin-like activity has not been detected in myxinoid pituitary extracts, either by bioassay or immunological techniques (Aler *et al.*, 1971).

Evidence for MSH-like activity in the hagfish pituitary is similarly unconvincing. Unlike the lamprey, the body colour of *Myxine* is not affected by hypophysectomy (Matty and Falkmer, 1965b). Extracts of hagfish pituitaries when tested on the melanophore of frogs produced only variable and slight effects which could have been due to the presence of a melanophore dispersing substance from some source other than the adenohypophysis (Holmberg, 1972). Furthermore, no cytological differences have been noted in the adenohypophyses of dark or light-adapted animals and the glandular tissue apparently contains no cells that give a positive response to lead haematoxylin staining (Fernholm and Olsson, 1969). On the other hand, an earlier report recorded strong melanophore dispersing effects of hagfish pituitary extracts on the melanophores of crabs from which the eye stalks had been removed and this activity seemed to be concentrated in the posterior part of the adenohypophysis (Carlisle and Olsson, 1965). However, it is difficult to visualize any function for a melanophore dispersing agent at least in *Myxine*, which in the most recent experiments has failed to show any signs of pigmentary changes when subjected to illumination on black or white backgrounds (Holmberg, 1968).

10.3 The neurohypophysis

The vertebrate neurohypophysis develops as an extension of the hypothalamic floor (saccus infundibuli) in the region immediately behind the optic chiasma, which establishes contact with the aboral part of Rathke's pouch. In amniotes, the pars nervosa or neural lobe differentiates from the dorsal side of the infundibular sac. This is a region characterized by its content of neurosecretory material, passing into it through the axons of hypothalamic neurosecretory cells. Part of the hypothalamic floor lying immediately behind the optic chiasma is differentiated in the higher vertebrates as the median eminence; a neurohaemal area, where axons from other hypothalamic neurosecretory axons end in contact with capillaries. These drain towards the adenohypophysis, forming a hypophysial portal system through which the activity of the glandular cells can be controlled by humoral factors.

10.3.1 Lampreys

The cyclostome neurohypophysis has rightly been regarded as the most primitive vertebrate type, representing only a slight modification of the hypothalamic floor. This is especially true of the condition in the ammocoete, where the infundibulum shows no marked evagination and is recognizable only by the concentration in this area of stainable neurosecretory material. This is particularly marked in the region of the infundibular floor lying immediately above the meta-adenohypophysis.

On the infundibular floor of the adult lamprey, two distinct areas may be recognized (Sterba, 1972). Above the meta-adenohypophysis and separated from it only by thin strands of connective tissue and a capillary plexus, is a thickened region, rich in neurosecretory materials. This is regarded as the parallel of the neural lobe or pars nervosa of higher vertebrates (Fig. 10.6). In front of this region and separated from it by the infundibular commissure, is an area where the infundibular floor shows comparatively little modification and which contains relatively small amounts of neurosecretory material. However, on the basis of its fine structure, this whole area of the anterior neurohypophysis has been divided into anterior and posterior regions, differing in structure and in the relative content of neurosecretory material. Thus, the more anterior section lying over the pro-adenohypophysis and separated from it by a thinner zone of connective tissue, contains greater amounts of stainable neurosecretion than the more posterior areas (Fig. 10.7).

Most of the axons carrying the neurosecretory material to the posterior neurohypophysis or neural lobe originate from neurones in the pre-optic nucleus. These extend in a broad arc (Fig. 10.2) from the pre-optic recess to the

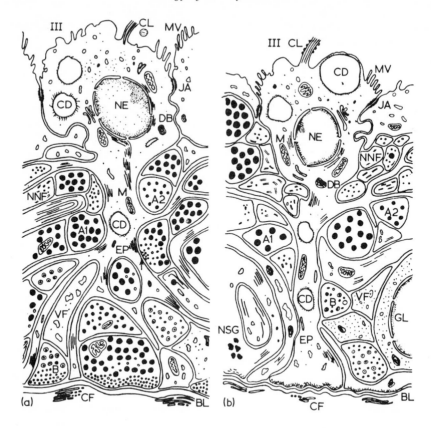

Fig. 10.7 Structure of the anterior neurohypophysis of the lamprey. From Tsuneki and Gorbman, 1975a.

(a) rostral region. (b) caudal region.

A1 axons with large granules; A2 axons with intermediate size granules; B axons with small granules; BL, basal lamina; CD. colloid droplets; CF. collagen fibril; CL. cilium; DB dense body; EP, ependymal processes; JA. junctional apparatus; M. mitochondria; MV. microvilli; NE. nucleus of ependymal cell; NNF. non-neurosecretory fibre; VF. vesicular formation; 111. third ventricle; Gl. glial cell; NSG. neurosecretory granules in ependymal process.

lateral walls of the posterior hypothalamus at the level of the meta-adenohypophysis. A majority of these neurosecretory axons enter the neurohypophysis at its anterior end, passing in two bundles from the ventral and lateral regions of the hypothalamus. Others run to the anterior end of the neurohypophysis in the areas where the connective tissue zone is attenuated.

10.3.2 Myxinoids

The infundibular process of the hagfish is much more developed than that of the lamprey and has been aptly described as like a shoe, with the toe pointing backwards (Fig. 10.1). The inner layer has a lining of ependymal cells whose elongated processes extend towards the external surface. Most of the neurosecretory fibres terminate in the outer layer, mainly in the dorsal wall of the infundibular process, where the neurosecretory substances are chiefly concentrated. These axons contain a variety of secretory granules ranging in diameter from about 65–400 nm. As in the lamprey, the main centre for the hypothalamic neurosecretory cells is the pre-optic nucleus, from which the axons pass through the neck of the infundibular process to reach the dorsal wall of the neurohypophysis. In addition, the hagfish has a unique vascular route for the passage of neurosecretions to the storage areas of the neural lobe. In this system, axons from some of the pre-optic neurons follow a shorter course, ending in a region just behind the optic chiasma, where there is a pre-hypophysial capillary plexus, from which portal vessels drain towards the neurohypophysis (Gorbman, 1965).

10.3.3 Functional differentiation of the neurohypophysis

Although there is general agreement on those areas of the cyclostome neurohypophysis that are functionally comparable to the storage area of the neural lobe of higher vertebrates, the significance of those regions that contain less stainable fuchsinophilic neurosecretory material is far from clear. This applies particularly to discussion of those areas of the cyclostome neurohypophysis that might be regarded as functionally equivalent to the tetrapod median eminence. Of course, if this latter structure is defined in terms of a vascular link between the hypothalamus and the adenohypophysis a strict parallel cannot be made, since we have no evidence that the cyclostomes possess a common vascular supply for the neuro- and adenohypophysis of the same pattern that exists in the hypophysial portal system of the higher vertebrates.

(a) *Lampreys*

In spite of the absence of vascular links between the hypothalamus and the adenohypophysis, ultrastructural studies have raised the possibility that humoral factors may pass towards the glandular elements via the connective tissues (Tsuneki and Gorbman, 1975a). This applies particularly to that part of the anterior neurohypophysis overlying the pro-adenohypophysis. Both this area and the more posterior regions have the same basic structure, consisting of an inner layer of ependymal cells lining the third ventricle and below this a fibre layer containing the processes of the ependymal cells, intermingled with axons and axonal endings. These axons contain secretory granules of three different size classes, 140–220, 95–140 and 65–100 nm. Especially in the anterior zone (Fig. 10.7), some of the axons that contain the larger granules make synaptic-like contacts with the branching processes of the ependymal cells. Functionally, the most significant difference between the anterior and posterior areas is that in the former, the axonal endings and the end feet of the ependymal cells actually make contact with the basal lamina of the connective tissue separating the neurohypophysis from the pro-adenohypophysis, whereas in the posterior region, the axonal endings are separated from it by a layer of end feet and the occasional glial cell processes (Fig. 10.7). In this posterior region therefore, direct passage of neurosecretions from neurohypophysis to meso-adenohypophysis would be less likely, although it could conceivably be transmitted via the cytoplasm of the ependymal cells. These structural and functional differences between the anterior and posterior regions have been regarded as comparable with the median eminence and neural lobe of higher vertebrates even though the cyclostomes lack the same vascular organization as the tetrapods.

The median eminence of the tetrapods contains high concentrations of monoamines, believed to be implicated in the regulation of the secretory activity of the adenohypophysis. In the anterior neurohypophysis of the lamprey, *L. japonica*, strong monoamine oxidase activity has been detected in that part of the neurohypophysis overlying the pro- and meso-adenohypophysis, whereas the activity of this enzyme was much weaker in the posterior neural lobe area over the meta-adenohypophysis (Tsuneki, 1974). In addition, in the anterior neurohypophysis of *L. tridentata*, axons are present that contain granules differing in size from those containing neurohypophysial peptide hormones and which might represent releasing hormones or monoamines. Unfortunately, this clear picture of the parallels between the anterior neurohypophysis and the median eminence has not been sustained by fluorescence studies on the actual distribution of monoamines themselves, which appear to be present in high concentrations in the rostral region of the posterior neurohypophysis, but in only low concentration in the anterior region (Tsuneki *et al.*, 1975). Nevertheless, it is possible that the weaker monoamine fluorescence in the anterior neurohypophysis may reflect the higher turnover of these compounds in an area where monoamine oxidase activity is most intense. The posterior neural lobe area of the lamprey neurohypophysis consists of an inner zone containing the cell bodies of the ependymal cells (tanicytes), a middle zone of neurosecretory fibres and an external layer of ependymal end feet and the enlarged terminals of neurosecretory fibres (Tsuneki and Gorbman, 1975b; Polenov *et al.*, 1974; Belen'kii, 1975). In *L. fluviatilis*, three types of nerve fibres have been distinguished; peptidergic A^1 and A^2 fibres with granules of 160–340 nm and 120–220 nm and monoaminergic B fibres with small granules 80–100 nm in diameter. The presence of two types of peptidergic granules is interesting in view of the fact that only a single neurohypophysial hormone, arginine vasotocin has so far been identified in the lamprey neurohypophysis (Rurak and Perks, 1976, 1977), although it should be noted that the A^1 fibres are said to be rare and the concentrations of their secretions may be very low (Belen'kii, 1975). The neural elements are separated by a basement membrane from the connective tissue zone that lies between the neural lobe and the meta-adenohypophysis. This connective tissue contains pericapillary spaces in those areas where the wide sinusoidal capillaries of the mantle plexus are found. From the structure of this region of the neurohypophysis several routes for the neurosecretion would appear to be possible. Peptides or monoamines might be passed into the capillaries and thus reach the general circulation, or alternatively they could enter the cerebrospinal fluid of the third ventricle. Belen'kii however, believes that as in other lower vertebrates, the meta-adenohypophysis is under the dual control of peptidergic and monoaminergic neurosecretory fibres from the hypothalamus, whose secretions reach the secretory tissues by diffusing through the connective tissue boundary layer.

(b) *Myxinoids*
By combining fine structural observations with histochemical techniques attempts have been made to elucidate the significance of regional differences in the organisation of the hagfish neurohypophysis. From this point of view, four areas may be distinguished: the anterior and posterior dorsal walls in front of, and behind the infundibular stem and the corresponding anterior and ventral walls of the infundibular process (Fig. 10.8). At an early stage, the fact that the stainable neurosecretory materials were seen to be concentrated largely in the posterior dorsal wall, led to the designation of this area as the neural lobe, while the existence of the pre-hypophysial capillary plexus, suggested that this anterior region might function as a median eminence. On the other hand, in its topography and fine structure it has been claimed that the anterior ventral region which is adjacent to the adenohypophysis offers a closer parallel to the tetrapod median eminence (Kobayashi and Uemura, 1972). In this region three types of neurosecretory axons were distinguished by these authors in *E. burgeri*, with small granules of mean diameters, 65, 80 and 110 nm which were thought to contain monoamines. As in the lamprey neurohypophysis, these type B axons in some instances end on the ependymal cell processes of the outer layer adjacent to the connective tissue separating the neurohypophysis from the adenohypophysis and, although in this species there are said to be some capillaries present in this connective tissue, it has generally been assumed that the transport of neurosecretions takes place by diffusion; a process for which there are many precedents in other cyclostome endocrine tissues. That this type of transport is practicable has been demonstrated by injecting peroxidase into the ventricle and following its passage through the connective tissue between neurohypophysis and adeno-hypophysis and even as far as the fibrous septa that surround the adenohy-pophysial cell nests (Nosaki *et al.*, 1975).

Although Henderson (1972) had been able to distinguish only one type of

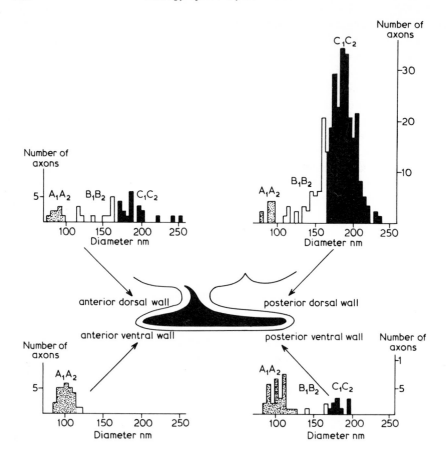

Fig. 10.8 Distribution of various types of neurosecretory axons in the neurohypophysis of the hagfish, *Eptatretus burgeri*. Adapted from data of Tsuneki *et al.*, 1976. Vertical axes: numbers of axons; horizontal axes – mean granule diameters Black areas – $C_1 C_2$ type granules; unshaded areas – $B_1 B_2$; stippled areas – $A_1 A_2$.

neurosecretory axon in the ventral wall of the neurohypophysis of *E. stouti* and two in the dorsal wall, no less than six types have been identified (by the application of statistical techniques) in the neurohypophysis of *E. burgeri* (Tsuneki *et al.*, 1976). These contained granules with mean diameters of 78–87, 98–112, 113–129, 130–164, 165–182, and 183–206 nm and were referred to as A_1, A_2, B_1, B_2, C_1 and C_2 fibres respectively. The various areas of the neurohypophysis show marked differences in the distribution of these axons (Fig. 10.8) and, in particular, the posterior dorsal wall is especially rich in types C_1 and C_2 with the largest granules, believed to contain the neurohypophysial octapeptide hormones, thus strengthening the homology between this region and the neural lobe. So far only one neurohypophysial hormone – arginine vasotocin – has been identified in the hagfish, but since this showed two peaks in chromatograms (Rurak and Perks, 1974) it has been suggested that the same octapeptide may be represented by the granules in C_1 and C_2. In the ventral wall regions the predominant axons are types A_1 and A_2 with the smallest granules. The A_1 granules at least are of a size which is usually considered to represent monoamines. This view is supported by histochemical tests for monoamine oxidase which has shown this enzyme activity to be located in the ventral wall and also in the anterior dorsal region (Tsuneki *et al.*, 1974). The remaining axon types may contain releasing hormones, although in view of the relatively inactive state of the hagfish adenohypophysis, their role in controlling the release of adenohypophysial hormones would appear to be severely restricted. In addition to monoamine oxidase activity, the ventral wall, and the anterior dorsal wall have been shown to exhibit acetylcholinesterase activity. In both these respects, a parallel with the tetrapod median eminence, based on histochemical criteria, would have to embrace the dorsal wall anterior to the infundibular stem in addition to the ventral walls, which more especially have been claimed to show the fine structural features of a median eminence. Whether or not these comparisons with the tetrapod median eminence turn out to be justified, there seems little doubt that, in spite of its primitive phylogenetic status, the neurohypophysis of the hagfish shows a surprising degree of organizational complexity and a well-developed neurosecretory repertoire.

10.3.4 Neuroendocrine functions

Direct experimental evidence for hypothalamic control over the adenohypophysis in the cyclostomes is almost completely lacking. Because of their relatively stable environment, changes in light or temperature could hardly be expected to play an important role in influencing the activities of hagfishes, although in these respects *E. burgeri* with its breeding season and breeding migration may prove to be the exception. At least in *Myxine*, no cytological changes have been observed when adenohypophysial tissues have been cultured in isolation from their normal connections with the hypothalamus (Fernholm, 1972a).

On the other hand, in the case of the lamprey, the seasonal incidence of many episodes in the life cycle provide at least *a priori* grounds for believing that some of these events may be triggered off by environmental factors and mediated through the hypothalamic-hypophysial system. Thus, the onset of metamorphosis occurs in all species of lampreys at the same season of the year and within a comparatively restricted period, the timing of the upstream migration is seasonal and the onset and duration of the spawning season is certainly affected by temperature changes (Section 12.2.1). In spite of this, experimental evidence for hypothalamic control is still lacking. No marked differences in sexual maturation have been observed in *L. fluviatilis* after the pro- and meso-adenohypophysis have been transplanted to sites away from their normal association with the neurohypophysis (heterotopic transplants) as compared with control animals in which the pituitary tissue had been transplanted to its normal site (orthotransplants) (Jørgensen and Larsen, 1967). Neither were there any obvious differences in the cytology of the pituitary tissues in the two groups (Larsen, 1973). In view of these negative findings it is of interest to note that in the non-parasitic species, *L. planeri*, a peptidergic neurosecretory system has been described in the torus semicircularis of the mid-brain, producing cysteine-rich protein secretions (Sterba *et al.*, 1973). Significantly, this system has been found only in the female and even there it does not appear until metamorphosis and after the eruption of the paired eyes. Both the numbers of neurosecretory cells and the intensity of their granulation increase with the progress of sexual maturation, suggesting that this system may be involved in the control of gonadal development. Some indication that the hypothalamo-hypophysial system may also be involved in the control of gonadal processes has come from the inhibition of the neurosecretory system accompanied by degranulation of the presumptive gonadotrophs of the adenohypophysis following the administration of chlorpromazine (Molnár and Szabo, 1975).

There are also other observations that suggest that adenohypophysial tissue may be influenced by the proximity of the neurohypophysis. For example, the pro-adenohypophysis which lies immediately below the 'median eminence-like' area of the anterior neurohypophysis is the first region of the adenohypophysis to differentiate in the larval lamprey and the basophils of this area are the first to show signs of secretory activity. In addition, as several authors have pointed out, pro-adenohypophysial cells in the dorsal regions adjacent to the neurohypophysis may show cytological differences from those in more ventral locations. Thus, in *L. tridentata*, the most deeply staining basophils are those at the most extreme tip of the pro-adenohypophysis (Tsuneki and Gorbman, 1975a) and, in *L. fluviatilis*, the dorsal cells are larger and more active in those regions that are the first to differentiate in the ammocoete (Båge and Fernholm, 1975).

The presence of a common capillary bed between the neural lobe of the lamprey and the meta-adenohypophysis would at least make possible a direct influence of the neurohypophysial hormones on the adenohypophysial tissues. However, apart from such local effects, these hormones may enter the general circulation and exert a wider systemic effect on the organs concerned with salt and water metabolism (Section 9.5). The earliest observations that were made on the effects on mammals of cyclostome pituitary extracts, demonstrated the existence of both oxytocic and vasopressor activity and such effects could be important in relation to the role of the gills in ionic regulation. In the hagfish, the endogenous neurohypophysial hormone, arginine vasotocin, produces contractions in the smooth muscles of the ventral aorta, even in very low concentrations, but the corresponding tissues of the dorsal aorta are much less sensitive (Somlyo and Somlyo, 1968). This has prompted the interesting suggestion that this hormone could be concerned in controlling the gill circulation. In the sea lamprey, injections of arginine vasotocin have resulted in elevations of the plasma free fatty acid levels, which could be an important factor in the mobilization of the lipid stores of the spawning migrant (John *et al.*, 1977).

10.4 Adenohypophysial functions

Some aspects of pituitary function are discussed in the chapters dealing with reproduction, osmotic regulation and the various peripheral endocrine tissues. In addition, the reader may be referred to recent reviews of the lamprey pituitary by Larsen and Rothwell (1972) and the more general account of cyclostome endocrinology by Falkmer *et al.* (1974).

10.4.1 Reproduction

The control and integration of reproductive processes is probably the oldest and original function of the vertebrate pituitary. This control is exercised through the gonadotrophic hormones, acting on both germinal and somatic tissues of the gonads. Through these hormones, the adenohypophysis is able to influence gametogenesis, culminating in the liberation of sperms and eggs, as well as controlling the production of steroid sex hormones by the somatic gonadal tissues and thus regulating the development of secondary sex characters and specific forms of breeding behaviour.

The characteristic development of the basophils of the pre- and meso-adenohypophysis of the lamprey during sexual maturation suggests that some or all of these may be gonadotrophs, but the results of hypophysectomy are not easy to reconcile with the cytological observations. In the higher vertebrates, pituitary ablation is usually followed by atrophy of the testis and atresia in the ovary, while castration results in characteristic changes in the pituitary gonadotrophs (castration cells), due to the interference with the normal feedback between gonadal steroids, hypothalamus and adenohypophysis. None of these effects have been seen in hypophysectomised lampreys, but it should be remembered that pituitary removal has so far only been practised on the sexually maturing spawning migrant stages. In the male lamprey, hypophysectomy early in the migratory period does not inhibit spermatogenesis, although the production of mature sperm is generally delayed. On the other hand, spermiation, involving the breakdown of the testis lobule and the liberation of the sperms into the body cavity, is usually inhibited. When the operation has been carried out in early autumn or winter, secondary sex characters generally fail to appear and if hypophysectomy is delayed until after these structures have already begun to develop, they may undergo regression (Dodd *et al.*, 1960; Evennett and Dodd, 1963, Dodd, 1972; Larsen, 1965, 1969b, 1973).

Early hypophysectomy in the female lamprey retards the growth of the oocytes and inhibits ovulation. The effects on the granulosa cells of the ovarian follicle are particularly marked. These fail to show their normal development and the accumulation of mucopolysaccharide secretions (Larsen, 1970). During normal sexual maturation, variable numbers of oocytes are subject to atresia, but there are no indications that its incidence is increased by removal of the adenohypophysis. In the female as in the male, an intact pituitary is essential for the development of secondary sex characters (Dodd *et al.*, 1960; Evennett and Dodd, 1963). What these experiments have shown therefore, is a marked pituitary influence on the gonadal changes that precede the liberation of the gametes and on the output of sex hormones from the somatic tissues of the gonads.

Earlier work involving partial hypophysectomy suggested that the meso-adenohypophysis was alone responsible for these influences on sexual maturation (Evennett and Dodd, 1963); a conclusion difficult to reconcile with the obvious development of the pro-adenohypophysical basophils during the spawning migration. However, subsequent work involving carefully controlled partial hypophysectomy has implicated the pro-adenohypophysis as well as the meso-adenohypophysis in sexual maturation (Larsen, 1965). For example, extirpation of the pro-adenohypophysis alone did not prevent the attainment of sexual maturity, but removal of the meso-adenohypophysis gave variable results, from complete inhibition to a lack of effect. After examining the pituitary regions of the experimental animals, it was concluded that complete inhibition of sexual maturity only occurred when pro-adenohypophysial tissue had disappeared after operation and that an intact pro- and meso-adenohypophysis are required for normal reproductive development. The involvement of both lobes would be consistent with the view that these may contain distinct types of basophil, producing two types of gonadotrophin with different target organs.

So far we have no conclusive evidence that hagfish pituitaries produce a gonadotrophic factor, or that they contain cells with the usual cytological characteristics of vertebrate gonadotrophs. Complete hypophysectomy in *E. stouti*, involving removal of both neurohypophysis and adenohypophysis has not provided any clear evidence for pituitary gonadotrophic activity (Matty *et al.*, 1976). In the operated animals the number of abnormal testis follicles was greater than in controls, but the progress of spermatogenesis from spermatogonia to spermatozoa was not obviously impeded, neither were any regressive changes noted in the somatic tissues of the gonads. Analyses of plasma testosterone concentration showed lower mean values in the operated animals, but the differences between these and the control values were not statistically significant. In the ovaries, although some abnormal features were noted in the mesovarium, oogenesis appeared to proceed normally in the hypophysectomised hagfish and the proportion of atretic follicles was not increased. No effect of hypophysectomy on oestradiol concentrations was detected (Matty *et al.*, 1976). On the other hand, in *E. burgeri*, the only hagfish known to have a definite breeding season, there have been some indications that gonadal growth and spermatogenesis may be retarded after hypophysectomy (Patzner and Ichikawa, 1977). However, in spite of earlier reports that hagfish pituitary extracts showed slight gonadotrophic activity when tested in a mouse bioassay (Strahan, 1959), these experiments fail to show any absolute dependence of the gonads on pituitary hormones, although the possibility of a more general and less obligatory influence has not been entirely excluded.

Experiments involving the administration of sex steroids to cyclostomes have so far failed to provide any convincing evidence for the existence of feedback mechanisms operating through the pituitary-gonadal axis. Sex hormones applied to castrated river lampreys, *L. fluviatilis* produced no decisive changes in the cytology of the adenohypophysis (Larsen, 1974), although some increase was noted in the intensity of staining of the pro-adenohypophysial basophils in animals that had been treated with testosterone. In similar experiments carried out on intact hagfishes, *E. burgeri*, oestradiol treatment resulted in a marked degeneration of large oocytes (Tsuneki, 1976). However there was no indication that the secretory activity of any of the granular cell types in the adenohypophysis had been inhibited and the effect of the hormone on the oocytes was thought to be pharmacological. Perhaps as a result of a direct stimulatory influence on spermatogenesis, testosterone produced a high proportion of mature sperm in the treated males and although there appeared to be some accumulation of granules in the type 1 adenohypophysial cells (Section 10.2.), there was too great an individual variability in this character in the control animals to justify the conclusion that a feedback control system exists.

10.4.2 Metabolism

Investigations of pituitary influences on metabolism have so far been limited to the migratory period in the life cycle of the lamprey; a time when the animals are no longer feeding and when their main source of energy is the lipid accumulated, either during the larval stage (in non-parasitic forms) or in the phase of parasitic feeding. Moreover, at this same period, the picture is further complicated by the diversion of metabolites to the rapidly maturing gonads. These conditions are reflected in the atrophy of the intestine, reduction in body length and weight, the depletion of lipids in the body wall musculature and the decrease in the liver stores of glycogen.

Hypophysectomy appears to have a retarding effect on many of these changes and diminishes the rate at which energy reserves are depleted. Thus, atrophy of the intestine occurs more slowly after early hypophysectomy and the process may even be reversed when the gonads are removed (Larsen, 1972,

1973). In normal animals the reduction that occurs in body length throughout this migratory period can be equated with the reduction in the tissues of the body wall, providing a rough indication of metabolic depletion. Throughout the early migratory period in autumn and winter, there is a slow reduction in body length which is replaced by a rapid shrinkage with the approach of sexual maturation in the spring (Larsen, 1965, 1969a). Hypophysectomy in the early stages does not affect the total length reduction that the animal has suffered by the time that it dies, but because the operated animals survive longer, the rate of length reduction is reduced. Hypophysectomy carried out in the spring has tended to prevent the final phase of very rapid shrinkage and in many instances has resulted in very prolonged survival. However, the precise mechanism whereby the removal of the pituitary reduces the mobilization of the tissues remains obscure. One possibility is that this effect might be related to a reduction in gonadotrophin output and thus indirectly to retarded gonadal growth. Alternatively, a more general effect on the metabolic rate may be involved. The latter possibility seems to be supported by experiments involving the transplantation of the pro- and meso-adenohypophysis to heterotopic sites. In some instances, these animals became sexually mature, but their survival was prolonged and the metabolic effects were similar to those seen in hypophysectomised animals (Larsen, 1973). This might indicate that it is the separation of the pituitary from the CNS that is responsible for a decreased rate of tissue mobilization.

One of the most puzzling and striking aspects of hypophysectomy is its effect on the life span of the lamprey. These animals normally die very soon after spawning, but operated animals may survive for months after reaching sexual maturity and one lamprey lived for nearly a year beyond its normal breeding season (Larsen, 1969a). Since this prolonged survival may occur after the animals have attained full sexual maturity, we may infer that the postponement of natural death is not simply a consequence of the removal of gonadotrophin secretion, but is rather related to the reduced rate of metabolism (Section 12.2.2).

Neither in lampreys nor hagfishes has there been decisive experimental evidence of a growth-hormone-like effect on blood sugar concentrations (Larsen, 1976a,b; Matty and Gorbman, 1978). Unlike the condition in the tetrapods, hypophysectomised lampreys or myxinoids show no trends towards reduced blood sugar levels, although in operated lampreys these concentrations may be less susceptible to elevation in certain forms of stress.

10.4.3 Metamorphosis

In view of the morphological and cytological changes in the pituitary at the time of metamorphosis, there have been suggestions that the adenohypophysis might, in some way, be involved in transformation, although there is as yet no very convincing evidence to support this belief. The rapid development and differentiation in the pro- and meso-adenohypophysis at the beginning of metamorphosis could be significant, but there is no way of knowing whether these are the result of the morphogenetic changes rather than a factor in their initiation.

An interesting feature of metamorphic development is the apparent correlation between the morphogenetic changes in the nasohypophysial stalk of the ammocoete and the growth and differentiation of the pro-adenohypophysis. In large ammocoetes of *L. planeri*, cavitation of the nasohypophysial cord appears to begin well in advance of the external metamorphic changes and at least in this species, proliferation from the cells of the nasohypophysial epithelium appears to contribute to the tissue of the pro-adenohypophysis. Thus, in some way the developments that occur at metamorphosis in the various parts of the nasohypophysial complex appear to be co-ordinated. This recalls the isolated cases of neoteny that were observed by Zanandrea (1956, 1958) in a local population of the North Italian brook lamprey, *L. zanandreai*, where a few female ammocoetes were observed with well-developed ovaries and mature eggs. Unfortunately we have no information on the condition of the adenohypophysis in these animals but it may be significant that a well-developed nasohypophysial sac was present.

10.4.4 Pigmentary control
Diurnal changes in the skin colour of lampreys are brought about by the dispersion or concentration of melanin granules in the pigment cells (melanophores) of the skin (Sections 4.2.1 and 8.2.4). After total hypophysectomy, involving the extirpation of the meta-adenohypophysis, the animals remain in a permanent state of pallor (Young, 1935; Larsen, 1965). Lamprey pituitary extracts are capable of causing melanophore dispersion in frog skin (Lanzing, 1954) and injections of mammalian intermedin produce similar effects in the ammocoete (Gorbman and Bern, 1962). The cytological, topographical and histochemical parallels between the meta-adenohypophysis of lampreys and the pars intermedia of higher vertebrates leaves little room for doubt that this region produces a hormone of the MSH type.

10.5 Evolutionary perspectives

The quite remarkable parallels between the nasohypophysial complex of lampreys and cephalaspids almost compel us to believe that this correspondence would probably have extended to the pituitary itself. The differentiation, already apparent in the cephalaspids, between the openings into the nasal and hypophysial parts of this complex, emphasizes the confusion so often introduced into comparative discussions by referring to the cyclostome nasopharyngeal opening and tract as a nasal opening and nasal tract. In the past, where doubts have been raised on a homology between the nasohypophysial tract and Rathke's pouch these have usually been based on the view that the tubular and persistent hypophysial canal of the cyclostome should be regarded simply as an extension of the nasal sac. In addition, such views have been influenced by misconceptions over the role of the nasohypophysial complex in relation to the origin of the adenohypophysis, which was thought to be derived from the nasohypophysial cord only in embryonic or early larval life. In fact, further contributions are made to the glandular tissue from the epithelium of the naso-hypophysial canal during metamorphosis and it is possible that these processes may extend even further into adult life. While it is true that aspiration of the olfactory sac is a function of the extended nasohypophysial sac of the adult lamprey, this is almost certainly a secondary development and there are good grounds for the belief that this function could not have been carried out by the nasohypophysial complex of the cephalaspids (Section 3.3). Finally, the presence in the ammocoete of a solid cell cord rather than a hollow structure is hardly relevant to the question of its homology, since among fishes it is only the embryonic elasmobranchs that show a typical hollow Rathke's pouch and in teleosts, the adenohypophysial tissue proliferates from a cellular cord in which cavities subsequently develop as schizocoeles (Wingstrand, 1966).

The close association between olfactory and hypophysial components that we see in the myxinoids and lampreys may be reflection of relationships that were established at very early stages in the evolution of the vertebrate pituitary, although the precise manner in which their functions are now carried out may bear little relation to the ancestral condition. If we accept an homology between the nasohypophysial cord or canal of the cyclostomes and Rathke's pouch, what is the explanation for this close association with the olfactory organ? Perhaps, as has been suggested, the early precursor of the pituitary may have been concerned with reproductive control and co-ordination, responding to the presence in the water of sexual products by secreting hormonal substances to effect the liberation of the gametes. If, in the ancestral condition, the pituitary was primarily concerned with reproductive processes, an association between chemosensory and glandular functions would hardly be surprising. It has been widely assumed that the adenohypophysis originated in a patch of glandular epithelium on the roof of the stomodaeum or immediate pre-oral region, which was subsequently invaginated to form an exocrine gland secreting through a duct into the mouth cavity. This is in harmony with the presence in the primitive polypteroids (and in the earlier ontogenetic stages of other actinopterygian fishes) of a persistent hypophysial duct from the pro-adenohypophysis (rostral pars distalis) to the oral cavity. This may be compared with the widespread occurrence in the cyclostomes of cyst-like cavities in the adenohypophysis, some of which at least may communicate with the lumen of the nasohypophysial tract through duct-like openings. What is more, as seems to be the case in the follicular cavities of the actinopterygian pars distalis described by Olssen (1968), the cells associated with these cysts are similar both structurally and in the nature of their secretions, to the mucoid cells of the nasohypophysial or oral epithelium.

At this early duct stage in the evolution of the hypophysis, Olssen considered that the gland would have been concerned with the production of a 'skin function' prolactin-type hormone, involved in the chemical and mechanical protection of the gut epithelium. This hypothesis may be in line with the localization of prolactin cells in the rostral pars distalis of the actinopterygian adenohypophysis, but fails to take account of the absence of any evidence for prolactin-like hormones or lactotrophs in the cyclostome pituitary. The apparent dominance in the pro-adenohypophysis of the lampreys, of glycoprotein-secreting elements believed to be gonadotrophs, suggests that reproductive control may have been the main function of the pituitary in the primitive vertebrates.

On the other hand a quite different view has been adopted by Fernholm (1972a) in relation to the hagfish pituitary. Here, he considered it likely that the two granular cell types produce hormones different to those associated with the gnathostome pituitary and perhaps akin to the ancestral molecules that might have existed in the Cambrian or Pre-Cambrian common vertebrate ancestor. For example, the type 1 cell with the smallest granules might be related to the primitive adenohypophysial cell, from which in the course of vertebrate evolution were differentiated the cells producing ACTH and MSH, while the type 2 cell could be the forerunner of the differentiated cells eventually producing growth hormone and prolactin respectively. Since the mucous materials produced by the PAS-positive cells were not considered to have any hormonal significance, this interpretation would exclude the existence of any cells corresponding to gonadotrophs in the early vertebrates.

In the absence of any division into separate lobes, the lack of a differentiated meta-adenohypophysis, the scarcity of granular cells and the sparseness of their secretory granules, the hagfish pituitary contrasts sharply with that of the lamprey. Were these conditions to be regarded as truly primitive, it would be difficult to imagine what selective advantages could accrue from the development of a structure with such apparently limited functional capacity. From this point of view, we are almost compelled to believe that this poor differentiation of the hagfish adenohypophysis is a secondary condition, resulting from its restricted habitat and unadventurous mode of life, more especially the general lack of seasonally controlled reproductive rhythms. However, these differences in the adenohypophysis of the two groups of cyclostomes are not parallelled in the structure of the neurohypophysis and indeed, in this respect, the simple infundibular sac of the lamprey may be considered to be more primitive than the well-developed infundibular process of the myxinoids. The fact that the hagfish neurohypophysis has not followed the apparent regression of the adenohypophysis could be related to the more basic metabolic functions of the hypothalamus and neurohypophysis. The widespread occurrence among the invertebrates of neurosecretory systems tends to suggest that the hypothalamic-neurohypophysial system of the vertebrates may be of greater phylogenetic antiquity than the adenohypophysis and for this reason might have been less susceptible to the kind of regressive changes that appear to have overtaken the myxinoid adenohypophysis (Holmes and Ball, 1974).

11
The peripheral endocrine tissues

11.1 The thyroid and endostyle

11.1.1 Embryonic development

Throughout the vertebrates the thyroid is a derivative of the mid-ventral floor of the embryonic pharynx. The cyclostome pattern differs from that of the gnathostomes only in that the thyroid tissue develops along almost the entire length of the pharynx rather than from a relatively short anterior region. In the earliest myxinoid embryos that have been examined, the thyroid appears to have arisen through the closing off of a ventral pharyngeal groove and not from an endostyle-like structure of the ammocoete type. From this embryonic rudiment, a continuous chain of cell cords is produced, passing upwards towards the oesophagus in the adipose tissue surrounding the gill pouches. These cell groups, at first solid, eventually hollow out to form thyroid follicles (Fernholm, 1969). In lampreys, the definitive thyroid tissue does not appear until metamorphosis, when it develops from certain epithelial elements that appear to survive the otherwise complete breakdown of the larval endostyle.

11.1.2 The endostyle

The large number of morphological and physiological investigations that have been carried out on the endostyle are an indication of the significance that has been attributed to this organ in relation to the evolution of thyroidal functions. Like the tunicates and cephalochordates, the ammocoete is a microphagous feeder, using its mucous secretions to trap minute food particles suspended in the water and, in spite of its much greater complexity, there can be little doubt that the specialized, closed endostyle of the ammocoete is morphologically homologous with the simpler organ of the protochordates. The significance of this correspondence was of course greatly heightened by the disclosure of remarkable parallels in the iodine metabolism of the protochordate and lamprey endostyles. Both share a common capacity for iodine concentration and iodine binding, with the formation of the usual

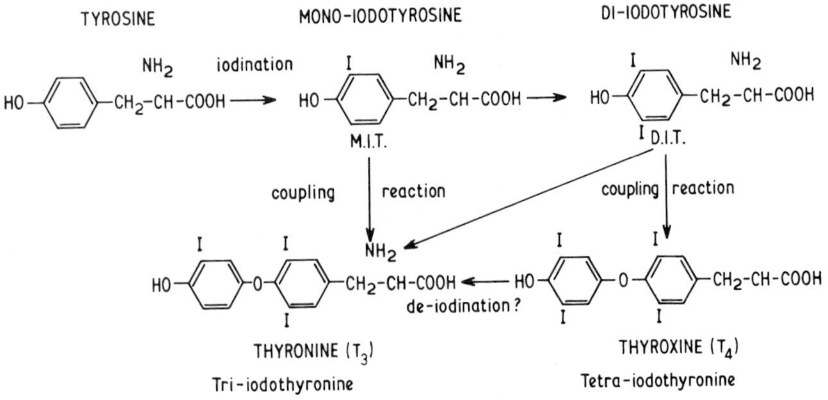

Fig. 11.1 Biosynthesis of thyroidal compounds.

range of iodo-tyrosine derivatives – mono- and di-iodotyrosine (MIT, DIT), thyronine (T_3) and thyroxine (T_4) (Fig. 11.1). These biosynthetic activities are concentrated within the endostyle, although at least in the protochordates, perhaps not exclusively so (Barrington, 1974).

The ammocoete endostyle consists essentially of paired, tubular lateral chambers (Fig. 11.2.), which in their anterior sections are completely separated by a median septum. In the posterior regions this septum is incomplete and the two chambers are in communication with one another. These tubes are also completely cut off from the pharyngeal lumen except at one point. Here, the endostylar duct opens into the pharynx dorsally, while ventrally it leads into a median chamber directed backwards from the point where the duct enters. This median chamber is coiled upwards in its posterior sections. In addition to the endostyle, the pharyngeal floor has a deep ciliated groove and, close to the exit of the endostylar duct, this is joined by two extensions of the lateral chambers, running forwards and upwards along the hyoid arches as ciliated, pseudobranchial grooves. This arrangement presumably serves to spread out the mucous sheets immediately behind the velum, where they would be in a position to trap food particles entering in the respiratory water current.

Within each chamber of the endostyle are four large tracts of glandular type I cells (Fig. 11.2). These groups are approximately triangular in section and the cells show ultrastructural and histochemical resemblances to the mucous

Fig. 11.2 (a) General view of endostyle. Redrawn from Barrington and Sage, 1972.
    (b) Transverse section of the right anterior chamber of the endostyle to show the distribution of the various cell types. From Sterba, 1953.

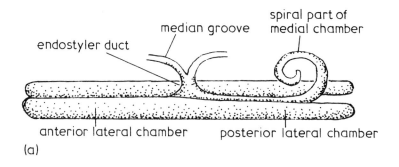

endostyler duct

median groove

spiral part of medial chamber

anterior lateral chamber

posterior lateral chamber

(a)

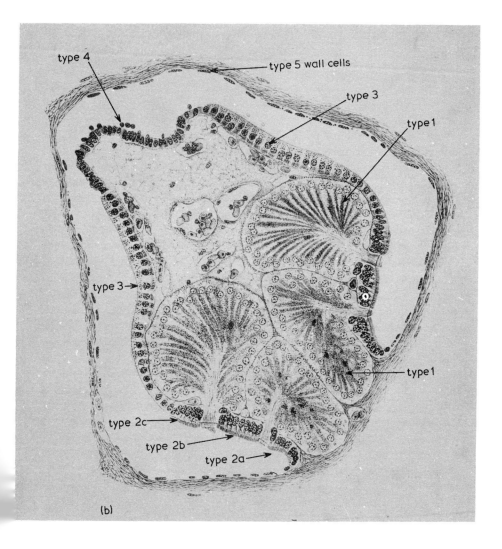

type 4

type 5 wall cells

type 3

type 1

type 3

type 1

type 2c

type 2b

type 2a

(b)

cells of the protochordate endostyle. These type I cells do not take part in iodine binding, but their highly developed granular endoplasmic reticulum intimately associated with the Golgi, indicates that they are concerned in the synthesis and transport of protein materials. (Fujita and Honma, 1968; Hoheisel, 1969). From the histochemical evidence it is assumed that the secretions of these cells contribute to the mucous feeding mechanisms. A protease is present in endostylar extracts and phosphomonoesterase and non-specific esterase have also been located in the glandular tracts. Earlier observations had suggested that proteolytic digestion took place within the pharynx and diatoms are said to undergo lysis in this part of the gut (Schroll, 1959). On the other hand, it is difficult to determine whether this digestive activity is attributable to the endostylar secretions or to the activity of the general pharyngeal epithelium (Sterba, 1953).

Other cell types in the endostylar epithelium are not sharply demarcated from one another in their cytological or physiological characteristics and for this reason, Barrington and Franchi (1956) preferred to distinguish two main fields of cellular activity; an alimentary field centred on the type I tracts and a thyroidal field around the type 3 cells that are prominent in iodine binding. This latter cell is ciliated and columnar, with a well-developed endoplasmic reticulum and vesicular Golgi, within which are electron-dense materials, tending to concentrate towards the apex of the cell forming multivesicular bodies. These have been interpreted as storage sites for iodinated products (Barrington and Sage, 1972). Granules originating in the smooth endoplasmic reticulum are thought to be lysosomes, containing non-specific phosphatase and esterase, whose distribution coincides with that of bound iodine in autoradiographs, after injections of ^{131}I.

Type 2 cells have been further subdivided into types 2a, 2b and 2c; the latter adjacent to the type 3 zone (Fig. 11.2). Both types 2 and 3 are ciliated, with microvilli on their apical surfaces but, in type 2c, the rough endoplasmic reticulum and Golgi are better developed and the lysosomes more conspicuous. Although bound iodine can be demonstrated in all three cell types, it is much more conspicuous at the surface of the type 2c cell. Type 4 cells have no granular endoplasmic reticulum and ribosomes are scanty. Traces of bound iodine have been seen in these cells, but they are considered to have little ability to bind this element or to synthesize protein. Type 5 cells, lining the endostylar lumen show active pinocytosis and are regarded as mainly absorptive elements, although their response to goitrogens might be regarded as evidence for some synthetic capacity (Barrington and Sage, 1963a,b).

11.1.3 Metamorphosis and the lamprey thyroid

In the adult lamprey, the thyroid lies beneath the tongue musculature and extends from the second to the sixth gill pouch. It consists of follicles of varying diameter, in which the cells vary considerably in height and width

(Fig.11.3). Both ciliated and non-ciliated cells are present (Fujita and Honma, 1966) and these show the usual hallmarks of protein-secreting cells – a well-developed granular endoplasmic reticulum and Golgi – although neither of these organelles is as well-developed as in the thyroid cells of higher vertebrates. Small vesicles of varying degrees of electron density are present in the apical regions, together with lysosomal bodies that may be derived directly from the endostylar epithelium (Hoheisel, 1970). In addition to the normal follicles, which may contain variable amounts of a loose-textured, PAS-positive colloid, parafollicles have been distinguished in which the epithelial cells are of a squamous type and in which the follicular lumen is devoid of colloid (Lanzing, 1959). Cell nests with no central lumen may also be present between the follicles. Compared to the thyroid of higher vertebrates, the lamprey gland is poorly vascularised and the capillaries lack fenestrations.

During metamorphosis, the structure of the endostyle is completely broken down by a combination of cytolysis and phagocytosis. These processes have made it very difficult to follow the fate of different parts of the endostylar

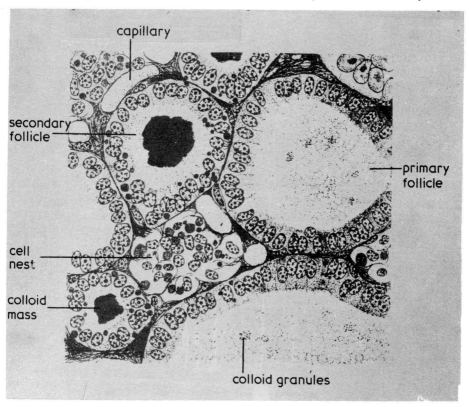

Fig. 11.3 Histology of the thyroid tissue of the lamprey. From Sterba, 1953.

epithelium at the critical periods when the adult thyroid follicles are being formed. In view of these difficulties, it is not surprising that there should still be uncertainty as to which cell types in the endostyle contribute to the adult follicles. Sterba (1953) distinguished two types of thyroid follicles; primary follicles containing a remnant of the original endostylar lumen and secondary follicles in which the lumen was only formed during their metamorphic development. The persistence of the endostylar lumen has also been suggested by Wright and Youson (1976), who observed a proliferation of type 2 and type 5 cells at the angle of the endostylar lumen during the metamorphosis of *P. marinus*. Most of the earlier authors who studied metamorphic changes in the endostyle believed that the greatest contribution to the adult thyroid follicles was made by type 4 cells, although the participation of other cell types has not been excluded. Sterba for example, suggested that the primary follicles were formed mainly by the type 4 cells, but that the secondary follicles originated in type 3 cells. Gorbunova (1975) considered that types 4 and 5 are the least specialized of the endostylar elements and that these persisted throughout metamorphosis, subsequently differentiating to form the definitive thyroid follicles. Wright and Youson found that iodine binding continued throughout metamorphosis, first in the transforming endostyle and later in the reconstructed thyroid follicles. Since the type 4 cells play little or no part in iodine metabolism they considered that other cells, such as types 2 or 3 which are more active in iodine binding, must participate in the development of the adult thyroid follicles. Subsequently, Wright *et al.* (1978b) have been able to detect the presence of thyroglobulin in cell types 2c and 3, as well as in some type 5 cells, by the use of immunocytochemical techniques.

11.1.4 The thyroid gland of the myxinoids

The hagfish thyroid consists of diffuse tissue extending over a wide area of the pharyngeal floor, along the course of the ventral aorta as far back as the last branchial pouch (Gorbman, 1963; Henderson and Gorbman, 1971). The follicles, which may occur either singly or in groups, are embedded in connective tissue and show great variability in size. For example, in *Myxine* their diameters range from 0.18–1.0 mm. As in lamprey thyroids, the cells vary in height, but are more uniform within an individual follicle and, perhaps because of the absence of an endostylar stage, no ciliated cells are present. The follicular lumen may contain a coarsely granular material, quite unlike the colloid of higher vertebrates. Capillaries are relatively rare and like those of the lamprey are without fenestrations, but pinocytotic vesicles have been described within the endothelial cytoplasm (Fujita and Shinkawa, 1975). As in other poorly vascularised cyclostome endocrine tissues, it is presumed that secretion occurs through the basal lamina and the connective tissues.

In their ultrastructure there are distinct parallels between the thyroid cells of hagfishes and lampreys. In the myxinoid thyroid, only one cell type has been described with short apical microvilli, rough endoplasmic reticulum and Golgi field, although like the lamprey thyroid cell, these are not as well developed as in the thyroid epithelia of higher vertebrates. Like the endostylar and thyroid cells of lampreys, the hagfish cells contain dense bodies which have been regarded as secondary lysosomes containing condensed colloid.

11.1.5 Interpretations of thyroidal mechanisms

In all vertebrates, thyroidal activity involves the same basic sequence of biosynthetic processes:

(1) The accumulation of iodine from the environment and its oxidation
(2) Iodination of tyrosine and the formation of MIT and DIT (Fig. 11.4)
(3) The coupling of these compounds to form T_3 and T_4 in thyroglobulins
(4) The hydrolysis of thyroglobulin and the liberation of the active hormones, T_3 and T_4.

After the injection of radio-iodine ^{125}I or ^{131}I or its addition to the ambient water, it is taken up by the thyroid or endostyle cell, but in comparison with higher vertebrates the rate of accumulation in the cyclostome is slow, perhaps because of the relatively small volume of the tissue and its poor vascularisation. In the lamprey, even greater concentrations of iodine may occur in the notochord or gonad. For example in *L. fluviatilis*, only 0.2–1.8% of the injected dose of ^{131}I appeared in the thyroid, compared to 10% in the notochord and about 3% in the ovary (Larsen and Rosenkilde, 1971). However, in these extra-thyroidal tissues, the iodine exists mainly in the inorganic form as iodide, whereas in the endostyle or thyroid tissue, a higher proportion is organically bound to protein.

By combining the ultrastructural and autoradiographic evidence, it is possible to outline some of the basic cellular processes that may be involved in the thyroidal activity of cyclostome follicle cells (Fig. 11.4). Within a short period of its administration, radio-iodine is concentrated around the apical surfaces of the cells and, to a more variable extent, within the follicular lumen, suggesting that these are the sites where iodination mainly occurs. Thyroglobulin or its protein precursors, are presumably synthesized within the rough endoplasmic reticulum and processed within the Golgi, where secretory vesicles are formed. These then pass towards the cell surface, where their contents are extruded into the lumen and iodination occurs. Judging by the relative intensity of the luminal contents as seen in autoradiographs, the storage functions of the colloid are much less important in the cyclostomes than in higher vertebrates, although in the sea lamprey, *P. marinus*, thyroglobulin has been detected immunocytochemically both in the luminal colloid and in the colloid droplets within the thyroid cells (Wright *et al.*, 1978a). The iodinated products are re-absorbed into the cells by pinocytosis and the pinocytotic vesicles are thought to associate with lysosomes in the

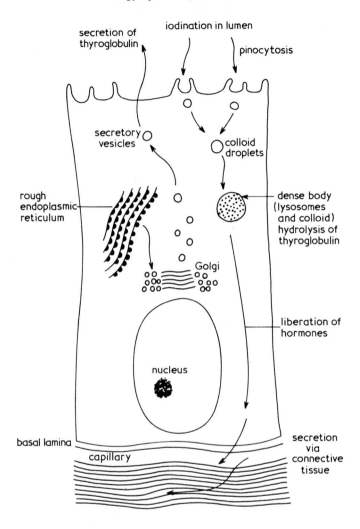

secretion of
thyroglobulin

iodination in lumen

pinocytosis

secretory
vesicles

colloid
droplets

rough
endoplasmic-
reticulum

dense body
(lysosomes
and colloid)
hydrolysis of
thyroglobulin

Golgi

liberation of
hormones

nucleus

basal lamina

capillary

secretion
via
connective
tissue

Fig. 11.4 Ultrastructural features of the cyclostome thyroid cell and interpretations of
the biosynthetic sequence.

dense bodies. In both lamprey and hagfish thyroid cells, X-ray microanalysis
has shown that these structures contain relatively high levels of iodine
although, in the case of the hagfish, the relative concentrations in the
cytoplasm and luminal colloid were too small to be measured by this
technique (Fujita, 1975). Presumably, hydrolysis takes place within the
secondary lysosomes, containing the highly condensed thyroglobulin and the
hormonal products are liberated to pass through the basal lamina.

In the case of the endostyle, the secretory processes must differ from those of the thyroid gland, in so far as there can be no extracellular storage of the iodinated products. In autoradiographs, bound iodine has been seen on the cell surfaces of type 2c and 3 cells within 30 minutes of the administration of radio-iodine (Fujita and Honma, 1969). After 2 hours, labelling was seen in vesicles at the cell apices and after a further 2 hours it was evident in the multivesicular bodies. These organelles were considered to be intracellular storage sites for the iodinated products. The distribution of the lysosomes appears to correspond to those regions where iodine binding is most intense, suggesting that these organelles may either be involved in biosynthesis or in the liberation by hydrolysis of the hormonal products (Barrington and Sage, 1972). At the same time we cannot exclude the possibility that hydrolysis may be extracellular and that hormonal materials may be liberated by digestion in the gut or, more probably, through the agency of the protease that has been identified in endostylar extracts.

Where analyses have been made on endostylar or thyroid tissue following the administration of radio-iodine, the amounts of the isotope incorporated into thyroxine has usually represented only a small fraction of the total radioactivity, most of which is present as iodide, with smaller amounts in the form of MIT and DIT. Thyronine T_3, which appears to be metabolized more rapidly, has either not been detected or has been present in only very low concentrations. The distribution within the iodoproteins of the various iodinated products is quite similar in cyclostomes (Table 11.1) with DIT as the major constituent, followed by MIT. T_3 was not detected in either sample in this instance and T_4 represented only 2.5% in the hagfish and 2.9% in the lamprey. Quite similar results have also obtained for the ammocoete and adult of *L. planeri*, where the iodinated products represented only 0.002% of the total protein. Together, T_3 and T_4 made up 3–5% and there were larger

Table 11.1 Iodoamino acid composition of hagfish and lamprey iodoproteins and of dogfish thyroglobulin. From Susuki *et al.*, 1975

	^{125}I content g 100 g^{-1}	I$^-$	MIT	% DIT	T_3	T_4
Hagfish iodoprotein* labelled with ^{125}I for 4 days	0.0057†	6.4	15.1	70.4	0	2.5
Lamprey iodoprotein labelled with ^{125}I for 4 days	0.0022	13.0	20.3	53.2	0	2.9
Dogfish thyroglobulin.	0.50	10.0	12.5	26.6	0	50.9

*Cited from Susuki and Gorbman, 1973.
†Cited from Tong *et al.*, 1961.

amounts of MIT and DIT (Monaco *et al.*, 1978). Compared to the thyroglobulin of the dogfish, the most significant features of cyclostome iodoproteins are therefore their low iodine content and lower hormonal levels.

Although it has been said that the cyclostomes are characterized by their low levels of circulating thyroid hormones (Fontaine and Leloup, 1950), this is hardly borne out by the information in Table 11.2. This shows that the concentration of T_4 in cyclostome serum is of the same order as in two teleost species reported by Refetoff *et al.* (1970) (4.3–4.5 μg 100 ml^{-1}) and generally higher than in the trout, where a serum concentration of 1.1 μg 100 ml^{-1} has been recorded (Jacoby and Hickman, 1966). Even lower values have been reported in a number of freshwater and marine teleosts by Leloup and Hardy (1976). Whereas in freshwater forms, T_4 levels (0.085–1.5 μg 100 ml^{-1} were always higher than those of T_3, the reverse was true in sea water, where values for the latter (0.36–3.6 μg 100 ml^{-1}) were said to be higher than those of mammalian species.

For the sea lamprey we have information on T_4 levels at all stages of the life cycle with the exception of the parasitic phase. Surprisingly, thyroxine concentrations in the blood of the ammocoete are very much higher than in the upstream migrant stages of either *P. marinus* or *L. tridentata* and are comparable with the exceptionally high concentrations in the blood of spawning female sea lampreys. During the early stages of metamorphosis, there is a dramatic fall in hormone levels (Fig. 11.5) and this decline continues

Table 11.2 Thyroxine concentrations in the serum of cyclostomes.

Species	T_4 μg 100 ml^{-1}	Authors
Lampreys		
P. marinus		
Ammocoetes	7.35	Wright and Youson, 1977
Prometamorphic	3.78	Wright and Youson, 1977
Macrophthalmia	0.46	Wright and Youson, 1977
Upstream migrants	0.087–0.23	Leloup and Hardy, 1976
Sexually mature males	0–1.8	Hornsey, 1977
Sexually mature females	5.4–10.8	Hornsey, 1977
L. tridentata		
Upstream migrants	0.5	Packard *et al.*, 1976
Hagfishes		
M. glutinosa		
After 1–2 months in laboratory	6.26–7.79	Henderson, 1975
E. stouti	2.4	Henderson and Lorschneider, 1975
Anaesthetised and sham operated	2.2–4.9	Matty *et al.*, 1976
	3.4	Packard *et al.*, 1976
E. burgeri	3.7	Tsuneki and Fernholm, 1975

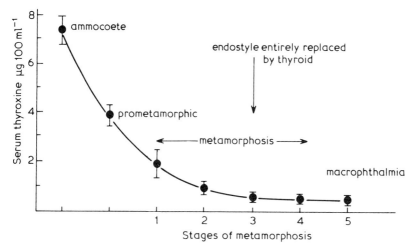

Fig. 11.5 Serum levels of thyroxine (T_4) during the metamorphosis of the sea lamprey, *Petromyzon marinus*. Redrawn from Wright and Youson, 1978.

right through to the macrophthalmia stage at the completion of transformation (Wright and Youson, 1977). Unfortunately, we have no information on T_4 levels in the period of adult life that intervenes between the end of metamorphosis and their entry into the rivers on their upstream migration, but since the values are somewhat similar in macrophthalmia and early upstream migrants it is quite possible that these low levels may be characteristic of the parasitic phase. Suggestions that the low levels of T_4 in early migrants may be due to the cessation of feeding have not been borne out in the ammocoete, where starvation did not result in decreased hormone levels. The pattern of events in the metamorphosis of lampreys is obviously in direct contrast to the situation in the frog tadpole, where an initially inactive thyroid begins to release increasing amounts of thyroid hormones during metamorphosis. On the contrary, the ammocoete endostyle shows the maximum biosynthetic activity followed by a rapid decline during transformation as this organ degenerates and is replaced by the developing adult thyroid follicles. This is parallelled by the activity of endostylar oxidative enzymes, which is at its maximum in the earliest stages of metamorphosis, decreasing when the first thyroid follicles appear (Pajor, 1977). Although it would be premature to postulate a causal relationship between the declining T_4 levels and metamorphosis in the ammocoete, it may be recalled that some metamorphic changes have been observed in larval lampreys after treatment with potassium perchlorate; an agent that inhibits thyroidal biosynthesis (Hoheisel and Sterba, 1963). Thyronine (T_3), although in only low concentrations, has been detected throughout the migratory period of the sea lamprey (Hornsey, 1977), but T_4 was only present in measurable amounts when the animals were already spawning. Indications of an inverse relationship between T_3 and T_4 concentrations suggest that as may be the case in higher animals, the former may be the more active hormonal compound. The sharp increase in T_4 levels, that seems to occur in the spawning sea lamprey, may appear to conflict with observations made by Pickering (1972, 1976a) on upstream migrant river lampreys (*L. fluviatilis*). In these animals, the rate of uptake of ^{125}I by isolated thyroid tissue showed a consistently downward trend from the beginning of the migration in August at least up to November. Moreover, as judged by the decreasing height of the thyroid epithelial cells, the gland was considered to be less active as the animals approached sexual maturity, when compared to those that had just embarked on their upstream movement in the autumn. Nevertheless, a declining level of thyroidal activity during the migratory period would not necessarily be incompatible with a rise in blood hormone concentrations in the brief final period when the lampreys are spawning. Indeed, in view of the well known effects of thyroid hormones on the vertebrate central nervous system, it is conceivable that such an increase might help to explain the marked changes in metabolic rhythms and behaviour patterns that occur in breeding lampreys, when restless and continuous daytime activity replaces the nocturnal habits of the upstream migrants (Section 4.2.3). However, speculation apart, it must be admitted that we are totally ignorant of any possible role for thyroid hormones in the life of the cyclostomes. Based on analogies with the amphibians, attempts to accelerate the transformation of the ammocoete by the administration of thyroxine proved completely unsuccessful (Stokes, 1939; Leach, 1946) and the removal of the endostyle has not inhibited metamorphosis (Sterba, 1953). A metabolic effect of thyroxine comparable with that of higher vertebrates has not been demonstrated (Leach, 1946; Matty, 1966) and, as we have seen, oxygen consumption rises continuously throughout metamorphosis at a period when T_4 concentrations are undergoing a continuous and marked decline.

11.1.6 Pituitary regulation of thyroid activity

In higher vertebrates, the activity of the thyroid is influenced by thyroid-stimulating hormone (TSH) secreted by basophils in the adenohypophysis. Earlier studies on the lamprey pituitary had suggested that the comparatively sparse basophils of the mesoadenohypophysis might represent such thyro-trophs (Chapter 10). Some support for this idea came from experiments involving prolonged treatment with the goitrogen, thiouracil, which resulted in cytological changes in these mesoadenohypophysial basophils (Båge, 1969), although it is possible that these could have been due to the toxic qualities of this compound. In *L. fluviatilis*, surgical removal of the pituitary has failed to produce any changes in iodine uptake, thyroxine levels or in the cytology of the thyroid epithelium (Larsen and Rosenkilde, 1971; Pickering, 1972). Similar conclusions were also reached by Pickering (1976a) who cultured thyroid tissue from the same species together with lamprey pituitary extracts or mammalian TSH. Neither treatment produced significant changes in the uptake of ^{125}I by the gland tissue or in the distribution of the radio-activity in the various iodine compounds recovered from the culture medium. Pickering also reported similar negative results after measuring the iodine uptake of isolated thyroid tissue from animals that had been hypophysecto-mised two months previously.

In ammocoetes, removal of the pituitary has not produced significant cytological changes in the endostyle, nor has this operation affected the rate of organic iodine binding (Pickering, 1972). On the other hand, a number of earlier authors had reported cytological changes or alterations in iodine metabolism following the administration of mammalian TSH to the am-mocoete, although such experiments cannot prove that the animal produces its own TSH. In this connection, experiments involving the application of goitrogens are of some interest. These compounds inhibit iodine binding and by so doing reduce the levels of circulating thyroid hormones. The existence in the higher vertebrates of a negative feedback system between the thyroid and the pituitary thus leads to an increased output of TSH from the adenohy-pophysis following the administration of goitrogens. After such treatments, ammocoetes show certain changes in the endostylar epithelium that would be consistent with hypersecretory activity, although curiously, these changes were not confined to those areas that are known to be primarily concerned in iodine metabolism and even affected the type I glandular cells (Barrington and Sage, 1963a,b). Even more importantly, these hypersecretory responses could be elicited in hypophysectomised ammocoetes, thus excluding the possibility that they are due to a pituitary-endostylar feedback system (Barrington and Sage, 1966). These authors were inclined to regard these paradoxical results as due to a feedback mechanism (at a purely intracellular level) on those cellular systems normally involved in the synthesis of proteins materials participating in iodine binding.

If the situation regarding pituitary control of the lamprey thyroid is equivocal, it has been equally difficult to find decisive evidence for pituitary thyrotrophins in the myxinoids from some of the earlier experimental work involving hypophysectomy. These failed to show significant changes either in the cytology of the thyroid tissue or in the plasma thyroxine levels of the operated animals (Falkmer and Matty, 1966; Gorbman and Tsuneki, 1975; Henderson, 1975; Henderson and Lorschneider, 1975). On the other hand, some evidence for the existence of a thyrotrophic factor in hagfish pituitary extracts has been claimed as a result of bioassays and more recently the uptake of ^{131}I by *E. stouti* was found to be nearly doubled after the administration of mammalian TSH (Kerkoff *et al.*, 1973). The treated animals also showed a lower percentage of labelled MIT and higher proportions of DIT and T_4 compared with the control animals. Matty *et al.*, (1976) have also found that transplantations of hagfish pituitary tissue into the cranial sub-cutaneous sinuses of *E. stouti* resulted in elevated T_4 levels and a similar result was obtained after the *in vitro* culture of pituitary and thyroid tissue of the same species (Dickhoff and Gorbman, 1977). It seems therefore, that there is rather better evidence for a thyrotrophin in the hagfish than in the lamprey and this is all the more surprising when we recall the undifferentiated condition of the myxinoid adenohypophysis and the rarity of granular or basophilic elements (Chapter 9).

In view of the uncertainty surrounding the existence in the lamprey pituitary of a TSH factor, it is interesting to find that thyrotrophin releasing factor (TRH) has been identified in the brain tissue of the adult lamprey, *P. marinus*, in the brain and pituitary of the ammocoete and also in the head region of *Amphioxus*. This has led to a suggestion that this tripeptide – Glu-His-Pro – has an ancient chordate lineage and may have been first developed as a central neurotransmitter (perhaps in connection with the olfactory system), before it became adapted as part of the thyroid control system (Jackson and Reichlin, 1974; Nicoll, 1977). Certainly, it does not appear to have this latter function in the hagfish, since injections of synthetic TRH have had no effects on the adenohypophysis or thyroid tissues, nor on serum T_4 levels of *E. burgeri* (Tsuneki and Fernholm, 1975). Furthermore, TRH has not increased the production of T_4 by cultured pituitary-thyroid tissue in *E. stouti* although the pituitary tissue itself increased the output of T_4 from the thyroid tissue by over 50 % (Dickhoff *et al.*, 1978). TRH is also present in high concentrations in the amphibian hypothalamus, but here again is said to have no effects on pituitary TSH. On the other hand, synthetic TRH has been shown to stimulate the release of α-MSH from the neuro-intermediate lobe of the frog pituitary, suggesting that its normal function may be to act as an MSH – releasing hormone (Trochard *et al.*, 1977).

11.1.7 Cyclostomes and the evolution of the thyroid

That the vertebrate thyroid gland is morphologically homologous with the endostyle of ammocoetes and protochordates has rarely been challenged by zoologists, although there is no direct evidence that this latter structure was present in the fossil agnathans. Many of these animals are thought to have been microphagous feeders, but the manner of their mucus production and the pattern of their ciliary feeding tracts is completely unknown. Undoubtedly, the transformation of the larval endostyle to the closed thyroid follicles within the life cycle of the lamprey has great intellectual attraction, in portraying a simple picture of the phylogenetic history of the gland as originating in the mucus feeding mechanism of some protochordate-like vertebrate ancestor. On the other hand, when the origin and significance of iodine metabolism are considered, the picture is less clear, especially in view of the widespread occurrence of iodine binding in both protostome and deuterostome invertebrates. Furthermore, there is some evidence to suggest that iodine uptake may be an inherent property of the entire vertebrate endoderm, but that this potency is suppressed in the course of ontogeny, except in those special areas from which the thyroid gland is subsequently developed. Thus, iodine incorporation has been demonstrated in the salivary glands, bronchi and even in the duodenum of the embryo rat (Csaba and Nagy, 1975).

An interpretation of the evolution of thyroidal activity that has gained wide currency is the one originally proposed by Gorbman (1958). This draws attention to the presence of iodoproteins in the exoskeleton and pharyngeal teeth of invertebrates, suggesting that at some point in time these compounds assumed a metabolic role and that, as in the protochordate endostyle, the pharyngeal site became dominant. At this stage, hydrolysis of the protein would have occurred in the gut, leading to the liberation of the active iodinated compounds. In *Amphioxus* and in the ammocoete, the iodoproteins are no longer skeletal proteins and specialized cells have appeared in the endostyle that are capable of synthesizing T_3 and T_4. However, these compounds may still be liberated in the gut by hydrolysis after being conveyed there in the mucus stream. Finally, with the replacement of microphagy by the macrophagous feeding habits of the vertebrates and the loss of the mucus carrier, the iodinated proteins were confined to the closed thyroid follicles, where hydrolysis occurred, setting free the hormonal compounds to enter the circulation.

In what may appear to traditional zoologists as a somewhat heretical approach to the problems of thyroid evolution, Etkin and Goa (1974) have questioned the hallowed position that the endostyle has hitherto occupied in the phylogeny of the vertebrate thyroid gland. In this connection it is interesting to note that thirty years earlier, similar doubts had been expressed by Leach (1944) who remarked that 'the usual statement that the endostyle of ammocoetes is phylogenetically intermediate between the acraniate endostyle and the thyroid of true vertebrates is to be seriously questioned.' Accepting the widespread and 'accidental' nature of iodine binding as a consequence of the chemical affinity between iodine and tyrosine, Etkin and Goa suggest that in the early stages of thyroidal evolution, the iodinated proteins may have had some structural or enzymatic role and that iodine metabolism may have been developed by vertebrates invading freshwater as a way of sequestering and storing iodine. The presence in blood or skeletal tissues of proteins capable of binding iodine would have facilitated the coupling of iodotyrosines to form T_4, but at the stage which might be represented by the protochordates, the presence of such compounds should not be assumed to have any hormonal significance. Although in the vertebrate ancestors, iodine-metabolizing cells may have tended to concentrate towards the floor of the pharyngeal region, it would be unnecessary to assume that they were associated with endostylar function. As these authors point out, iodine metabolism in the protochordates is not confined to the endostyle and even in the ammocoete the glandular tracts have no thyroidal activity. The presence of cells showing such activity in the lamprey endostyle could be interpreted as a result of the suppression of thyroid differentiation consequent on the introduction of a prolonged larval stage. Such an interpretation is in harmony with the views expressed elsewhere in this volume (Chapters 3, 12) on the place of the larval stage in the evolution of lampreys. The presence of an embryonic thyroid primordium or 'rest' whose final differentiation has been delayed by the extension of larval life has its counterpart in other groups of animals with specialized larval forms, such as the heterometabolous insects or the amphibians.

As explained earlier, we are still uncertain as to the exact contribution made to the adult thyroid follicles by the various cell types that have been identified within the larval endostyle, although several authors have considered that this may involve the type 4 cell, which plays little or no part in iodine binding. In this connection it is interesting to note a comment made by Barrington and Sage (1972) who, in reference to the contribution made by this cell type to the adult thyroid, remarked 'that this could be readily understood if they were genetically programmed for activity in the adult'.

11.2 Hormones of the pancreas and gut

11.2.1 The organization of the exocrine and endocrine pancreas

Unlike the higher vertebrates, the cyclostomes have no discrete pancreatic organ lying outside the gut and connected to it by pancreatic ducts. Nevertheless, both components of the vertebrate pancreas can be recognized; exocrine zymogen cells secreting their digestive enzymes directly into the gut lumen and endocrine tissue representing the islets of Langerhans of the higher vertebrates.

The distribution of zymogen cells in the anterior intestine and intestinal caecae of ammocoetes and adult lampreys have already been described in Chapter 5 and, at metamorphosis, there are important changes in this region of the gut. The most posterior segment of the ammocoete oesophagus persists in the adult, but the remainder of the definitive oesophagus is a neoformation, developed from a dorsal lamina on the roof of the larval pharynx. In Northern Hemisphere lampreys, the adult intestine forms a blind caecum over the posterior end of the oesophagus, where it breaks up into a complex of branching diverticula surrounded by islet tissue (Fig. 11.6). In the Southern genera, the larval caecae are reduced in size at metamorphosis, but in *Mordacia* a small structure persists ventral rather than dorsal to the gut as in the Northern forms. This difference like those between the ammocoete caecae of the various genera need not necessarily be of profound phylogenetic

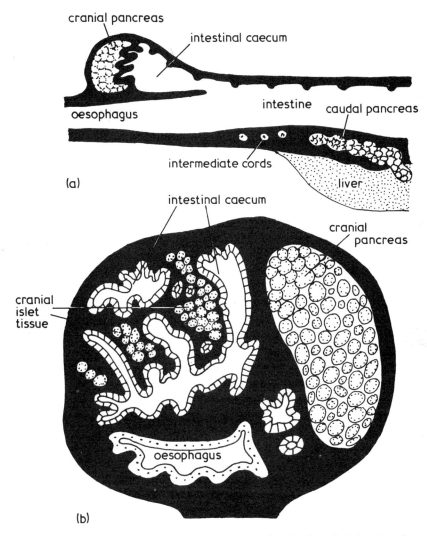

Fig. 11.6 The islet organ of the adult lamprey (a) Sagittal section through the junction of oesophagus and intestine. (b) Transverse section through the region of the cranial pancreas. Based on Barrington, 1972.

significance and both are probably a result of variations in the torsion that takes place in the intestine during early larval and metamorphosing stages. In so far as they may foreshadow the evolutionary history of the gnathostome exocrine pancreas, these intestinal evaginations are of considerable interest, pointing towards a closer affinity between the petromyzonts and the gnathostomes.

Equally significant in relation to the evolution of the vertebrate pancreas is the condition of the cyclostome endocrine tissue. First described by Langerhans himself in 1873 and later to be known as the follicles of Langerhans, the islet tissue of the ammocoete is formed by the proliferation of cells at the base of the intestinal submucosa in the region around the junction of the oesophagus and intestine. These cells were later shown to be involved in the control of blood sugar concentrations and therefore comparable with the β cells of the pancreatic islet tissue (Barrington, 1942). In adult lampreys, this tissue is concentrated in two main areas; a cranial pancreas, lying above the intestinal caecae and a caudal pancreas situated below the intestine where it passes over the surface of the liver (Fig. 11.6). Between these two areas, further more scattered groups of islet tissue occur along the course of the degenerating bile duct.

In myxinoids, the separation of exocrine and endocrine tissues is even more complete and in this respect might be regarded as more primitive. The exocrine component is represented by diffuse zymogen cells in three main areas; an intra-hepatic region around the major liver vessels, in the mesenteries between the liver and intestine and finally within the mid-gut epithelium. The endocrine tissue forms a discrete, compact mass – the islet

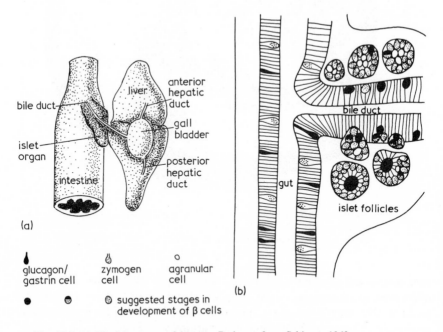

Fig. 11.7 (a) The islet organ of *Myxine*. Redrawn from Schirner, 1963.
(b) Distribution of endocrine cells in the intestine and bile duct of *Myxine* and the formation of islet follicles. From Östberg, 1977.

organ – attached as a small nodule to the bile duct, close to the point where it enters the gut (Fig. 11.7).

11.2.2 The islet tissues

Both in their general structure and in their cellular components, there are some quite striking parallels between the islet tissues of lampreys and hagfishes and these show significant divergencies from the typical gnathostome pattern. The islet lobules of *Myxine* tend to form hollow follicle-like structures, containing cavities 25–100 μm in diameter but these are less frequent in the islet tissues of *Eptatretus* species (Winbladh, 1976). In *Myxine*, these follicles sometimes enlarge to form cyst-like tumours (hamartomas) which may occupy almost the entire islet organ (Falkmer and Patent, 1972; Östberg *et al.*, 1972; Falkmer and Winbladh, 1964). The functional significance of the normal follicular structure of the islet organ of *Myxine* is not understood but, at least in some cases, the lining cells show microvilli and coated vesicles, suggesting that they may be involved in the absorption of protein materials from the luminal contents (Winbladh and Horstedt, 1975). The islet tissue of larval lampreys often contains follicular cell groups, but in the adult the islet cords or lobules are more usually solid structures. However, in the cranial pancreas of upstream migrant river lampreys, *L. fluviatilis*, follicular structures and islet cell tumours closely resembling those of the *Myxine* islets, have been found in a considerable proportion of the animals examined and appear to increase in frequency towards the end of the migratory period (Hardisty, 1976).

In both groups of cyclostomes, a feature of the islet tissue is its relatively poor vascularisation and this is most marked in the hagfish, where capillaries are said to be absent. As in the thyroid gland, secretion is believed to take place by emiocytosis into the surrounding connective tissues. Judging by their cytochemical and immunological reactions it is generally believed that apart from some agranular cells and connective tissue elements, the islet organs of cyclostomes consist only of β cells, reacting to antisera raised against mammalian insulin (Noorden and Pearse, 1974; Östberg *et al.*, 1975; Östberg *et al.*, 1976a). In the islet tissue of *Myxine*, a rare additional type of cell has been described, which in its ultrastructure appears to show some of the characteristics normally associated with A cells (Thomas *et al.*, 1973). The same cell type has been found in the bile duct epithelium where it has been regarded as a precursor of the functional β cell (Östberg *et al.*, 1976b) (Fig.11.7). The islet tissues of lampreys contain both light and dark cells, probably representing phases in the secretory cycle of the β cell (Hardisty *et al.*, 1976). More difficult to evaluate are the four types of granular cell described by Brinn and Epple (1976) in the islet tissue of *P. marinus* and although these may be derived directly or indirectly from the β cells, these authors did not exclude the possibility that they are functionally distinct cell types.

11.2.3 Endocrine cells of the gut and bile duct

The vertebrate gut and its various appendages, produces a whole spectrum of polypeptide hormones concerned in the integration of the digestive processes, including the secretory activities of the glands and the passage of food through the tract. Amongst the better known of these hormones are glucagon, produced by the A cells of pancreatic islets, gastrin secreted in the pyloric glands of the stomach, secretin and cholecystokinin-pancreozymin (CCK).

Investigations into the endocrine cells of the cyclostome gut have relied mainly on immunological techniques, based on antibodies raised against mammalian polypeptides. These have been supplemented by electron microscopy and by studies of the effects of cyclostome gut extracts on mammalian tissues. In the bile duct epithelium of the hagfish *Myxine*, there are scattered cells which react to mammalian insulin antisera, in addition to the rare granular cell already referred to. These latter cells have been seen to proliferate within the bile duct mucosa and are believed to contribute to the formation of fresh islet tissue (Fig. 11.7). Eventually, after acquiring an envelope of connective tissue, groups of these cells separate from the bile duct epithelium and enter the lobules of the islet organ (Östberg *et al.*, 1976b). Although these accounts have established the immediate origin of the β cells from the bile duct mucosa of the adult hagfish, they do not of course resolve the problem of their more remote embryological history.

A number of detailed studies have been made on the ontogeny of the islet tissues of larval lampreys (Boenig, 1927, 1928, 1929; Morris and Islam, 1969a; Ermisch, 1966). These larval follicles arise from cells lying at the base of the oesophageal epithelium at its junction with the intestine, but further contributions are also made by the epithelium of the intestine and the bile duct. From these parent cells, follicles are formed which then pass from the epithelium to the underlying sub-mucosa, where they are surrounded by connective tissues and blood vessels (Barrington, 1972). After a central lumen has developed within the follicles, the contents of the granular cells, together with cell debris may be discharged into the lumen, which may also contain blood cells from the surrounding capillaries. Throughout larval life, further follicles continue to develop, both by proliferation of the epithelium as well as by amitotic divisions of existing follicle cells.

The cranial pancreas of the adult lamprey which develops at the posterior end of the persistent larval oesophagus, is produced directly from existing larval follicles in the same region. These tend to lose their central lumen to form the more compact cords and lobules of the adult islet tissue. However, with the degeneration of the bile duct during metamorphosis, further cords are proliferated from its epithelium to form the caudal pancreas. Thus, the larval islet tissue originates mainly from the gut epithelium, but as in the hagfish, some of the endocrine tissue of the adult is derived from the bile duct epithelium.

The intestinal epithelium of *Myxine* contains scattered endocrine cells, which judged from the distribution of their secretory granules and the fact that these are not passed into the gut lumen, have been described as 'open' basal granulated cells. (Fujita and Kobayashi, 1974). These elements, with their base resting on the epithelial basal lamina, extend throughout the entire depth of the gut epithelium, with their apices bearing microvilli, projecting into the intestinal lumen (Östberg and Boquist, 1976). From these morphological characteristics it is clear that these cells pass their secretions through the basal lamina into the underlying connective tissue of the gut wall (Fig. 11.7). In animals that have been starved it has been noticed that the numbers of secretory granules increases, suggesting that their liberation is linked to feeding and digestion. Despite their uniform ultrastructure, these cells nevertheless show immunofluorescent reactions to antisera to porcine glucagon and also (although less intensely) to antisera against gastrin, the terminal C-peptide of gastrin (pentagastrin) and to caerulein, a peptide with biological activity characteristic of cholecystokinin-pancreozymin (CCK) (Fig. 11.10). These cells have shown no reactivity to either insulin or secretin antisera (Östberg *et al.*, 1975, 1976a).

Using similar techniques, parallel investigations have also been carried out on the pancreas and gut of larval and adult lampreys (Van Noorden *et al.*, 1972). In the ammocoete, endocrine cells in the intestinal epithelium are scattered along the length of the gut in the larger animals, although they tended to occur more frequently in the anterior third. Similar cells in the adult lamprey (*L. fluviatilis*), were more numerous and larger than those of the ammocoete and occurred either singly or in groups throughout the entire length of the intestine from the liver to the cloaca although, as in the larval stage, they were more abundant in the anterior sections. These endocrine cells are oval or flask-shaped, with a wider base resting on the basal lamina and a slender process extending upwards to reach the luminal surface of the epithelium.

No immunoreactivity to glucagon or secretin antisera has been observed in the islet tissue of larval or adult lampreys and the islet cells respond only to anti-insulin sera, supporting the view that the β cells are the only functional cell type in the endocrine pancreas (Van Noorden *et al.*, 1972; Van Noorden and Pearse, 1974). On the other hand, in the intestine, cells believed to be identical with those seen in electron micrographs were found to be reactive to glucagon, gastrin and caerulein antisera. These reactions were always more intense in adults than in ammocoetes and are believed to occur within the same cell. There is thus a general agreement in the immunological characteristics of intestinal endocrine cells in lampreys and hagfishes.

Intestinal extracts from both lampreys and hagfishes exert a CCK-like activity when tested on the mammalian pancreas (Nilsson, 1973; Barrington and Dockray, 1970). The same authors also found some evidence for secretin-like activity in their intestinal extracts, in spite of the apparent absence of cells with secretin-like immunoreactivity in either group of cyclostomes. Glucagon immunoreactivity has also been observed in lamprey intestinal extracts as well as in extracts from the cranial pancreas, although in the latter case the response is probably due to glucagon-secreting cells in the intestinal diverticula which penetrate the islet tissue (Zelnik *et al.*, 1977).

In view of the possibility of cross-reaction and the absence of specificity, results based on immunological responses to mammalian antisera must be interpreted with caution. What these observations show is that the intestinal endocrine cells of the cyclostomes contain molecules sharing certain immunological properties and reactive sites with the hormones used in the preparation of the antisera.

11.2.4 Physiology of cyclostome islet tissues

Early investigations on the islet tissues of lampreys (described in detail by Barrington (1972)) were directed mainly towards structural and functional comparisons with the endocrine pancreas of higher vertebrates. Historically, the presence in cyclostome islet tissue of insulin-like activity and its essential role in blood sugar regulation was first demonstrated in larval lampreys, where destruction of the follicles of Langerhans by cautery was followed by an elevation of blood sugar concentrations. Conversely, after glucose loading, the vacuolisation of the islet tissue gave clear indications of hypersecretion (Barrington 1942; Morris and Islam 1969b). Subsequent studies on the adult lamprey have confirmed that as in other vertebrates, the insulin produced by the β cells is involved in the maintenance of relatively constant levels of circulating glucose and in the utilization and storage of carbohydrate reserves.

Surgical removal of the islet tissue of the adult lamprey (pancreotectomy) has been followed by decreased levels of insulin-like activity in the serum, associated with an increase in blood sugar levels to concentrations some five times higher than those of normal animals (Zelnik *et al.*, 1977; Hardisty *et al.*, 1975). Comparisons of their glucose tolerance curves showed that the operated animals had completely lost their ability to regulate blood sugar concentrations following a single glucose injection (Fig. 11.8).

Earlier experiments on *Myxine* gave no evidence of hyperglycaemia in isletectomised animals, even after force feeding, and their glucose tolerance curves showed no indications that their capacity for blood sugar homeostasis had been impaired (Falkmer and Matty, 1966). Furthermore, in contrast to the lamprey, where anti-insulin sera have provoked hyperglycaemia and decreased glycogen concentrations in liver and heart muscle (Plisetskaya and Leibush, 1972), this procedure did not result in elevations of blood sugar levels

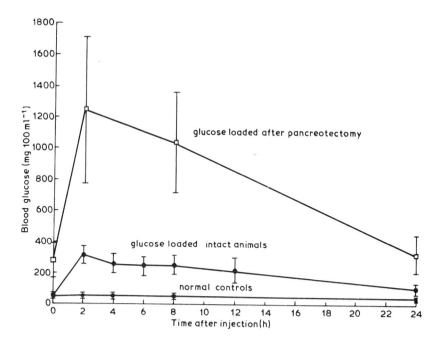

Fig. 11.8 Effect of removal of the islet tissue on the blood sugar homeostasis of the river
lamprey, *Lampetra fluviatilis*. From Hardisty *et al.*, 1976.

in *Myxine* (Falkmer and Wilson, 1967). More recently however, some
evidence of hyperglycaemia has been found in isletectomised *E. stouti*,
subjected to 10 weeks starvation (Matty and Gorbman, 1978), but in this as in
other divergencies between the myxinoid species in their carbohydrate
metabolism and its hormonal control, it is difficult to decide whether these
should be attributed to differences in handling or experimental methods
rather than to real species differences. For example, Falkmer and Matty
(1966) had commented that *Myxine* was able to withstand prolonged
starvation without displaying a reduction in blood sugar levels, whereas in *E.
stouti* these declined from 28 mg 100 ml^{-1} in freshly caught specimens to 11
mg 100 ml^{-1} after 10 weeks without food (Matty and Gorbman, 1978).

A common characteristic of both lampreys and hagfishes is the slow rate at
which a load of injected insulin is cleared from the circulation and the delay in
the establishment of normal blood sugar levels (Falkmer and Matty, 1966;
Hardisty *et al.*, 1976) after glucose loading. For example in *Myxine*, blood
sugar levels following glucose injection, showed a rapid rise within the first few
hours, succeeded by a slow return to normal values over a period of 1–2 days.
Similarly slow responses have been observed in lampreys and hagfishes after
injections of mammalian insulins (Falkmer and Matty, 1966; Leibson and

Plisetskaya, 1969). This hormone produces a long-lasting hypoglycaemia, which may take many days to return to baseline levels. While similar slow responses are characteristic of the lower vertebrates, they are certainly more marked in the lamprey than in teleosts (Fig. 11.9). The Pacific hagfish, *E. stouti* in contrast to *Myxine* requires only small dosages of bovine insulin (0.1–0.5 IU kg^{-1}) to induce hypoglycaemia and its sensitivity to the exogenous hormone is said to be some 1000 times greater than that of the Atlantic species. After a dose of 10 IU kg^{-1}, *E. stouti* still showed depressed blood sugar concentrations 15 days after the injection (Inui and Gorbman, 1977). Convulsions have rarely been seen in ammocoetes or adult lampreys, and never in the hagfish, after the administration of insulin and this has been related to the high glycogen content of the brain and its low levels of glucose-6-phosphatase activity. When administered to *Myxine*, mammalian insulin has been reported to increase the glycogen content of the muscles and decrease carbohydrate levels in the liver (Matty and Falkmer, 1965a), but these effects were not observed in *E. stouti*. However, in the latter species, insulin appears to be involved in protein synthesis, since its administration has been followed by depression of plasma nitrogen and increased incorporation into muscle protein of ^{14}C-labelled glycine (Inui and Gorbman, 1977, 1978). Exogenous insulin appears to have no effect on the glycogen content of the body wall

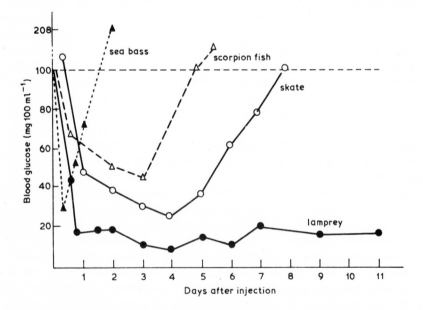

Fig. 11.9 The effects of an insulin injection (30–60 IU kg^{-1}) on the blood sugar levels of lampreys and teleosts. Redrawn from data of Leibson and Plisetskaya, 1968.

muscles of the lamprey (Bentley and Follett, 1965) and reports of its influence on liver glycogen have been unconvincing or even conflicting.

Lampreys are quite remarkable in the extent to which they are able to maintain their blood sugar concentrations during the long periods of fasting that are a normal feature of their life cycles; the first occurring during and for some time after metamorphosis, and the second during the upstream spawning migration. Coinciding with these changes from feeding to fasting, there are considerable changes in the orientation of their metabolism. Thus, in spite of the cessation of feeding, the metamorphosing sea lamprey, *P. marinus* shows a marked rise of blood sugar levels, with indications that the glycogen content of the muscles and liver diminish (O'Boyle and Beamish, 1977). These changes could indicate a drop in insulin output at a period when the islet tissue is undergoing its transition from the larval to the adult condition.

In the upstream migration of the river lamprey, *L. fluviatilis*, blood sugar concentrations appear to remain relatively constant throughout a period of 6–8 months without feeding, although a slight rise has been observed at spawning time (Larsen, 1976a). During the winter and early spring, the glycogen content of the liver and heart muscle decrease, whereas the carbohydrate reserves of the muscles of the body wall remains unchanged and are only exhausted during the final burst of activity that accompanies spawning (Plisetskaya *et al.*, 1976). This remarkable ability of the lamprey to maintain its blood sugar levels and the carbohydrate content of the somatic muscles throughout long periods of starvation, argues for the existence of efficient means for synthesizing carbohydrates from amino acids and above all, from the extensive lipid reserves that were built up during the preceding period of parasitic feeding. Both the somatic and heart muscles have been shown to contain highly active glycerol-3-phosphate dehydrogenase and glycerol kinase and rapidly incorporate ^{14}C-labelled glycerol. On the other hand, in the liver this gluconeogenic activity was very weak, confirming that the muscles are the main site of glycogen synthesis during the upstream migration (Savina and Plisetskaya, 1976; Savina and Wojtczak, 1977).

In lamprey muscle it has been shown that insulin increases the activity of glycogen-synthesizing enzymes (Plisetskaya and Leibson, 1973; Leibson *et al.*, 1976). This has led to the view that this hormone serves to maintain the carbohydrate content of the muscles throughout the period of winter starvation; an idea that is strengthened by the effect on the lamprey of antisera to mammalian insulin. This has resulted in hyperglycaemia and in a decrease in the glycogen concentrations of heart muscle and liver; parallelling the changes observed in the upstream migrant.

Changes in glycogen levels during the migratory period have been correlated with the levels of insulin reactivity in the blood (Plisetskaya *et al.*, 1976). When the first migrants enter the rivers in October, these insulin levels are high, but throughout the months following until February, they decline in parallel with the decreasing glycogen content of the liver and heart muscle. A rise in insulin levels occurs in the spring as the animals are approaching sexual maturity, followed by a dramatic fall at spawning time. The exhaustion of the carbohydrate reserves of certain organs has been linked by Plisetskaya to chronic insulin deficiency, developing at a time when the animals are no longer feeding. A similar fall is also said to occur in the anadromous sturgeon, although in this case the carbohydrate levels are restored when feeding is resumed.

Little is known of carbohydrate metabolism in the hagfish, but it appears that the liver is a relatively unimportant glycogen store and that, as in the lamprey, the muscular tissues represent the major proportion of the body carbohydrate. Unlike the situation in higher vertebrates, where removal of the liver is followed rapidly by hypoglycaemia and death, the hepatectomised hagfish has been found to survive for periods up to 30 days, without showing reductions in its blood sugar levels. Judging by their relative glucose-6-phosphatase activity, the kidney and intestine are probably major factors in the maintenance of blood sugar concentrations, and enzyme activity in the latter organ is about half that of the liver tissue (Inui and Gorbman, 1978). Similar conditions have also been observed in the lamprey, where hepatectomy has had little effect on blood sugar homeostasis. On the other hand, the importance of the liver in the elaboration of yolk proteins during sexual maturation was shown by the almost complete inhibition of vitellogenesis in the female lamprey (Larsen, 1978).

11.2.5 The evolution of islet tissue and gut hormones

The mode of origin of cyclostome islet cells from the epithelium of the gut, or its derivative the bile duct, points towards an ancestral condition in which there may have been a widespread and diffuse system of insulin-secreting cells within the digestive tract. This is apparently true of many invertebrates, where insulin immunoreactive cells have been detected throughout the digestive tract and its associated glands (Falkmer, 1972; Falkmer *et al.*, 1973). Evidence for such 'insulin cells' has now been found amongst molluscs, crustaceans, echinoderms and tunicates, whereas the existence of glucagon-like materials has yet to be established decisively. On the other hand, immunoreactivity to glucagon as well as to gastrin and insulin antisera has been reported in the gut of *Amphioxus* (van Noorden and Pearse, 1976). In this protochordate, no secretin-like activity has been detected, but, as in the cyclostomes, both the glucagon and gastrin-like activities appear to be located within the same cell type.

While the immediate derivation of vertebrate islet tissue from the gut endoderm is not in question, the ultimate and early embryological source of these cells is still disputed. A recent hypothesis has postulated the origin of islet cells together with other polypeptide secreting cells of the thyroid, adenohypophysis and gastro-intestinal tract from the neural crest or neurectoderm of the vertebrate embryo (Pearse, 1968). In this respect, this view has grouped together in one family, a wide range of endocrine elements, including the corticotrophs of the adenohypophysis, the C-cells of the thyroid, the chromaffin cells and the endocrine cells of the gut; all of which share a number of cytochemical features in common, above all the ability to take up and decarboxylate certain precursors of biogenic amines, believed to be concerned in the processing and secretion of cell granules. For this reason this family of endocrine cells has been termed the APUD series (amine precursor uptake and decarboxylation). However, there is now some doubt whether the origin of all these cells can in fact be traced back in ontogeny to the neural crest. For example, in bird embryos, transplanted primordia of the neural crest were found to give rise to parasympathetic ganglia, but not to cells with the cytochemical characteristics of pancreatic endocrine elements (Fontaine *et al.*, 1977) and in rat embryos from which neural crest precursors had been removed, functional β cells were subsequently identified (Pictet *et al.*, 1976; Phelps, 1975).

At what evolutionary stage, or in what sequence, the close association of exocrine and endocrine pancreatic tissues occurred is open to a number of alternative interpretations (Epple and Lewis, 1973). In the lamprey, and to a lesser extent in the hagfish, the islet tissues have already left their ancestral site within the intestinal epithelium to form masses of endocrine tissue, separated from the zymogen cells which still remain within the gut mucosa. Is this condition representative of an intermediate stage in the phylogeny of the gnathostome pancreas? If so, the condition in the lampreys with their so-called 'protopancreas' would be regarded as less primitive than that of the hagfish. At the same time, the segregation of the islet tissue from the gut epithelium in both groups might indicate that this migration preceded phylogenetically, the extra-mural concentration of exocrine elements in the history of the gnathostome pancreas. Such a view is not without its difficulties. The absence in cyclostome islets of A cells or of other differentiated elements of the gnathostome endocrine pancreas could only be explained on the assumption that the movement of the β cells out of the gut epithelium preceded that of the A cells and that the latter entered the islet tissue later, at a time when a distinct exocrine pancreas was already developing. An alternative hypothesis would involve the simultaneous migration from the intestinal wall of both endocrine and exocrine elements, but this would imply either a secondary regression of the exocrine pancreas in the cyclostomes, which seems highly improbable, or an independent evolution in agnathans and gnathostomes of distinct types of pancreatic organization. Finally, a third possibility would involve the origin of exocrine and endocrine tissues from a common anlage, which at first produced only exocrine tissue, but from which, at a later stage, endocrine elements were differentiated. This hypothesis, which has no support from our knowledge of the relation between zymogen and endocrine cells in cyclostomes would, like the second alternative, place the condition in these animals outside the phylogenetic sequence followed in the gnathostomes.

The first indications that the cyclostomes might possess gut hormones with some of the properties of those of higher vertebrates, came from observations on the physiological effects in mammals of intestinal extracts from lampreys and hagfishes. When tested on rats, intestinal extracts from the lamprey increased both the flow and the enzyme content of the pancreatic juice and in rabbits produced contraction of the gall bladder (Barrington, 1969; Barrington and Dockray, 1970). In mammals, these physiological activities are controlled by gastro-intestinal hormones; secretin promoting the flow of pancreatic secretion and cholecystokinin-pancreozymin, the release of pancreatic enzymes and stimulation of the gall bladder. Secretin-like activity has also been detected in intestinal extracts of *Myxine* (Nilsson and Fänge, 1970) and immunoreactivity to glucagon in similar extracts from lampreys (Zelnik *et al.*, 1977).

How can these findings be related to the ultrastructural and immunoreactive characteristics of the endocrine cells in the intestinal mucosa of the cyclostomes? Neither in the hagfish nor in the lamprey has secretin-like immunoreactivity been observed, but glucagon reactivity has been obtained from both hagfish and lamprey intestinal extracts and from their endocrine cells. The presence of secretin-like biological activity in intestinal extracts can probably be explained by the existence of a glucagon-type polypeptide, whose molecular structure is sufficiently like that of secretin for it to exert the appropriate physiological effects on the mammalian pancreas. In view of the absence in the cyclostomes of a discrete pancreas, it is difficult to visualise the necessity for secretin-like control of the diffuse zymogen cells, which are themselves in direct contact with the contents of the gut. This interpretation is in line with the close similarities that exist in the molecular structure of glucagon and secretin (Fig. 11.10) and which has led to the suggestion that both have been evolved from a common parent molecule.

How are we to regard the evidence for cholecystokinin-pancreozymin (CCK) activity in the intestinal extracts? The endocrine cells of the hagfish intestine, in addition to their immunoreactivity to glucagon antisera, also reacted less strongly with antisera to gastrin, pentagastrin and to caerulein and similar reactions occur in the endocrine cells of lampreys. Caerulein, a decapeptide isolated from frog skin and a more potent stimulator of gastric acid secretion than gastrin, shares with the latter and with CCK an identical C-terminal pentapeptide (Fig. 11.10) and even its last eight amino acid residues differ from those of CCK at only one position. This explains why antisera to caerulein can cross-react with CCK and gastrin in radioimmunoassays. The presence in both lamprey and hagfish intestinal extracts of

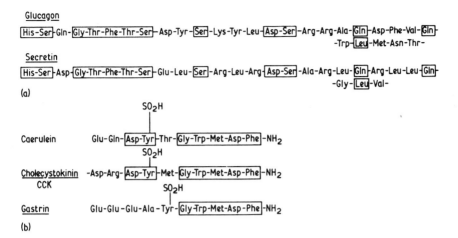

Fig. 11.10 Amino acid sequences of polypeptide hormones.
(a) Porcine glucagon and secretin.
(b) The C-terminal pentapeptide sequences of caerulein, gastrin and CCK.

CCK biological activity could thus equally well be attributed to the presence of a gastrin-like polypeptide. This has led to a suggestion that the same cyclostome endocrine cell may produce two separate polypeptides; one belonging to the secretin-glucagon family and the other to the gastrin-CCK group, each perhaps released in response to different conditions in the intestinal environment. Many of the organs or tissues which in the higher vertebrates are targets for their differentiated gastro-intestinal hormones, are either absent altogether in cyclostomes or are present in only a less developed form. Thus, there is no stomach or discrete exocrine pancreas and, in adult lampreys, even the gall bladder and bile duct have disappeared. One way out of these difficulties would be to assume that the target organs for the cyclostome hormones have been modified during the evolution of the higher vertebrates; a process for which the history of vertebrate endocrine systems provides ample precedents. In addition, the biological properties of the undifferentiated cyclostome polypeptides may differ considerably from those of the related mammalian hormones. For example, it has been suggested that hagfish endocrine cells may be involved in controlling the islet tissues, since in mammals both CCK and enteric glucagon are said to act as signals for insulin release (Östberg, 1976).

Unfortunately, one of the limitations inherent in the immunological techniques that have been used in the study of cyclostome polypeptides is that they do not always enable us to decide whether we are dealing with separate hormones or with different manifestations of a single larger molecule. Thus, those who cling to the traditional doctrine of 'one cell one hormone' may prefer to believe that the intestinal endocrine cell of the cyclostome is in reality producing a single type of polypeptide. For example, Weinstein (1972) had proposed that the gastro-intestinal hormones of higher vertebrates were originally derived from a primitive ancestral molecule which he termed 'prosecgastrin' with sequences at one end of the chain resembling secretin and at the other end, like gastrin. Elaborating on this concept, van Noorden and Pearse (1974) suggested on the basis of their observations on the lamprey intestine, that a corresponding parent molecule 'proglucgastrin' might exist, having as one part of the chain sequences similar to both glucagon or secretin and another region with a gastrin-like sequence. Such a molecule might then be expected to show both the secretin and CCK-type of biological activity that has been found in cyclostome gut extracts. If indeed the cyclostomes should be found to possess such a common parent molecule, the splitting up of this polypeptide might have been brought about by successive gene duplications producing first a glucagon-secretin and a gastrin-CCK peptide, followed later by the evolution of the separate hormones and accompanied by the increasing differentiation and the alimentary tract and its associated glands. With regard to the timing of these events, Weinstein (1968) has calculated from a comparison of the sequences of secretin and glucagon, that these two polypeptides were separated by gene duplication during the Mesozoic, some 200 million years ago; a figure that agrees broadly with the distribution of glucagon in amphibians, birds and mammals. Judging from the distribution in various vertebrate groups of separate gastrin and CCK polypeptides, it would appear that these may have been separated somewhat more recently than the secretin-glucagon gene duplication. Thus, in mammals birds and in reptiles it has been found that gastrin-like and CCK-like molecules occur within separate cell systems whereas, in amphibians and in teleosts, the same cell appears to be responsible for the production of both types of activity (Larssen and Rehfeld, 1977). These authors believe that gastrin and CCK have both evolved from a common caerulein-like ancestral molecule, which could in its turn have shown the kind of glucagon-like immunoreactivity that is found in the cyclostomes. Although it is possible that some of our views on the evolution of these molecules may have to be modified when the position of the protochordate intestinal polypeptides has been clarified, there seem little doubt that the cyclostomes occupy a primitive and key position in relation to the phylogeny of the vertebrate enteric hormones.

11.3 The interrenal and chromaffin tissues

The mammalian adrenal consists of two distinct functional components, differing fundamentally in their embryological origins and in the type of hormonal compounds that they produce; the adrenal cortex elaborating a range of steroid hormones and the medulla producing catecholamines. In the cyclostomes, these two tissues although recognizable, are not combined to form a discrete organ, but are loosely distributed over wide areas of the body. The tissue that is regarded as homologous with the adrenocortical tissue of the higher vertebrates is commonly referred to as the interrenal, while the counterpart of the adrenal medulla is the system of chromaffin cells.

11.3.1 The localization and structure of the interrenal
First recognized in the lamprey embryo shortly before hatching, the origin of the interrenal has been traced to a proliferation and thickening of the coelomic epithelium at the angle between the mesentery and the somatopleur at the level of the second pronephric tubule. After separating from the epithelium, the interrenal cells migrate dorsally away from the coelomic surface, eventually

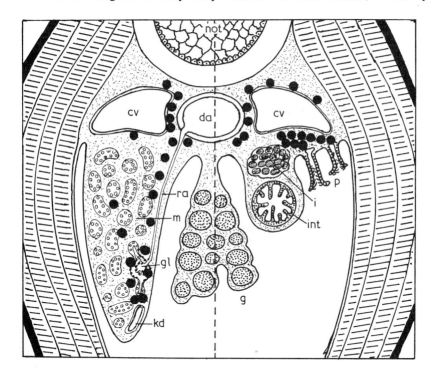

Fig. 11.11 The distribution of interrenal tissue (black circles) in a transverse section through the pericardial region of an adult lamprey. To the left of the dotted line the section is at the level of the mesonephros; to the right it is further forwards in the pronephric region.
cv. cardinal veins; da. dorsal aorta; gl. glomerulus; i. islet tissue; g. gonad; m. mesonephric kidney; kd. kidney duct; p. pronephric funnels; ra. renal arteries, not notochord; int. intestine.

concentrating around the ventral sides of the great vessels in the vicinity of the pronephric tubules (Sterba, 1955).

This tissue was first described in *L. planeri*, where it appeared .that the largest masses of interrenal cells remained within the pronephric region, although it was known that smaller groups subsequently extended posteriorly throughout the greater part of the trunk. However, more recent studies on ammocoetes and adult sea lampreys *P. marinus*, have shown that much greater numbers of interrenal cells may occur in association with the mesonephric kidney. In the pericardial regions of *Lampetra* larvae, small groups of interrenal cells are found on the ventral and lateral surfaces of the dorsal aorta, around the cardinal veins and among the pronephric tubules, but during the later stages of metamorphosis and in the adult, the largest groups are usually to be found at the base of the persistent pronephric funnels

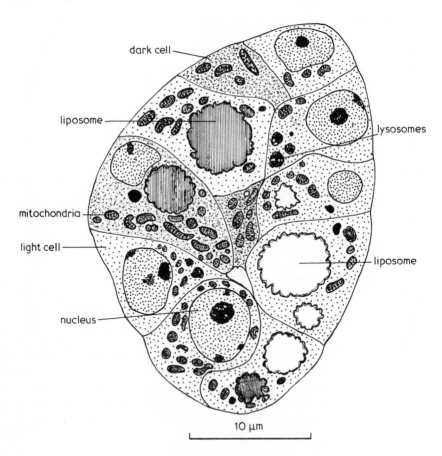

Fig. 11.12 A group of interrenal cells from the lamprey.

(Hardisty, 1972a,b) (Fig. 11.11). In the sea lamprey, the interrenal cells of the trunk region are mainly associated with the arterial supply of the kidney (Youson, 1972) where they have presumably been carried in the walls of the renal arteries during the development of the mesonephros. Entering the kidney along these arteries, the interrenal cells occur along the course of both efferent and afferent arterioles of the glomus as well as within the renal capsule itself, where they are separated only by the basement membrane from the capillaries and visceral epithelium. Other groups of cells are found between the kidney tubules (Fig. 11.12). In *L. planeri*, the total mass of interrenal tissue has been put at about 0.04 mg 100 g^{-1} body weight^{-1}; a minute figure by mammalian standards, where for several species the adrenal tissue represents from 10–126 mg 100 g^{-1} (Hardisty, 1972b).

The larger groups of interrenal cells are often arranged in cords or follicles, divided internally by thin connective tissue septa. Characteristic of these cells are the large liposomes, often exceeding in diameter the size of the nucleus and containing lipid and cholesterol (Fig. 11.12). Ultrastructurally, they have all the hallmarks of other vertebrate steroid-secreting tissues (Hardisty, 1972b; Youson, 1972). The smooth endoplasmic reticulum is well-developed, the mitochondria have tubular or vesicular cristae and the cytoplasm contains large numbers of free ribosomes or polysomes. Furthermore, the relationships between the mitochondria, lipid vacuoles and endoplasmic reticulum parallel the patterns seen in other steroidogenic cells (Hardisty and Baines, 1971).

Until recently, the interrenal tissue of myxinoids had not been identified. By using ^{14}C-labelled isoaxole, an inhibitor of the steroid enzyme $\triangle^5$,3β-hydroxysteroid dehydrogenase, it has been possible to locate sites where the enzyme may be concentrated (Idler and Burton, 1976). Since this is a key enzyme in steroid biosynthesis (Fig. 11.13), its presence should be regarded as evidence that the tissue concerned is involved in steroid secretion. In *Myxine*, the activity of the enzyme inhibitor was found to be concentrated in the pericardial and pronephric regions, but there was no evidence for its presence in either the mesonephros or posterior cardinal veins. Histological examination of the pronephros revealed cells with some of the characteristics of interrenal cells, situated around the bases of the pronephric tubules and in the central mass. Equally significant was the hypertrophy and hyperplasia that occurred in the central mass of the pronephros after injections of saline solutions or of mammalian corticotrophin (ACTH). In addition, antibodies to mammalian ACTH were concentrated in the same areas. Should the identity of this hagfish tissue be confirmed, its distribution would be similar to that of lampreys in early ontogenetic stages although, unlike the latter group, it apparently remains confined to the pronephric and pericardial regions.

11.3.2 Biosynthetic activity and pituitary control
Although the cytological, histochemical and embryological evidence points

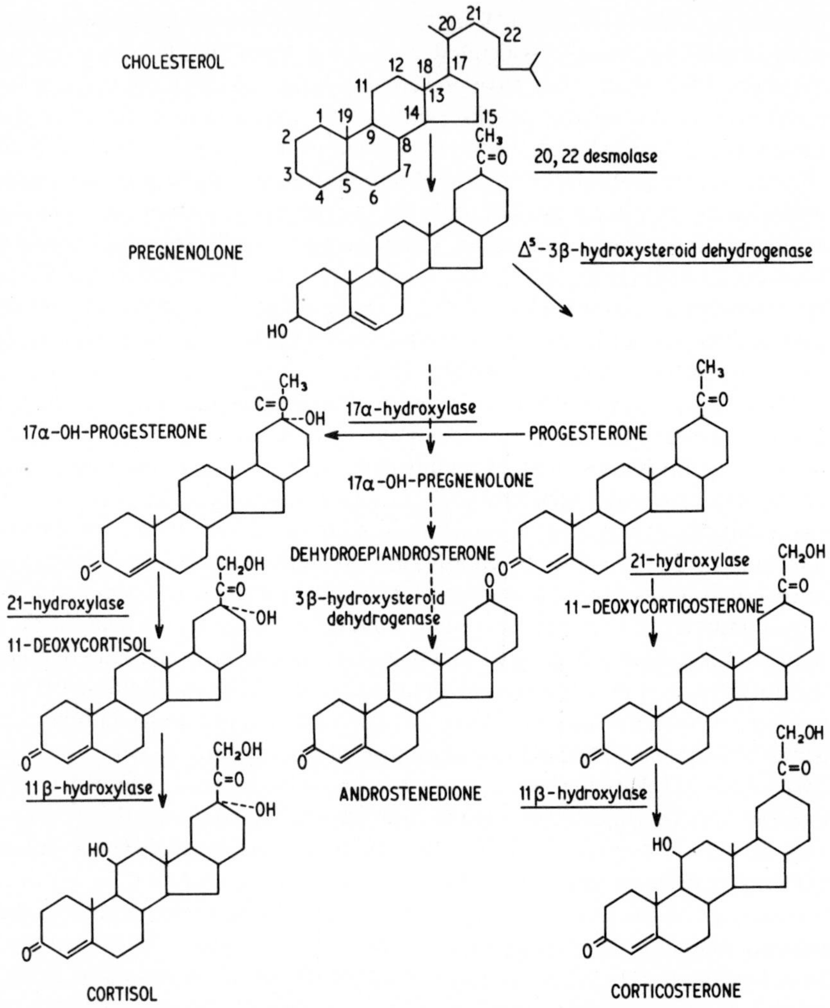

Fig. 11.13 Some of the pathways and enzymes involved in the biosynthesis of steroid hormones.

very clearly towards an homology between the lamprey interrenals and the adrenocortical tissues of higher vertebrates (Seiler *et al.*, 1970), the nature and extent of their biosynthetic activities are still obscure. On the other hand, in the myxinoids, where the plasma contains some of the usual vertebrate corticosteroids, their cellular source remains to be firmly established. In lampreys, the minute volumes of active interrenal tissue and the consequently low concentrations of hormones or enzymes might be at least partly responsible for some of the difficulties encountered in biochemical studies, such as the repeated failure to localize the enzyme $\triangle^5,3\beta$, hydroxysteroid dehydrogenase by histochemical methods, although some evidence of its activity has come from the application of sphectophotometric techniques using tissue homogenates (Weisbart *et al.*, 1978).

Attempts to establish the existence of corticosteroids in lampreys have involved blood analyses as well as *in vitro* investigations on tissue containing interrenal cells. Cholesterol, the parent precursor in steroid metabolism has been shown to be present in the liposomes of interrenal cells and these take up ^{3}H-labelled cholesterol (Youson, 1975a) thus strengthening their homology with adrenocortical tissue. However, in spite of the use of highly sensitive methods of analysis, the presence of corticosteroids in lamprey blood has not yet been demonstrated unequivocally (Weisbart and Idler, 1970; Buus and Larsen, 1975) and even if such compounds are present, their concentrations must be so low that it would be difficult to imagine that they have any physiological significance. On the other hand, experiments involving the incubation of kidney tissue containing interrenal cells with the appropriate ^{14}C-labelled steroid precursors, indicate that this tissue is capable of forming compounds which are intermediates in the synthesis of vertebrate steroid hormones (Fig. 11.13) and that it can produce some of the enzymes involved in steroid metabolism. In spawning migrant sea lampreys, tissue from the kidney was incubated with ^{14}C-labelled cholesterol, 11-deoxycorticosterone, 11-deoxycortisol and progesterone. As a result, neither corticosterone, cortisol, cortisone or deoxycorticosterone were formed, but 17α-hydroxyprogesterone was synthesized from cholesterol, indicating the presence of the enzyme 17α-hydroxylase (Weisbart, 1975). A further series of experiments involved kidney and interrenal tissue from ammocoetes and feeding stage sea lampreys. After incubation with ^{14}C-progesterone, again no corticosterone, cortisone or cortisol were produced, but in this case 11-deoxycortisol, 17α-hydroxyprogesterone and androstenedione were detected (Weisbart and Youson, 1975). These results indicate the presence of 17α- and 21-hydroxylase activity in addition to 20-desmolase and, when considered in relation to the observations on spawning migrants, they suggest that there may be some changes in the biosynthetic capacities of the interrenal cells when the animals abandon parasitic feeding to embark on their spawning run. The ability of feeding stage animals to form androstenedione is of particular interest in view of the ability of other vertebrate adrenocortical tissue to produce C-19 steroids. Curiously, *in vivo* experiments on parasitic phase sea lampreys involving the intracardiac administration of ^{3}H-labelled progesterone, have only provided evidence for 21-hydroxylase activity and the formation of 11-deoxycorticosterone (Weisbart and Youson, 1977).

In the blood of *Myxine*, many of the normal vertebrate corticosteroids have been detected after prior treatment of the animals with mammalian ACTH

(Idler and Truscott, 1972; Idler *et al.*, 1971). In this animal, cortisol, cortisone, corticosterone and 11-deoxycortisol were present and, although their source is not known, the fact that their production was stimulated by ACTH suggests that this may have been interrenal tissue. Moreover, this possibility is enhanced by the fact that, although the gonadal tissue of a hagfish has been shown to be capable of producing a range of steroid compounds *in vitro* (Hirose *et al.*, 1975), none of identified metabolites (with the exception of 11-deoxycortisol) include the steroids that were found in the blood of *Myxine*. Thus it seems likely that these corticosteroids were synthesized in some tissue other than the gonads and that this tissue, like the adrenocortical tissues of higher vertebrates can be stimulated by the pituitary ACTH.

Evidence for pituitary control of the lamprey interrenal is at present far from conclusive. Although various cell types in the adenohypophysis have been nominated as possible corticotrophs, their identification has not been finally settled (Chapter 10). Hypophysectomy does not result in any obvious changes in the morphology of interrenal cells at least at the level of the light microscope but, on the other hand, injections of mammalian ACTH have resulted in nuclear enlargement and hyperplasia of the interrenal cells of both ammocoete and adult lamprey (Sterba, 1955; Hardisty, 1972; Youson, 1973a). Much more significant however, are the ultrastructural changes that have been seen in sea lamprey interrenal tissue after ACTH injections (Youson, 1973b). These include alterations in nuclear shape, increased numbers of mitochondria, development of the Golgi and smooth endoplasmic reticulum, reduction in the numbers of lipid droplets and a folding of the plasma membrane adjacent to the perivascular spaces. All of these are reminiscent of the changes that have been seen in the adrenal cortex of the higher vertebrates following the administration of ACTH.

11.3.3 The chromaffin cells
The chromaffin cells of the adrenal medulla of the higher vertebrates originate from the embryonic neural crest which also gives rise to the sympathetic ganglia. These cells are therefore regarded as modified post-ganglionic neurones, which in the course of their differentiation have lost the characteristics of nerve cells and have assumed glandular functions. Throughout the vertebrates, cells similar in their origins and cytochemical characteristics to those of the adrenal medulla occur in other sites throughout the body and are classified as chromaffin cells. Although we have no direct evidence of their origin from the neural crest, the chromaffin cells of the cyclostomes resemble those of other vertebrates in their cytochemistry, fine structure and the presence of storage granules containing catecholamines.

In lampreys, apart from the chromaffin cells of the heart wall referred to in Section 6.3.4, the greatest concentrations of these elements are to be found in the pronephric region. Here they are usually located in the tunica adventitia of

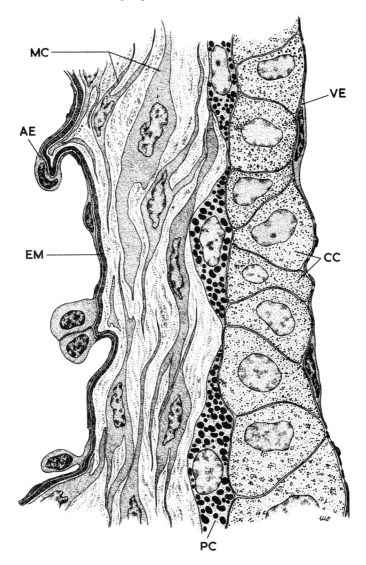

Fig. 11.14 Chromaffin cells from the lamprey. From Paiement and McMillan, 1976. CC. chromaffin cells; PC. pigment cells. MC. smooth muscle; EM. elastic membrane; AE. endothelial cells making up tunica intima of artery together with elastic membrane; VE. single layer of endothelial cells in wall of vein.

the major arteries, particularly the dorsal aorta and the proximal part of the mesenteric artery, and separated from the lumen of the corresponding veins only by a thin endothelial layer (Fig. 11.14). They are especially numerous in the walls of the cardinal veins and the sinus venosus and are often closely associated with nests of interrenal cells. In hagfishes, their distribution has not been described in detail, but they are said to be particularly numerous in the branchial and portal hearts and in the cardinal veins.

In the ammocoete, the chromaffin cells measuring from $12–35\,\mu m$ in diameter, are often stellate in form with processes extending between neighbouring cells and separated from the underlying connective tissue of the arterial wall by a thin basal lamina (Paiement and MacMillan, 1975). The cytoplasm is filled with granules about 130 nm in diameter, surrounded by a translucent halo and associated with the tubules of the smooth endoplasmic reticulum. Although nerve fibres may be present in the underlying connective tissue, there is no evidence that the chromaffin cells are innervated. From their cytochemical reactions it is believed that the extra-cardiac chromaffin cells of larval lampreys contain a primary catecholamine, either noradrenalin or dopamine.

The chromaffin cells of the lamprey heart are said to contain mainly adrenalin, whereas noradrenalin predominates in the circulating blood (Plisetskaya and Prozorovskaya, 1971) and is also present in the chromaffin cells of the cardinal veins and the sinus venosus (Fänge, 1972). These differences in the tissue distribution of the two catecholamines are in general agreement with biochemical analyses, which have shown that adrenalin accounts for most of the catecholamine content of the atrium and ventricle, whereas in the exceptionally high catecholamine concentrations of the sinus venosus,the ratio of noradrenalin to adrenalin is about 3:1. Differences in the distribution of catecholamines in the hearts of lampreys and hagfishes have been described in Section 3.3.4.

11.3.4 The physiological role of the interrenal and chromaffin tissues
Activation of the interrenals and/or the chromaffin tissues has been implicated in many of the short-term responses of lampreys to various stresses. These include the diuresis that occurs when larval or adult lampreys are handled and the rise in blood sugar concentrations that is seen when the animals are subjected to reduced oxygen tensions or surgical treatments (Morris and Islam, 1969b; Morris, 1972; Leibson and Plisetskaya, 1969). Hyperglycaemia has also been observed in *Myxine* in various forms of injection or surgical trauma. Under conditions of hypoxia, hypertrophy of the chromaffin cells occurs in the lamprey (Hardisty *et al.*, 1976) and increased levels of blood catecholamines have been observed after animals have been maintained in poorly oxygenated water, or after forced swimming or surgical trauma (Stabrovsky, 1967; Mazeaud, 1969).

In their responses to adrenalin injections, the two groups of cyclostome show a curious difference. In lampreys, the effect is a marked but transient hyperglycaemia, which has not been observed in *Myxine* (Falkmer and Matty,

1966). It is difficult to accept the suggestion that this is due to the high levels of endogenous catecholamines in the hagfish, since this must also be true of the lamprey. As in the mammal, adrenalin reduces the glycogen synthetase activity of the liver and muscles of lampreys, thus tending to decrease the rate at which glycogen is produced and stored in these organs (Plisetskaya and Zheludkova, 1973). At the same time it promotes the degradation of glycogen to glucose in these tissues by increasing the level of activity of amylolytic enzymes.

In higher vertebrates, stress responses mediated in the first place by catecholamines, are maintained over longer periods by the pituitary-adrenocortical system and the secretion of corticosteroids. Whether a similar mechanism exists in cyclostomes is unknown. A possible intervention of the interrenal of lampreys in certain forms of stress, has been suggested by the hypertrophy and hyperplasia of this tissue, particularly after surgical interventions. Evidence that the interrenal might also be involved in carbohydrate metabolism rests solely on the effects of a single dose of injected cortisol, which is said to increase the liver stores of glycogen and raise blood sugar concentrations (Bentley and Follett, 1965).

By analogy with other vertebrates it might be anticipated that the interrenal would be implicated in water and electrolyte metabolism, but the indications we have are conflicting. Some evidence of interrenal hyperplasia and hypertrophy has been reported after saline injections in lampreys (Hardisty, 1972) but, on the other hand, in young sea lampreys, acclimated to full strength sea water, the nuclear and cell volumes of the interrenal are said to show a significant decrease (McKeown and Hazlett, 1975). The common embryological origin of adrenocortical and nephrogenic tissue from the mesonephric blastema has been advanced as a rationale for the involvement of corticosteroids in water and salt balance and the control of kidney function, and especially in lampreys, the close association between the interrenal cells and the vessels of the kidney or glomus might suggest some functional relationship.

The trend in higher vertebrates towards an increasingly close association of the adrenocortical and chromaffin tissues has yet to receive an entirely satisfactory explanation. In the past, this close proximity of the two tissues has been attributed to the dependence of catecholamine biosynthesis on corticosteroids. Thus, the conversion of noradrenalin to adrenalin involves the transfer of a methyl group catalysed by the enzyme, phenylmethylanolamine-N-methyl transferase (PNMT), whose synthesis and activity are promoted by corticosteroids. Since the levels of these hormones required for this activity are thought to be much in excess of their normal concentration in the circulation, a close contact between the two tissues might be physiologically advantageous. However, there are considerable doubts whether this explanation is generally applicable to the lower vertebrates and, in lampreys, there is some evidence that the methylation of noradrenalin may be independent of corticosteroids (Mazeaud, 1972). Nevertheless, even in these animals these two tissues, although widely dispersed, are frequently closely associated and the two cells types quite often occur within a common connective tissue envelope.

11.4 The endocrine tissues of the gonads

11.4.1 The endocrine tissues of the testis
In teleosts where the organization of the testis is similar to that of lampreys, the steroid-secreting tissues are either located in the walls of the lobules (lobule boundary cells) or, like the Leydig cells of higher vertebrates, in the interstitial tissues between the lobules (Lofts and Bern, 1972).

In sexually immature river lampreys, *L. fluviatilis*, scattered undifferentiated cells of a fibroblast type are found in the interstices between the testis lobules and below the fibrous coat covering the surface of the gonad (Hardisty *et al.*, 1967). By the late winter or early spring, these cells have divided to produce small islets of polygonal cells whose cytoplasm now contains numbers of mitochondria and small granular lipid inclusions (Fig. 11.15). With further progress towards sexual maturity and the appearance of spermatids and spermatozoa, these interstitial cells show large cholesterol-rich, lipid masses (liposomes) like those of the interrenal cells (Section 11.3.1), the smooth endoplasmic reticulum is well-developed and the mitochondria

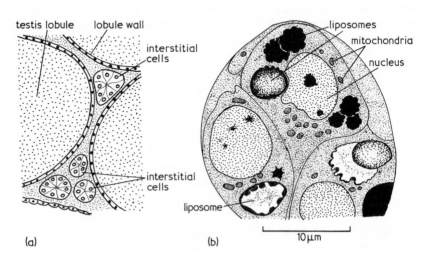

Fig. 11.15 The interstitial cells of the lamprey testis.
(a) The interlobular location of the tissue.
(b) High power view of a group of active interstitial cells.

show a dense matrix and tubular cristae, characteristic of steroid-secreting cells (Chieffi and Botte, 1962; Follenius, 1964; Barnes and Hardisty, 1972). Furthermore, their implication in steroid biosynthesis has been confirmed by their positive reaction to the techniques used for the histochemical demonstration of 3β-hydroxysteroid dehydrogenase. Shortly before the sperm are set free into the body cavity of the sexually mature animal, significant changes occur in the lobule wall cells, suggestive of their involvement in some form of steroid metabolism. These undergo a rapid and synchronous swelling, lipid masses appear in the cytoplasm, cytoplasmic vesicles increase in size and number and the mitochondria develop structural characteristics similar to those of the interstitial cells.

The lobules of the hagfish testis are surrounded by a continuous layer of somatic cells (Sertoli cells) and in the testis of *E. stouti* these show occasional small lipid droplets, while some of the mitochondria may develop tubular cristae (Tsuneki and Gorbman, 1977a). However, these features were not sufficiently developed to justify a conclusion as to their possible steroidogenic activity. Earlier light microscopists (Nansen, 1888; Schreiner, 1955) considered that the Sertoli cells of *Myxine* had a nutritive function although it is clear from the latter author's descriptions that they undergo marked changes in the course of the spermatogenic cycle and that, at least at certain periods, they develop large lipid accumulations. Furthermore, Schreiner himself had envisaged a secretory function for these elements, suggesting that they might produce substances that inhibit the development of adjacent ovarian tissue in the mixed gonads of *Myxine*. In the testis of *E. stouti*, the interstitial tissue contains two types of synthetically active cells; one of which has large numbers of mitochondria (a majority with tubular cristae) and elements of both smooth and rough endoplasmic reticulum. This was regarded by Tsuneki and Gorbman as a probable homologue of the interstitial cells of the lamprey and the absence of conspicuous lipid masses could be a reflection of the low synthetic activity of the hagfish gonad.

11.4.2 The somatic elements of the cyclostome ovary
As in the gnathostomes, the ovarian follicles of the cyclostome consist of an inner layer of follicle cells or granulosa, derived from the peritoneal epithelium (germinal epithelium) on the surface of the developing ovary. This is surrounded by a thecal layer derived from modified fibroblasts.

In the ovarian follicles of the lamprey, the granulosa at first consists of fusiform cells with spindle-shaped nuclei but, as vitellogenesis proceeds, these become prismatic and their lateral and apical surfaces adjacent to the oocyte develop short microvilli (Fig. 11.16). Initially, the granulosa covers most of the vegetative surface of the oocyte and the cells increase in height, particularly at the vegetal pole. This growth reaches its peak at the spawning season, shortly before ovulation, when the individual cells begin to separate.

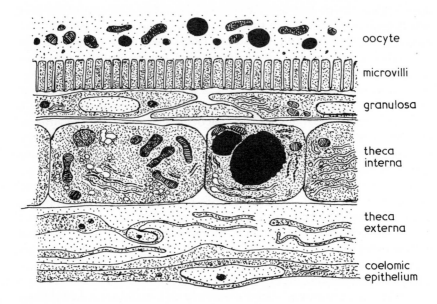

oocyte

microvilli

granulosa

theca interna

theca externa

coelomic epithelium

Fig. 11.16 A section through the ovarian follicle of a lamprey at the onset of vitellogenesis. Drawn from an electron micrograph by Busson-Mabillot, 1967.

After ovulation, the remnants of these cells remain attached to the egg, forming a sticky coat that enables it to adhere to the sand grains in the nest. Mucopolysaccharides are an important element in the secretions of the granulosa cells and are thought to play a significant role in the final rupture of the follicle at the time of ovulation (Larsen, 1970). Judging from their ultrastructural features, it seems more likely that the thecal tissues are responsible for the steroid secretions of the ovary and it is significant that these undergo a marked differentiation during the breeding season. At first slender and fusiform in shape, they become polygonal, developing a smooth endoplasmic reticulum and mitochondria with tubular cristae (Busson-Mabillot, 1967).

In the ovaries of *Myxine*, Fernholm (1972b) was unable to detect the presence of the enzyme 3 β-hydroxysteroid dehydrogenase by histochemical techniques and no cellular elements with the structural characteristics of steroidogenic tissues could be distinguished. In the course of oocyte growth the granulosa cells undergo changes similar to those that occur in the lamprey follicles and in the largest oocytes they form a single layer of high columnar cells with long processes extending to the surface of the oocyte. At this stage, the cytoplasm contains dense whorls of rough endoplasmic reticulum, vesicular smooth endoplasmic reticulum, Golgi, lipid droplets and mitochondria with lamellar or irregular cristae. As the oocyte increases in size the theca, which at first is composed of a single layer of modified fibroblasts, becomes multi-layered and divisible into a theca interna and externa, but although some of these cells may show a well-developed Golgi and rough endoplasmic reticulum, together with some lipid droplets, there are no indications that they are concerned in steroid biosynthesis (Tsuneki and Gorbman, 1977b). Thus, neither the granulosa nor the thecal elements show the typical hallmarks of active steroidogenic tissues and the former layer at least, is no doubt primarily concerned in the transport and processing of metabolites involved in the maturation of the oocyte. At the same time, as conditions in the lamprey gonad indicate, the assumption by somatic cells of the structural differen-tiations associated with active steroid metabolism may be relatively transient and sudden and in the case of the hagfish these might easily be overlooked without a more intensive investigation, taking into account the likelihood that there may be considerable changes in physiological state depending on precise phases of the spermatogenic or ovarian cycle.

11.4.3 The hormonal activities of cyclostome gonads

It has now been established that the vertebrate sex hormones, oestradiol and testosterone are present in the blood of cyclostomes or in gonadal extracts although in hagfishes the concentrations of these substances is extremely low (Botticelli *et al.*, 1963; Piavis *et al.*, 1975; Matty *et al.*, 1976). In the ovary, the precise cellular site of steroid synthesis has not been firmly established, but in the lamprey testis the involvement of the interstitial cells has been confirmed by the presence in these cells of the enzyme 3β-hydroxysteroid dehydrogenase (Hardisty and Barnes, 1968). The steroidogenic activity of the lamprey gonad has been further demonstrated by the failure of these animals to develop secondary sex characters after castration (Evennett and Dodd, 1963) and the reappearance of these structures when sex hormones are administered to the operated animals. That the sex hormones are specific to the development of the male and female secondary sex characters has been shown by experiments involving the administration of oestradiol to male lampreys and testosterone to females (Larsen, 1974). In some cases this led to the inhibition of the sex characters normal to the particular sex, but in other cases there appeared to be some simulation of the sex characters of the opposite sex. Although the mechanism is not understood, it is interesting to note that oestradiol tended to prolong the life span of the sexually mature lamprey.

Other than their involvement in the development of these sex characters, we have few indications of the wider role of sex steroids in the physiology of lampreys, although it is likely that as in other vertebrates, they may be involved in some aspects of gametogenesis and in the onset of spawning behaviour. In the female lamprey, oestrogens are believed to be implicated in the synthesis of calcium-rich phosphoproteins in the liver, and the administration of this steroid has resulted in increases in ovarian weight and oocyte volumes during the vitellogenic phase (Pickering, 1976b). This is in line with the processes of vitellogenesis in other oviparous vertebrates, where under the influence of oestrogens, yolk proteins are synthesized in the liver and transported to the oocytes.

A wider involvement of oestrogens in metabolic processes is also indicated by the effects of these compounds on the intestinal atrophy that occurs in the course of the spawning migration. Castration early in the migratory period of the river lamprey, *L. fluviatilis*, between August and November, prevented this intestinal degeneration (Larsen, 1969a). Furthermore, when the operation was delayed until the gut had already atrophied, castration resulted in some restoration of the gut tissues. On the other hand, when pellets of oestradiol or testosterone were implanted into the body cavity, the atrophy of the intestine was stimulated in both castrated and normal animals, but these treatments were ineffective either in normal animals late in the migratory season, or in cases where the gut had already hypertrophied as a result of castration (Larsen, 1974). As the removal of the pituitary early in the migratory period is also capable of preventing or retarding the atrophy of the gut, it may be assumed that this acts by virtue of its control of the secretory activity of the gonadal tissues (Chapter 10).

There is some evidence that steroidogenic activity in the gonads of lampreys does not become significant until comparatively late stages of sexual maturation. In the testis, for example, the differentiation of the interstitial cells does not seem to occur until spermatogenesis is well advanced and the concentrations of testosterone in the blood remain at very low levels until the spring, when the secondary sex characters appear (Hardisty, 1971). Similarly in the female sea lamprey, oestradiol concentrations in the plasma have been found to rise sharply at sexual maturation, declining even more rapidly after ovulation has occurred (Piavis *et al.*, 1975). This may explain the apparent failure to demonstrate the production of testosterone by testis tissue *in vitro* after incubation with ^{14}C-labelled progesterone (Weisbart and Youson, 1975). In these experiments, using the most refined methods of steroid analysis, it was only possible to establish the presence of 11-deoxycortico-sterone (and therefore of the enzyme 21-hydroxylase) in gonadal tissue from larval and feeding stage sea lampreys.

Although Fernholm (1972) was unable to find any indications of steroid biosynthesis or hydroxysteroid enzymic activity in ovarian tissues of *Myxine*, clear evidence of steroid secretion has been reported in the ovaries of *E. burgeri*, a species with a seasonal breeding season (Chapter 11). In experiments on ovarian tissue from this species, Hirose *et al.* (1975) cultured the tissue with ^{14}C-labelled pregnenolone as substrate and identified among the metabolites, 17 β-hydroxyprogesterone, androstenedione and 11-deoxycortisol. When 17 β-hydroxypregnenolone was used as the substrate, androstenedione was produced, suggesting that as in higher vertebrates, there may be two alternative biosynthetic pathways from pregnenolone to androstenedione. These authors believed the presence of 11-deoxycortisol to be of particular interest, since corticosteroids have been implicated in fish ovulation and 21-hydroxylase has also been found in the ovaries of teleosts and protochordates.

In another member of the same genus, *E. stouti*, measurements have been made of both testosterone and oestradiol concentrations in the blood (Matty *et al.*, 1976) in hypophysectomised, sham-operated and anaesthetised animals. As mentioned elsewhere (Section 10.4.1.) there was no evidence that the removal of the pituitary has any effect on the levels of circulating sex hormones and these concentrations appeared to decline the longer the animals were kept in captivity. In the male hagfish, there was no significant correlation between body length and plasma testosterone concentrations, but in the females there was a distinct tendency for hormone concentrations to decrease with increasing body length. If the latter is assumed to be an index of the degree of sexual maturation, this relationship might imply that, in smaller animals, the testicular tissues tend to be better developed (or to have undergone less regression) and therefore may retain a greater capacity for testosterone production. In a very high proportion of the hagfish examined, the oestradiol concentrations were below the sensitivity of the analytical technique and of the individuals with the highest hormone levels, five were sexually mature females and two were large males.

12
Reproduction and development

Although the breeding habits of hagfishes remain almost a complete mystery and we know very little about their reproductive physiology, the histology of their unpaired gonad indicates that the sexual differentiation of these animals is in a curious transitional state between gonochorism and hermaphroditism. Whether the myxinoids have a determinate life span is unknown, but they certainly produce a succession of eggs; each ovulation being followed by the renewed growth of a reserve of smaller oocytes, which are presumably replaced by the division of oogonial stem cells persisting in the ovary throughout their reproductive life. At least within the genus *Eptatretus*, ovulation may be restricted to a definite breeding season, but in *Myxine*, females containing mature eggs have been found throughout the year, although the interval between successive ovulations is not known.

As is the case in the eel and some species of salmon, the single breeding season of the lamprey marks the end of its life span. Like birds and mammals, the proliferation of the germ cells is confined to an early ontogenetic stage and in the lamprey is restricted to a relatively short period in early larval life. As a result they produce a finite stock of oocytes, whose development is more or less synchronous and of which the vast majority are destined to undergo degeneration (atresia) at various stages of their development. Those eggs that survive are finally shed during the short breeding season in spring or early summer. Thus, the two groups of cyclostomes have adopted quite different reproductive strategies; the myxinoids repeatedly ovulating a small number of large, yolky shelled eggs throughout their life, the lamprey producing very large numbers of small eggs which mature simultaneously and which are all shed together at the close of the life cycle.

12.1 Sex differentiation and gametogenesis

12.1.1 Myxinoids
The inherent 'bisexuality' of the gonads of *Myxine* had already attracted the

272

attention of 19th century zoologists, who remarked on the extreme rarity of 'true males' and the presence in the genital fold of both ovarian and testicular tissue; the former predominating in the cranial, and the latter in the more caudal areas of the gonad. Observations on the relative development of the two tissues in relation to body length suggested that these animals might be protandric hermaphrodites, functioning first as males and later as females (Nansen, 1888). However, as a result of more detailed histological studies, Schreiner (1955) was unable to detect any constant relationship between length (or age) and sexual status and he concluded that *Myxine* was in fact, a dioecious animal. In those individuals where the cranial or caudal segments contained exclusively male or female germ cells, sex differentiation was believed to have occurred at body lengths of 170–180 mm. At this time, either the cranial region became a functional ovary with rapidly developing oocytes or the caudal testicular region became dominant. In the latter case, the cranial ovarian segment developed only a few oocytes and these rarely grew to diameters of more than 1 mm before becoming atretic. Such animals were destined to become functional males. A third category consisted of animals in which both eggs and testis follicles occurred together in the cranial region and in this type, the direction of their future development would depend on the relative volumes of the testicular and ovarian tissues. Where the numbers of oocytes was small, the testicular tissue continued to develop, but the eggs did not progress far before undergoing atresia. Such animals would become functional males. On the other hand, if the volumes of male and female tissues were to be more equally balanced, both eventually degenerated resulting in the production of sterile individuals. These forms were surprisingly common and at body lengths of 25–29 cm accounted for about 13% of the 4000 hagfish examined by Schreiner. The picture that emerges is one of a delicately poised mechanism of sex differentiation in which the individuals in a population show a wide spectrum of sexuality and which probably vary widely in the length (or age) at which their sexual orientation is finally decided. From an initial common hermaphrodite stage the gonads of *Myxine* may develop in any one of three directions:

(1) The testicular region may continue to develop, while the ovarian region undergoes complete regression resulting in the development of pure males. Only 19 of this type were seen by Schreiner out of a total of 4000 animals.
(2) The testis may remain more or less static, while the ovary continues to develop, producing a functional female.
(3) The two areas develop together exercising mutual inhibition and resulting in a sterile gonad. Such animals were distinguished by Schreiner from others where sterility was attributable to a primary absence of germ cells.

In the *Eptatretidae* sexual status is more determinate than in *Myxine*. Early reports on *E. stouti*, indicated that sex ratios were approximately normal and that the two sexes did not differ in their range of body lengths. No female tissue occurred in the males and no testicular tissue in the females (Conel, 1931). Similarly, the male gonads of *E. burgeri* were said to show no traces of oocytes. Among 17 specimens of this species examined by Schreiner, there were five females showing some signs of rudimentary hermaphroditism, one animal was sterile and eleven were true males — a far higher proportion than he had observed in *Myxine*.

Information on the early development of the gonad of *Myxine* is limited to animals already over 100 mm in body length. In the smallest specimens described by Schreiner, germ cells were already present in the epithelium of the genital ridge (germinal epithelium). In the more caudal regions, germ cells (spermatogonia) were leaving the germinal epithelium to enter the underlying mesenchyme where they proliferated to produce testis follicles. Unlike the future female germ cells (oogonia) of the cranial region, which showed meiotic changes soon after they had left the germinal epithelium, the onset of meiotic prophase was delayed in the spermatogonia of the more caudal regions. Furthermore, the germ cells of the latter area invariably differentiated in a male direction, but in the cranial segments the further development of the germ cells was less constant. Here, they showed all transitional states between the two extremes of pure ovarian and pure testicular tissue.

The course of the ovarian cycle has been described by Patzner (1974) in comparatively young stages of *Myxine* with body lengths of 15 cm. Here, two regions of the genital fold could be distinguished (Fig. 12.1). The proximal part continuous with the mesovarium consists of a mesenchymal core covered by a flattened peritoneal epithelium, but at the distal end this germinal epithelium is several cells thick and contains the oogonia, from which successive waves of oocytes are produced. With the onset of the meiotic prophase, the oocytes now move below the surface accompanied by follicle cells, and grow to diameters of 1–2 mm while remaining within the central region of the genital fold. Further oocyte growth is then arrested until the previous generation of mature eggs has been shed. The largest and oldest eggs are ultimately found in a row at the boundary between the gut and the mesovarium and at this stage the ripe eggs hang below the free border of the ovary. Throughout oogenesis, a high proportion of the oocytes are lost by atresia and, where this occurs in the later stages, the atretic follicles progressively move away from the distal border towards the proximal regions of the ovary. Thus, the ovarian cycle involves a movement of oocytes towards the proximal region adjacent to the mesovarium and away from the germinal epithelium, but as they become more mature, they increasingly sink downwards below the distal border. Finally, when the egg has reached its maximum dimensions (in *E. burgeri* about 20×8 mm) the shell and the adhesive filaments are secreted by the follicle cells. The filaments have characteristically expanded and lobed heads and are arranged in tufts at each pole of the egg,

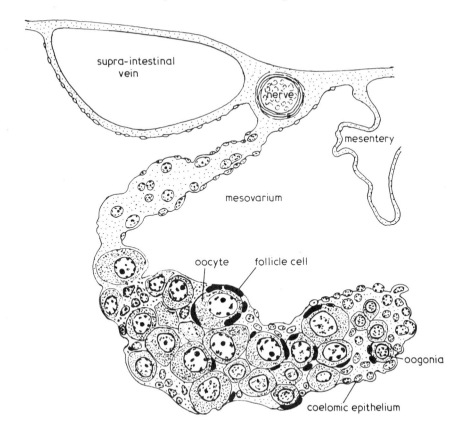

Fig. 12.1 The ovary of a young female *Myxine*. Redrawn from an electron micrograph of
Patzner, 1974.

anchoring them together in a string and presumably attaching them to the
substrate after ovulation. Below the animal pole the shell shows a thin suture
– the opercular ring – marking the line along which the shell is believed to
break open at hatching. At the animal pole and surrounded by the apical tuft
of anchor filaments is the micropylar funnel, formed by a cord of granulosa
cells. Electron micrographs reveal that at the bottom of this funnel, which is
about 0.2 mm in diameter, there is a honeycomb-like structure consisting of
some 3500 individual 'cells', only one of which was found to be open
(Fernholm, 1975). This implies that the effective diameter of the aperture
through which the sperm must pass is about 3.0–3.5 μm, or only twice the
diameter of the sperm head.

Within individual testis follicles, the male germ cells develop synchronously but, even in the same animal, different follicles may represent widely varying stages of spermatogenesis from spermatogonia to spermatozoa. Amongst over 1000 specimens of *Myxine* examined by Jesperson (1975), 200 were adult males but, of these, only one contained ripe sperms. On the other hand, a much higher proportion of ripe males were found in the several species of *Eptatretus*. In his ultrastructural examination of the sperm of *E. burgeri*, Jesperson remarked on the long middle piece (20 μm) consisting of the flagellum surrounded by 2–4 slightly twisted, irregular mitochondria; a type of structure said to be typical of animals practising internal fertilization.

12.1.2 Lampreys

Here the early formative stages of gonadal development and sex differentiation are already completed during the larval stages and it is within this period that the fecundity of the adult female is determined by the number of oocytes laid down in the ammocoete ovary. Thus, the proliferation of the germ cells is confined to an early phase of larval life, to be followed almost immediately by the onset of the meiotic prophase and oocyte growth (Fig. 12.2). By the time of metamorphosis, the gonad already contains the entire

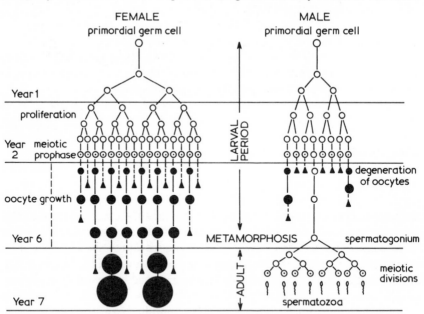

Fig. 12.2 Schematic representation of the course of gametogenesis in a non-parasitic lamprey, *L. planeri*.

The timing of events must only be regarded as approximate.

▲, degenerating germ cells; • ● ●, stages in oocyte growth; ⊙, meiotic germ cells.

stock of potential eggs but, because of the extensive degeneration that occurs throughout their development, the number of oocytes that survive to maturity represents only a small fraction of those originally present (Hardisty, 1965a, b). After a delay of up to a year, mitotic division in the small number of primordial germ cells originally present in the proammocoete results in the formation of an undifferentiated gonad containing isolated germ cells and cell nests. There are indications that this proliferative phase may begin earlier and be more intense in the presumptive female ammocoetes. The onset of the meiotic prophase, followed by oocyte growth occurs to some extent in all gonads, irrespective of their future sex, although it is more synchronous and extensive in the future females. Before sex differentiation is complete the larval gonads will therefore contain variable numbers of oocytes, both normal and atretic, as well as undifferentiated cells, either occurring singly or in small groups. Since these cell nests were formerly regarded as male elements (Okkelberg, 1921; D'Ancona, 1943, 1950), this stage of gonadal development has been considered as a form of juvenile or rudimentary hermaphroditism whereas, in reality, these larval cell nests are not direct forerunners of the spermatogonial lobules of the adult testis, and cannot be said to be 'male-orientated' (Hardisty, 1965b).

Although the presence in all ammocoete gonads of variable numbers of oocytes may prompt comparisons with the myxinoid gonad, the genital fold of the ammocoete shows no regional differences in their distribution. Even more importantly, unlike the male cells of *Myxine* which at an early stage proceed to form testis follicles, the vast majority of undifferentiated cells in the early ammocoete gonad show no such precocious male trends and either enter meiotic prophase and degenerate or temporarily differentiate as oocytes before undergoing atresia. In the lamprey, the differentiation of the male gonad is thus a 'negative process', consisting of the gradual elimination of growing oocytes and meiotic germ cells. These degenerative processes may be accompanied by some increase in the connective tissue stroma of the future testis but, because of the massive germinal destruction, the gonad may actually decrease in size at this stage of its development. The future male germ line can be traced to a very few stem cells, which for some obscure reason escape these regressive processes and their eventual proliferation to produce small testis lobules is delayed until late larval life and the approach of metamorphosis. In these lobules, the meiotic prophase and finally the production of secondary spermatocytes, spermatids and spermatozoa follow one another quite rapidly within a few months of the onset of spawning.

12.2 Reproductive biology

12.2.1 Lampreys
In their patterns of sexual maturation there are important quantitative

differences between the non-parasitic and the parasitic species of lampreys. In the latter, both the testis and the ovary undergo a slow and steady development throughout the phase of parasitic feeding and by the time and mature adults enter the rivers on their spawning migration, the testis lobules already contain masses of spermatogonia, while the oocytes in the female are in a comparatively advanced stage of vitellogenesis and growth. Lacking this interlude of adult feeding and growth, the non-parasitic lampreys enter the final period of sexual maturation from the lower baseline of gonadal development that they have attained at the end of their larval life and, within the six to nine months that elapses between metamorphosis and spawning, must be compressed the entire cycle of oocyte growth and spermatogenesis (Fig. 2.4).

During these final periods of sexual maturation, all lampreys, whatever their type of life cycle, undergo metabolic and morphological changes, most of which can be related directly or indirectly to the cessation of feeding and to the metabolic demands of the ripening gonads. These include atrophy of the gut and loss of its osmoregulatory functions, changes in the dentition, shrinkage in the length of the body and loss of weight, degenerative changes in the liver, depletion of lipid stores and liver glycogen and regression of the haemopoietic tissues. Finally, as the breeding season approaches, secondary sex characters develop; the urinogenital papilla of the male is everted to form a 'penis-like' structure and the male sea lamprey develops a rope-like thickening in front of the first dorsal fin. In the female, the leading edge of the second dorsal fin becomes swollen and oedematous and in addition, a small anal fin develops immediately behind the cloaca. In ripe animals of both sexes the fins become higher and more vascular, while the cloacal labia also become oedematous. In the southern hemisphere species, *G. australis* and *M. lapidica*, ripe males develop a large gular sac hanging down behind the oral disc and in both genera there is a remarkable increase in the size of this disc, accompanied by an extension of the snout region.

Shortly before spawning, the oocytes are liberated from the ovarian follicles and lie free in the body cavity. The mature egg is enclosed in a double-layered chorion formed during the third and final vitellogenic phase. Although the mechanisms of ovulation are not completely understood, it has been suggested that the mucopolysaccharide granules in the disintegrating granulosa cells may cause an osmotic uptake of water, contributing to the rupture of the follicle (Larsen, 1970). After ovulation, the remains of the granulosa cells still attached to the egg swell on contact with the water and assist in the adhesion of the eggs to the substrate of the nest. At the animal pole of the egg where the granulosa cells are absent is a tuft of mucoid fibres, which extends far into the water and is said to facilitate the movements of the sperms towards the egg, perhaps at the same time offering them some protection from this hostile osmotic environment (Killie, 1960).

As in the myxinoids, the sex products of the lamprey leave the body cavity through coelomic pores but, in this case, opening only just before spawning to connect the body cavity with the urinogenital sinus. This development is apparently controlled through the pituitary-gonadal axis, since injections of pituitary extracts have resulted in the opening of the pores in immature adults or even ammocoetes, while similar changes have occurred in the adult (but not in larvae) after the administration of steroid sex hormones (Knowles, 1939).

Under laboratory conditions, the rate of sexual maturation depends on the temperatures at which the lampreys are maintained (Larsen 1965), but is independent of light. Animals kept in permanent darkness have reached breeding condition at the same time as those subject to normal illumination (Larsen, 1973). However, under field conditions long-term temperature trends appear to influence both the onset and the duration of the spawning season and, once spawning has started, the behaviour of the spawning lampreys is markedly affected by relatively small transient changes in stream temperatures (Hardisty, 1961; Sjöberg, 1977). In the case of the two *Lampetra* species, *planeri* and *fluviatilis*, spawning usually begins when spring water temperatures rise rapidly to about 11°C, but the sea lamprey *P. marinus* spawns later in early summer at temperatures of around 15°C.

At least in some species, the males are believed to reach the spawning grounds first to begin the preliminary nest building activity. Here, when conditions are favourable the animals may gather in large numbers within relatively restricted areas which remain in use from one year to another. The possibility that these local assemblages are influenced by olfactory signals is raised by a technique used by French lamprey fishermen. This involves placing a completely ripe male in a conical wickerwork basket to attract the migrating females (Fontaine, 1938). In addition to the presumptive ion uptake cells which are found in the gills of both sexes of the river lamprey, *L. fluviatilis*, the interplatelet areas of the spawning male show a distinctive type of mitochondria-rich glandular cell, containing large lipid bodies with translucent centres (Pickering and Morris, 1977). Could these cells be responsible for the secretion of a pheromone-like sexual attractant?

The early stages of nesting activity involve the removal of larger stones from the nest site. Attaching themselves to the stones, the body is raised vertically while the tail makes vigorous vibratory movements. This has the effect of loosening the stones, which tend to be carried downstream with the current. However, even in sites where the current is slack, small stones lifted by the lampreys may be carried to the sides or even upstream from the nest. As a result of this activity, a depression is produced in the stream bed cleared of larger stones and varying in its diameter with the body size of the spawning lampreys and their numbers (Fig. 12.3). Because of the vortex created by the structure of the nest, the animals are usually able to move freely within it without being swept away by the current. The smaller non-parasitic lampreys

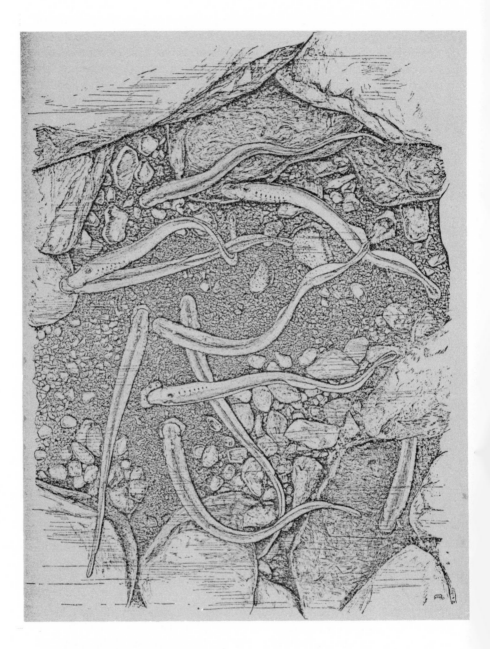

Fig. 12.3 Spawning brook lampreys engaged in 'stone-carrying' and pairing activities. From Dean and Sumner, 1898.

tend to be more gregarious than the larger parasitic species and it is by no means uncommon to find 30 or more animals associated with a single nest. The nests of the large sea lamprey, on the other hand are generally occupied by only a single pair and under these conditions, a form of territorial behaviour has been described, in which a resident male is said to drive off others by attaching with its sucker and making violent vibratory movements (Applegate, 1950; Hardisty and Potter, 1971b).

Lampreys practice a form of sexual congress in which the male attaches to the female usually on the head or branchial region, rapidly coiling his tail into a loop round the female so that the 'penis' is directed towards the cloaca. This is immediately followed by a short burst of intense vibratory activity, during which a few eggs are expelled. This pairing act has two functions. It assists in ovulation through the pressure exerted by the male on the pre-cloacal region of the female and it also increases the chances of successful fertilization by directing the jet of milt directly on to the eggs. This is important since the sperms are said to remain active for less than a minute once they have made contact with freshwater. Only a small number of eggs are extruded at each spawning act so that this must be repeated over and over again, perhaps for several days before the major part of the egg stock has been exhausted.

12.2.2 The death of lampreys after spawning

The failure of lampreys to survive more than a few weeks at the most, after their first and only spawning puts them in the same category as the eel and the Pacific salmon, whose death after the completion of a single reproductive effort has sometimes been cited as an example of a genetically programmed event – a description that does little to answer any of the significant questions that confront the gerontologist. Obviously, for those species like the lamprey in which no reserve of undifferentiated germ cells remains after their single breeding season, further survival would be biologically meaningless, but this does not explain the physiological mechanisms that result in such relatively synchronous natural death.

As described elsewhere (Sections 2.2 and 10.4.2), spawning is preceded by a variable, but prolonged period of starvation, during which the animal can be regarded as a closed system in which energy reserves built up over the parasitic feeding phase are in part diverted to the developing gonads or used to furnish the energy expended in the rigours of the upstream migration and the activities associated with nest-building and spawning. One of the manifestations of these processes is the slow and continuous reduction in body length and weight, followed in the terminal stages of sexual maturation by a phase of more rapid shrinkage (Larsen, 1965, 1969a). At this period, the lipid reserves of the spawning or spent lamprey have dropped to about 4 % of the wet weight of the body but, as Beamish *et al.*, (1978) have pointed out, it seems unlikely that this depletion is the sole or even the main factor in death since these terminal lipid levels are still higher than they are in the macrophthalmia stage, at a time when it is just beginning to feed after the metamorphic period. As in the spawning salmon, there is some evidence that when lipids have been almost exhausted, the lamprey may begin to catabolize its tissue proteins and perhaps it is at this stage that degenerative changes reach a critical and irreversible point in some vital tissues. Among such degenerative changes are the massive destruction of red cells (Larsen, 1973), the atrophy of the haemopoietic tissues (George and Beamish, 1974), a reduction in haemoglobin concentrations (Potter and Beamish, 1978) and degeneration of the liver. Nevertheless, not all tissues are equally affected by such regressive changes. This has been demonstrated in experiments involving the transplantation of eyes from spawning lampreys into ammocoetes, where the grafts remained viable for at least 6 months beyond the normal time of death of the donor animals and even then their further survival was only limited only by the termination of the experiments (Bytinski-Salz, 1956). In addition, the possibility of halting or even reversing degenerative changes has been illustrated by the effects of castration in inhibiting the atrophy of the intestine (Section 11.4.3).

The effects of endocrine factors on the time of death of the lamprey, *L. fluviatilis*, have been studied by Larsen (1973). The possibility that hyper-secretion of the adrenocortical tissues might be involved in the post-spawning mortality of Salmonid fishes has been widely discussed but, at least in lampreys, this can be discounted in view of the absence of evidence for the usual range of vertebrate corticosteroids, much less for their presence in high concentrations. In an extensive series of experiments, Larsen has shown that prolonged survival can be induced by maintaining the lampreys at constant, low temperatures, by castration and more especially by hypophysectomy. Thus, it would appear that any factor that interferes with the metabolic changes involved in sexual maturation or which retards their progress, is capable of prolonging the life span. Hypophysectomy is known to reduce the metabolic rate of the lamprey and it is probably this effect rather than a more specific influence exerted through the removal of gonadotrophins, that is responsible for delaying the time of natural death. Although in Larsen's experiments, some of the hypophysectomised lampreys survived for periods up to 10 months beyond the normal time of spawning, they eventually suffered a shrinkage similar to that of normal animals, although this had obviously occurred at a reduced rate. It would seem therefore, that death is ultimately due to degenerative changes induced by starvation and that, as Larsen has established, any procedures that slow down tissue mobilization are effective in prolonging their natural life span.

12.2.3 Myxinoids

In spite of the prize offered in 1854 by the Copenhagen Academy of Sciences for information on the reproductive habits of *Myxine* (which remains unclaimed), there has been little advance towards a resolution of the many problems posed by these enigmatic animals. How are we to explain the wide variations in the sex ratios of European populations and how can fertilization be achieved in areas where functional males are rare? Why have fertilized eggs so rarely been found, how are they fertilized and where are they deposited? Is breeding confined to some inaccessible spawning grounds?

Data collected towards the end of the last century suggested that *E. stouti* was markedly localized in its distribution off the Californian coasts and that males and females tended to occur in separate areas. This raises the possibility that the disparities in the numbers of male *Myxine* in different localities might reflect sex differences in habitat, feeding behaviour or patterns of movement. Furthermore, it is possible that information on numbers or distribution based on captures on baited lines or traps may be misleading. For example, appetite or feeding responses may well vary at different periods in the breeding cycle or as Holmgren (1946) suggested, the ripe hagfish may, like many other breeding fish, burrow deeply during spawning and so be out of the reach of trawls.

Fernholm's (1974) accounts of the seasonal movements of *E. burgeri* agree with earlier records for the same species. These indicated that these animals were not caught off the Japanese coast between the end of July and the end of October. Towards the start of this apparent breeding season, the proportion of males increased, but when the animals reappeared in October all the mature males were found to be spent. Similarly, in late autumn a high proportion of the females had empty follicles or corpora lutea, indicating recent ovulation. Although a number of earlier workers had postulated a seasonal breeding cycle for *Myxine*, this cannot be substantiated either by the state of the gonads or the times of the year when infertile eggs have been recovered in trawls, some from depths of 100–200 m (Holmgren, 1946). Living as it does in such deep water, the cyclic processes of oogenesis in *Myxine* probably owe more to endogenous rhythms than to seasonal factors.

In the ovary of *Myxine*, it is possible to recognize as many as four generations of spent follicles or corpora lutea in various degrees of involution (Walvig, 1963). These must represent successive ovulations, but at what time intervals is not known. If, as has been suggested in *E. burgeri*, they represent annual ovulations, an animal with four generations of regressing follicles must be at least four years of age, not allowing for the time that must have elapsed before the young animal had matured its first crop of ripe eggs. Even so, older animals still appear to retain a reserve of immature oocytes (and probably oogonia) which could provide for further reproductive activity and tagging experiments have given some indications that egg production may continue even when growth has slowed down or ceased altogether (Foss, 1963).

The absence in myxinoids of sex dimorphism or secondary sex characters is easily understood in the context of their relatively undifferentiated sexual status and the poverty of their sensory equipment. Although the ovarian tissue has the ability to produce a range of steroids (Section 11.4.3), it seems more likely that these may be involved in the regulation of gonadal activity, rather than in the development of breeding signals. Sex recognition, whatever the form it takes, is presumably dependent on olfactory stimuli.

12.3 Metamorphosis and the significance of the larval stage

At normal stream temperatures, the hatching of the lamprey egg occurs at 10–14 days after metamorphosis and the pro-ammocoete as it is called at this stage still retains considerable quantities of yolk, although the stomadaeal depression has broken through to the mouth cavity and the gills have become functional. The first movements to develop are lateral undulations involving almost the whole body, but with the increasing absorption of the yolk (now restricted to the gut lumen), typical swimming movements appear, although at this stage the pro-ammocoete does not yet exhibit light avoiding responses. When disturbed, they swim vigorously towards the surface, then subside towards the bottom where they lie motionless on their sides. Quite abruptly, this behaviour changes and by about the third week they begin to develop burrowing responses, ceasing their surface swimming and diving rapidly into a soft substrate. These successive phases of larval behaviour obviously ensure, first their emergence from the sand of the nest and their distribution downstream by the current and secondly, their establishment in suitable ammocoete habitats.

In relation to the embryonic development of the myxinoids, opinion has been sharply divided, between those who like Løvtrup (1977) maintain that the large yolky egg represents a primitive feature and others who believe that the absence of a larval stage represents a secondary consequence of an evolutionary increase in the volume of the hagfish egg and its yolk content. On the other hand, a majority of authors have tended to accept without question, the primitive nature and evolutionary significance of the larval stage of the lamprey, assuming a similar ammocoete stage to have already been present in some at least of the fossil agnathans. Watson (1954) for example, argued that the development of the upper lip and the resulting dorsal position of the nasohypophysial opening must have preceded the formation of the oral hood as part of the feeding mechanism of the ammocoete. Since the nasohypophysial complex occupied the same position in both cephalaspids and anaspids and was assumed to develop in the same way through the hypertrophy of a post-hypophysial fold, Watson inferred that these groups would have passed through an ammocoete stage. Because the myxinoids also show a post-hypophysial fold, although with no obvious functions, he considered that they too must once have possessed a larval stage, which was subsequently suppressed by their secondary evolution of large yolky eggs. On account of the paired nostrils opening into the mouth, Watson excluded only the heterostracans from this common possession of an ammocoete-like larval form. More recently however, Whiting (1972) in his comparisons of the morphology of the proammocoete and the fossil agnathans has concluded that both the heterostracans and the cephalaspids may have passed through a proammocoete or ammocoete stage.

Somewhat different conclusions were reached by Strahan (1958) in his analysis of probable growth patterns in the agnathan head and, if these are valid, Watson's argument would lose much of its force. Thus, Strahan suggests that the most direct transformation from embryo to adult lamprey must involve a stage in which an ammocoete-like upper lip is present. This structure was therefore regarded as an inevitable stage in the growth processes involved in the development of the oral funnel, rather than as a unique adaptation to larval feeding. For this reason he excluded the ammocoete stage from the life cycle of the cephalaspids, although he believed that it may have been present in the anaspids.

Certain features of the fossil *Mayomyzon* raise considerable doubts whether a larval stage was present in these Carboniferous lampreys. In a number of respects this fossil lamprey appears to have retained primitive characteristics which in present day lampreys are to be found only in embryonic or early larval stages. These include the anterior position of the branchial region, the comparatively short pre-oral region, the tubular shape of the snout, the probable absence of disc teeth, the less extensive piston mechanism and the lack of a direct connection between the sense capsules and the neurocranium. Moreover, the somewhat expanded annular cartilage suggests that the pre-oral region may have resembled the oral hood of the ammocoete. An even more convincing argument is the small size of *Mayomyzon* (30–60 mm). If these very small creatures were, in fact, mature adults, this would seem to imply that they had not yet evolved the parasitic method of feeding, that alone has made possible the much greater body size of present-day lampreys. It may of course be argued that these fossils represent recently metamorphosed or marcrophthalmia stages of a much larger animal. However, this would imply that even if a larval stage were present, it must have been of very short duration, since the smallest of these fossils is no larger than present day ammocoetes after little more than six months of larval life (Fig. 4.6). Moreover, if we assume that both the size of the eggs and the patterns of oogenesis in *Mayomyzon* were at all like those of present day lampreys, their small body size would imply very low levels of fecundity, quite incompatible with an extended life cycle like those of living forms.

Damas (1944) has suggested that in their biology and anatomy, the ancestors of the lampreys would have been intermediate between the present-day ammocoete and the adult and that these subsequently became more divergent and specialized. As de Beer (1952) has emphasized, it is a corollary of larval adaptation that the better suited it becomes to its mode of life, the greater will be the divergence between the form of larva and adult and the more violent the metamorphosis. This would be all the more so in the lamprey, where the larval stage has been prolonged to an exceptional extent and occupies such a large part of the life cycle. As the specialized larval adaptations are further developed, the future adult tissues and structures remain as undifferentiated embryonic primordia, whose further development is suspended until the time of metamorphosis. Thus, what was originally a continuous sequence of development from embryo to adult is now interrupted by the interpolation of a larval stage and morphogenetic processes previously confined to embryonic life, are now postponed for periods of several years. Examples of this kind of delayed differentiation are to be found in the persistence of mucocartilage (Section 7.2), the development of the eye, kidney, thyroid and adenohypophysis. At metamorphosis, the larval kidney regresses and is replaced by an adult mesonephros arising from undifferentiated nephrogenic tissue in the nephric fold of the ammocoete (Youson *et al.*, 1977). The curious history of the ammocoete eye involves the early appearance of visual cells forming the so-called retina A (Keibel, 1928). This is surrounded by other visual elements that do not differentiate until metamorphosis (retina B). As we have seen earlier, it is also possible that the adult thyroid follicles may be derived from relatively undifferentiated elements of the endostylar epithelium (Section 11.1), perhaps playing little or no part in the thyroidal biosynthesis of the ammocoete. The significance of these conditions had already been appreciated by Leach (1944) who noted that, whereas in a frog tadpole the thyroid and pituitary are well-developed embryonic structures, in the ammocoete both thyroid follicles and hypophysial development are concurrent with a metamorphosis taking place some seven years after the embryonic period. As further examples of delayed differentiation Leach cited the undifferentiated tubular muscles of the larval pharynx and the retention of a rudimentary eye.

No doubt it is true of the ammocoete, as it is of larval amphibians, that the timing of their metamorphosis is dependent on both age and size. Thus, within a particular larval population there may be a considerable range of ages amongst metamorphosing ammocoetes depending on differences in their growth rates. While therefore we can only speak of an average age at metamorphosis, this shows characteristic species differences that must be assumed to have a genetic basis. Although there are some conspicuous exceptions, parasitic species tend to have a shorter larval life than the non-parasitic forms and these differences are particularly evident among the members of paired species (Section 2.2.1). This ability to vary the duration of the larval phase must have been an important factor in the adaptability of the group.

We have as yet no firm evidence of hormonal involvement in the initiation or control of metamorphosis, although the switch that has been reported in larval metabolism towards lipid accumulation in the period before transformation (Lowe and Beamish, 1973; O'Boyle and Beamish, 1977) could be related to hormonal changes. That environmental factors are in some way implicated is shown by the fact that throughout the Northern Hemisphere, metamorphosis is generally only observed within a restricted period from June to September (Beamish and Potter, 1972; Potter *et al.*, 1978). This degree of synchrony strongly suggests the involvement of temperature or daylength, perhaps mediated through the pineal complex. However, whatever may the precise factors that are concerned in triggering transformation, these probably operate long before external morphological changes appear, perhaps early in the spring. There is now some evidence that certain internal changes may begin some time before the external signs of metamorphosis appear and this has led to attempts to define a pro-metamorphic period, analogous with that of the anuran tadpole (Youson, *et al.*, 1977). Among these precocious internal changes are the cavitation of the nasohypophysial stalk, the development of the accessory olfactory organ and the replacement of the larval mesonephros by that of the adult. Once metamorphosis has begun, it proceeds very rapidly and most of the external changes are completed within 3–5 weeks, although some of the physiological transformations like those of the haemoglobins may take much longer, especially in the non-parasitic forms.

Reference has been made earlier (Section 10.4.3) to the discovery in one locality of Northern Italy of neotenous female ammocoetes of the brook lamprey, *L. zanandreai*. At first sight, the occurrence of neoteny in a non-parasitic lamprey might hardly seem surprising. These species, although not dispensing with metamorphosis havé nevertheless sacrificed most, if not all of the biological advantages of adult life. Since they never feed, their body size cannot be increased beyond the larval stage and as a result they are unable to mature more than a very small proportion of the oocytes in the ammocoete gonad (in *L. planeri* only about 20 %). The brook lampreys therefore, suffer the additional and relatively heavy mortality associated with the hazards of adult life, but without compensatory gains in the shape of increased fecundity. Why then have these dwarf forms not more often become neotenous and abandoned altogether this apparently functionless adult stage? Here, the answer is to be sought in the physiological, morphological and behavioural characteristics of larval lampreys which would preclude their ability successfully to complete the reproductive cycle. Apart from its photophobic burrowing responses and low metabolic rates, ill-adapted to the vigorous free swimming activities associated with spawning, the cardiovascular system and low heart ratio of the ammocoete would be quite unable to cope with the sustained locomotory activity involved at the breeding season. Lacking the better developed fins of the adult or its paired eyes and rheotactic responses, it would be incapable of making its way upstream towards suitable spawning sites and, once there, possession of a suctorial oral disc would be essential for spawning activity, copulatory pairing and the ability to maintain station in the face of rapid currents.

12.4 Evolutionary aspects

12.4.1 The nature of myxinoid sexuality

In one of the earliest investigations carried out on the European hagfish, *Myxine glutinosa*, Nansen (1888) commented that this animal 'still seems to be seeking, without yet reaching that mode of reproduction most profitable for it in the struggle for existence.' This picturesque phrase aptly describes the major evolutionary problem that is raised by the sexual status of these animals. Are they evolving towards a hermaphrodite condition or moving towards dioecism from an originally hermaphrodite state? An answer to these questions might be easier if we had more complete information on other myxinoid species. There are some indications that a more stable dioecism may exist within the genus *Eptatretus*, where the proportion of pure males appears to be much higher than in *Myxine*. From the morphology and habits of the former genus, it has been argued that *Myxine* may represent a more aberrant (or apomorphic) genus and that the species of *Eptatretus* may be more primitive and closer to the ancestral myxinoids (Fernholm and Holmberg, 1974). On the other hand, Schreiner considered that the ancestors of the hagfishes were probably self-fertilizing hermaphrodites, subsequently evolving towards protandry and the postponement of ovarian function as the eggs increased in size and developed a higher yolk content. Throughout these trends, a mutual antagonism developed between male and female gonadal territories, finding its extreme expression in the production of sterile gonads. These sterile individuals might be regarded as a means of eliminating unfavourable genotypes that otherwise might upset the delicately balanced mechanisms of sex differentiation. This interpretation would imply that, in their reproductive physiology, the *Myxinidae* are the more primitive or plesiotypic forms and the *Eptatretidae* the more advanced or apotypic.

Hermaphroditism is most likely to develop in species with low mobility, for example in parasitic or sessile forms, where there will be little sex dimorphism and the male will not evolve mechanisms for seeking out and holding the female (Charnov *et al.*, 1976). Low mobility limits male reproductive success (measured by the numbers of eggs that can be fertilized competitively) and this will rise at less than a linear rate with the input of resources into male function. Schreiner's view was based on the rather dubious assumption that the ancestors of the myxinoids were parasitic and wide-ranging forms with limited opportunities for cross-fertilization. Later, they adopted what he called saprophytic modes of feeding, leading them to collect together in large numbers, favouring cross-fertilization and the development of protandric hermaphroditism. This change was, in turn, associated with a reduction in fecundity and the development of large and more yolky eggs. As evidence for these changes, Schreiner cited the presence in many specimens of *Myxine* of a rudimentary left genital fold and also pointed out the relationship between increased oocyte numbers and more intense atresia. However speculative these views may be, some relationship between the habits and ecology of hagfishes and their peculiar sexual condition seems highly probable. Although they are said to be present in considerable numbers within restricted areas, it is possible that these local populations may be isolated from one another and that individual hagfishes do not move very far from their 'home' territory. The absence of sexual dimorphism would also be consistent with a relatively recent escape from a condition of functional hermaphroditism.

12.4.2 Sex differentiation

The various inductive theories of sex differentiation stress the primary role of the somatic gonadal tissues in controlling the sexual orientation of the germ cells. Schreiner believed that there was evidence for this kind of inductive effect in the onset of meiosis in the oogonia of *Myxine* immediately they have acquired an envelope of follicle cells and migrated from their original site in the germinal epithelium. The somatic tissues were also held responsible for the antagonistic effects of male and female gonadal territories in the potentially hermaphrodite genital fold. A similar incompatibility also appears to be present in the early ammocoete gonad, where the oocytes that develop in the presumptive male gonads invariably degenerate before reaching the final stages of oocyte growth. It appears that with the progressive development of the somatic substrate, the cellular environment of the future testis may become increasingly hostile to the further development of germ cells that have differentiated in a female sense. This may be compared with the situation in hermaphrodite males of *Myxine*, where variable numbers of oocytes are present which fail to reach maturity. In both hagfish and lamprey, oocytes occur almost universally whatever the future sex of the gonad and, in *Myxine*, they are only absent in the rare so-called true males. In hermaphrodite males they appear in variable numbers and their further development and the size that they attain is related to the relative volumes of testicular and ovarian tissue.

In *Myxine*, it is appropriate to speak of a range of sex phenotypes ranging from pure males to pure females through a graded series of hermaphrodite gonads containing both ovarian and testicular tissue of which only one will become dominant and functional. This does not appear to differ fundamentally from the situation in ammocoete gonads before and during sex differentiation, except that, in this case, the potentially male germ cells remain as undifferentiated elements and never form recognizable testicular structures.

Nevertheless, in regard to the numbers of growing oocytes, the ammocoete gonads show a continuous spectrum in which the extremes of higher and lower numbers are believed to correspond to the future male and female gonads (Hardisty, 1965a,b). The conditions of the gonads in both groups of cyclostomes can probably best be explained by postulating a sex-determining mechanism involving a delicately poised and multifactorial system, capable of producing a wide range of sexual phenotypes. Such a system might be very susceptible to the kind of chromosomal variability that has been reported in a Swedish population of *Myxine* (Section 13.9) where enormous variability was observed in chromosome counts and the existence of haploid individuals and even parthenogenetic reproduction has been suggested (Nygten and Jahnke, 1972).

In lampreys, the appearance of a prolonged stage of sexual indeterminacy could be a consequence of the introduction of a larval stage. This could have the effect of spreading out over a longer period developmental processes that were originally completed during the embryonic phase. By delaying the differentiation of ovarian and testicular tissues, a slowing down of gonadal development might then give the appearance of a prolonged intersexual stage. It is interesting to find that a similar explanation was put forwards in the late 19th century to account for the hermaphrodite gonads of myxinoids (Dean, 1899). Unlike Schreiner, Dean supposed that the hagfishes were derived from dioecious ancestors with right and left genital folds, but that with the loss of the left gonad, development of the right gonad became precocious thus prolonging the undifferentiated stage common to all vertebrates, allowing its development to proceed to a point where both male and female tissues are already present when final sex differentiation took place.

12.4.3 Fecundity and survival

As relict groups, the cyclostomes have survived from the Palaeozoic by adopting completely contrasting survival strategies. The route followed by the myxinoids has involved the production of small numbers of large yolky and shelled eggs from which the young hatch in a relatively advanced stage of development. Lampreys, on the other hand, have maximized fecundity to a point where egg numbers are limited only by the body size that the female is able to attain; the small eggs hatching to produce an immature larval stage with habits and mode of feeding completely different from those of the adult lamprey.

Fecundity in lampreys varies widely in different species and is broadly correlated with body size. The smallest non-parasitic forms (in which the larval stage tends to be extended and the more vulnerable adult stage curtailed) have egg numbers varying from 400–9000, whereas in the largest parasitic species like the sea lamprey, egg numbers may reach several hundred thousand. Although the frequency of ovulation in hagfishes has not been determined, at least in *E. burgeri*, there are indications that only a single crop of eggs may be shed annually. Eight females of this species kept in cages on the sea bottom throughout the period when they would normally spawn i.e. from September to November, deposited a total of 178 eggs (Fernholm, 1975). Thus, the average number of 22 ovulated by each female corresponds quite closely with the number of mature eggs that are normally found in the ovaries of other species. In both hagfishes and lampreys a considerable proportion of the oocytes are lost by atresia. In *Myxine*, this has been put at about half all the oocytes between 2–23 mm in diameter and, in the non-parasitic lampreys, atresia may account for 60–90 % of all the oocytes in the ovaries of metamorphosing ammocoetes.

The enormous disparity in the fecundity of hagfishes and lampreys must of course imply the existence of corresponding differences in mortality rates. By virtue of their concealed and relatively inactive burrowing habits, the myxinoids are presumably subject to only low rates of mortality from predation and this is probably true also of the concealed larval stages of lampreys. In this respect, hagfishes conform to the pattern that characterizes other members of the deep sea community, where predation is thought not to be the main mechanism of population control (Grassle and Sanders, 1973). Thus, compared to shallower water species, these forms are said to be distinguished by small brood sizes and with slow growth rates and high longevity, show a higher proportion of adult to younger stages. Based on the annual cycle of oocyte growth, Patzner (1978) has calculated that the body length of *E. burgeri* only increases by about 4–5 cm a year. Assuming slow growth rates of this order, the largest hagfishes measuring up to a metre in length must presumably have attained a very considerable age. Low mortality and reproductive rates have been regarded as expedients adopted in populations where long term or evolutionary diversity is a product of past biological interaction within a physically stable environment such as that of the myxinoids (Wolff, 1977).

Such conditions are in sharp contrast to the life cycles of lampreys, where mortality rates are certain to be high during the more adventurous and exposed episodes of migration and spawning. Statistics from the albatross population of South Georgia have provided some idea of the severity of the losses that may be sustained by lampreys during their oceanic migration. On Bird Island, the lamprey *Geotria australis* (presumably originating from spawning populations in South American rivers) forms an important part of the diet that the Grey-headed Albatross supplies to its young. Analysis of the diet of the 5000 albatross chicks on this island has indicated that the lamprey accounts for about 10 % of their solid food, representing an annual consumption of 20 tons, or about 350 000 lampreys. If we extrapolate from these figures to the entire albatross population of South Georgia, the total loss through bird predation would work out at 120 tons or over two million lampreys (Potter *et al.*, 1978).

Although we are ignorant of the breeding habits of hagfishes, it is unlikely that these would rival the complex breeding patterns of lampreys or involve the same endocrine mechanisms of control and integration. At the same time, in spite of these more complex adaptations, lampreys do not appear to be conspicuously successful in regard to their breeding efficiency. Experiments conducted on female sea lampreys under natural conditions have shown that the number of ammocoetes successfully hatched may not exceed more than 5.3–7.8 % of all the eggs that the female produces. Whatever may be the mode of fertilization or egg protection adopted by hagfishes (and in spite of the apparent rarity in some areas of functional males), their very low reproductive rates must necessarily imply a vastly greater degree of reproductive efficiency.

13
Comparative biochemistry, immunology and cytogenetics

Discussion of the phylogenetic status of the cyclostomes has so far been very largely confined to morphological considerations, with little regard to comparative biochemistry and physiology. To a certain extent, the responsibility for this situation can be laid at the door of the biochemists themselves who have often shown some reluctance to follow the logic of their own data and have adopted an attitude reflected in the words of Florkin (1966) that 'we must adopt the methodological rule to be led by knowledge of phylogeny in our search for biochemical evolution, rather than be brought by biochemistry to the discovery of new aspects of phylogeny'. We must of course, recognize that when taken in isolation, the distribution of a particular molecule within or between groups of animals may be of little phylogenetic significance, although such information may become more important where we have more detailed information on biosynthetic pathways and enzyme systems. An example is the erratic distribution of creatine even within animals belonging to the same taxonomic group (Watts and Watts, 1974). This nitrogenous compound has been found in the tissues of cephalochordates, myxinoids and lampreys, although only the latter animals possess the necessary enzymes to synthesize it. In the other two chordate groups, as in some invertebrates, its presence may be due to the ability of these animals, either to recover it from their food or to absorb it from their aquatic environment. However, as Barrington (1974) has pointed out, even where we have information on biosynthetic pathways, their presence in different organisms or groups of animals may be open to several alternative explanations. While it may be evidence of a common ancestry, it is also possible that it is a result of parallel evolution or reflects biosynthetic abilities that are widely distributed among diverse animal groups, as exemplified in the occurrence of steroids or polypeptides such as insulin in a broad spectrum of animal groups.

These reservations, however, apply with much less force to the rapidly advancing techniques that are being developed in the field of molecular evolution. The increasing numbers of protein sequences now becoming available, which because they relate to the fine structure of the genome, constitute a rich source of genetic and evolutionary information which can supplement or even perhaps, in a few cases, supplant the more subjective judgements of the morphologist or palaeontologist.

Considerable precision can now be obtained from protein sequence data by the application of computer techniques for the construction of phylogenetic trees, in which the branch lengths are proportional to sequence differences, taking account of the probabilities of superimposed or back mutations. Providing the geological or palaeontological evidence is adequate to determine the age of the major points of divergence (nodal points) on such trees, it is then possible to estimate the rates of protein evolution, either in terms of accepted point mutation (PAM) 100 residues^{-1} 100 million years^{-1} or as nucleotide replacements (NR) 100 codons^{-1} 100 million years^{-1}. Although it may be true that we cannot at present interpret morphology in terms of molecular structure and that point to point information transfer from DNA stops at the level of the polypeptide (Hansen, 1977), to a quite surprising extent, phylogenetic trees for the best-known proteins tend to show agreement with the views of conventional taxonomists. This suggests that evolution at the level of the gene adequately reflects changes in the whole organism at the morphological level.

Unfortunately, sequence data for cyclostome proteins is still very incomplete and for none of the molecules discussed in this chapter do we have comparable information for both myxinoids and lampreys. For this reason, we are bound to rely on comparisons of amino acid analyses, although at best these can provide only crude approximations to the sequence changes that may have occurred. From time to time a number of statistical techniques, of varying degrees of sophistication, have been proposed for estimating the degree of homology between proteins from comparisons of amino acid composition (Metzger *et al.*, 1968; Marchalonis and Wellman, 1971; Harris and Teller, 1973; Dedman *et al.*, 1974). One of the simplest of these used extensively in this Chapter and devised by Marchalonis and Wellman is the difference index $S\Delta Q = (X_{i,j} - X_{k,j})^2$, where $X_{i,j}$ and $X_{k,j}$ represent the proportions of an amino acid j in the two proteins i and k. This index has been found to correlate quite closely with differences in amino acid sequences in cytochrome c, haemoglobin or immunoglobulin. For example, unrelated proteins show a much greater dispersion of $S\Delta Q$ values than homologous proteins; the median value for the former being 300 units, and for about 12% of the proteins examined the index was over 1000 units (Marchalonis, 1977). On the other hand in comparisons of related proteins the median $S\Delta Q$ values were only 80 for the haemoglobins, 30 for immunoglobulins and 20 for cytochrome c.

13.1 Cyclostome haemoglobins

13.1.1 Structure of lamprey and hagfish haemoglobins

The complete primary structure of one of the haemoglobins of the sea lamprey, *P. marinus*, is now known (Fig. 13.1), together with one of the two major haemoglobins of the river lamprey, *L. fluviatilis* (Braunitzer and Fujiki 1969; Li and Riggs, 1970). In addition, X-ray diffraction studies have demonstrated the essential similarity in the three-dimensional structures of lamprey and other vertebrate haemoglobin or myoglobin chains (Hendrickson and Love, 1971). Unfortunately, sequence data for the hagfish haemoglobin is limited to the first 30 residues at the N-terminal in one of the major components of *E. stouti* (Li and Riggs, 1972), but amino acid analyses are available for five haemoglobin components of this species, as well as for four haemoglobin species of *E. burgeri* (Bannai *et al.*, 1972) and three components in *Myxine* (Paleus and Liljeqvuist, 1972).

For the two species of lampreys belonging to the genera *Lampetra* and *Petromyzon*, the haemoglobin sequences show substitutions at only 12 positions. Five of these would involve single, and seven double nucleotide replacements. At positions 8–10 and 136–139 there are reversals of the sequence which would require two nucleotide replacements. In addition, the methionine at 141 in *fluviatilis* is absent in *marinus* and the latter also has an alanine at 119 which appears to be missing in *fluviatilis*.

The partial sequence for the hagfish shows some striking differences from the lamprey haemoglobins (Fig. 13.1) with no less than 17 substitutions out of the 30 residues. In 14 cases these would require single base changes and three would involve double nucleotide replacements. Both hagfish and lamprey share a unique additional segment of nine residues at the N-terminus and in this sequence five of the positions are occupied by the same amino acids in both groups. As Li and Riggs (1972) have pointed out, the fact that this agnathan segment has been preserved through the long period of evolutionary separation, suggests that it may have some functional significance peculiar to this group of vertebrates.

The tertiary structure of the lamprey haemoglobin shows the same predominantly α-helical structure of other globins, accounting for about 79% of all residues. These helices are arranged in basically the same pattern as in the monomeric myoglobin molecule, making it possible to use the same nomenclature that is applied to other globin chains. As in other haemoglobins, the haem group is situated in a haem pocket between the E and F helices, lined by hydrophobic residues and linked to the two histidine residues at E7 and F8.

In the absence of complete sequence data for hagfish haemoglobin, it is instructive to compare its amino acid composition with that of the lamprey and mammalian haemoglobins (Table 13.1). Although there have been some discrepancies in the molecular weights of hagfish haemoglobins as reported by

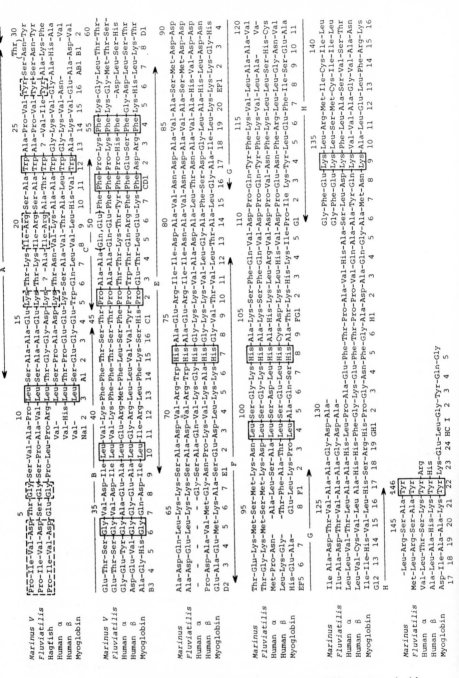

Fig. 13.1 Amino acid sequences of lamprey and hagfish haemoglobins compared with sperm whale myoglobin and the human α and β chains. Data from Riggs, 1972; Braunitzer and Fujiki, 1969; Li and Riggs, 1972.

Table 13.1 Values of the difference index $S \triangle Q$ for comparisons between lamprey and hagfish haemoglobins and human α and β chains. Dotted lines divide the parts of the matrix that are concerned with comparisons within and between hagfish species from the values that refer to comparisons between the two cyclostome groups and the human haemoglobin chains.

| | Lampreys | | Hagfishes | | | | | | | | | | | | Human | |
	LF	PM	A	B	C	D	E	F1	F2	F3	F4	M1	M2	M3	α	β
β	220	185	154	155	124	245	203	348	357	237	129	258	178	209	88	0
α	153	130	170	189	164	266	273	398	427	291	178	321	207	202	0	---
M3	93	93	54	38	54	42	60	72	82	69	60	42	46	0		
M2	81	71	23	15	27	43	44	100	100	62	30	57	0			
M1	167	169	61	52	73	37	52	45	47	41	66	0				
F4	72	62	8	10	7	51	60	116	126	67	0					
F3	165	166	61	52	66	25	63	55	54	0						
F2	215	223	109	82	114	56	63	2	0							
F1	200	206	101	75	105	48	62	0								
E	135	141	63	31	50	48	0									
D	115	114	34	25	40	0										
C	57	51	5	7	0											
B	68	60	7	0												
A	55	48	0													
PM	10	0														
LF	0															

Row groups: Human (β, α); *Myxine* (M3, M2); *E. burgeri* (M1, F4, F3, F2, F1); *E. stouti* (E, D, C, B, A); Lampreys (PM, LF). Column groups: *E. stouti* (A–E), *E. burgeri* (F1–F4), *Myxine* (M1–M3).

LF *Lampetra fluviatilis*
PM *Petromyzon marinus* Hb V

different authors, it is generally agreed that these are higher than in the
lamprey molecule and the difference is especially marked in the case of *Myxine*
haemoglobin. It has been suggested that this could be explained by assuming
that hagfish haemoglobin possesses the extra six residues that are found at the
carboxyl terminal of the myoglobin chain (Fig. 13.1), as well as the nine
residues that are missing from the lamprey chain between positions 130 and
131 (Li *et al.*, 1972).

Analysis of the difference indices for haemoglobins of the hagfish and
lamprey in Table 13.1 discloses a number of interesting features:

1. Whereas the haemoglobins of *L. fluviatilis* and *L. japonica* are identical in
their amino acid composition and show only relatively minor differences from
haemoglobin V of *Petromyzon marinus*, there are very marked divergencies in
the haemoglobins of the three hagfish species, where the comparisons between
the individual haemoglobins, yield mean values for $S\Delta Q$ of 45–65.

2. Both in comparisons within and between species, the various individual
haemoglobins of the hagfish show a very wide range of variability in amino
acid composition. For example, components A, B and C of *E. stouti* are
virtually identical ($S\Delta Q < 10$) and closely resemble F4 of *E. burgeri*
($S\Delta Q < 10$). There is also a considerable measure of agreement between these
Eptatretus haemoglobins and the M2 component of *Myxine*. In *E. burgeri*, F1
and F2 are almost identical, but differ widely from F3 and more especially,
F4 ($S\Delta Q$: 116, 126).

3. These intraspecific and interspecific divergencies in amino acid composition are apparently matched by differences in the physiological properties
of the haemoglobin components (Table 6.3). For example, components F1
and F2 have low oxygen affinities comparable with lamprey haemoglobins,
whereas F3 and F4 have the high oxygen affinities characteristic of myxinoids.
Likewise, in *E. stouti* the similarity in amino acid composition of A, B and C is
also reflected by similar oxygen affinities, even higher than those of the
haemoglobins F3 and F4 of *E. burgeri*.

4. Although comparisons between the lamprey haemoglobins and all the 12
hagfish components gives an overall mean $S\Delta Q$ value of about 120,
interpretation is complicated by the variability in the individual hagfish
haemoglobins. The closest agreement is between A, B and C of *E. stouti* ($S\Delta Q$:
48–68), F4 of *E. burgeri* (62–72) and M2 and M3 of *Myxine* (71–93). These
values may be compared with indices of 200–223 for comparisons between the
lamprey and F1 and F2 of *E. burgeri* and 114–141 in the case of components D
and E of *E. stouti*.

5. Similar difficulties arise in comparing cyclostome haemoglobins with
human α- and β-chains. Here again, the closest correspondence between the
various hagfish components and human haemoglobins is between A, B and C
of *E. stouti* ($S\Delta Q$: 124–266), F3 and F4 of *E. burgeri* ($S\Delta Q$: 129–291) and M2
and M3 of *Myxine* ($S\Delta Q$: 178–202). In contrast there is much greater disparity

in the case of D and E of *E. stouti* (203–273) and more especially, F1 and F2 of *E. burgeri* (348–427). For comparisons between lamprey and human chains the difference indices are of a similar order to the smallest values for some of the hagfish components (130–220).

6. In these comparisons with human α- and β-chains the two cyclostome groups display a striking divergence. Both lamprey haemoglobins show a distinctly closer approach to the human α-chain and a similar relationship has also been established in comparisons with the α- and β-chains of the rhesus monkey (Marchalonis, 1977). On the other hand, the position is reversed in the hagfishes where, with the sole exception of M3, every one of the 12 haemoglobins shows a closer similarity to the human β-chain. At least in the case of the lamprey, this situation can be confirmed from the known sequences. These indicate that 73 substitutions have occurred in relation to the human chain, compared to 78 in the β-chain.

Any attempt to draw phylogenetic conclusions from this haemoglobin data must obviously be highly speculative, bearing in mind the extent of the variability in the amino acid composition of the various hagfish haemoglobins, where differences both within and between the species are often as great or even greater than the divergence between hagfish and lamprey molecules. Unfortunately, much less information is available in the case of the lampreys, although the existence of only minor differences between three species representing two genera, seems to indicate that haemoglobin variability may be much less marked than it is in the myxinoids. In relation to the hagfishes, it may be significant that most of the observed variability in the two species of *Eptatretus* relates to F1 and F2 of *burgeri*, which together make up only 30% of the total haemoglobin and to D and E, which are only minor components in *stouti*. It might therefore, be argued that because these minor components make a less significant contribution to the physiological properties of the blood, they have been under less stringent selective constraints and have therefore, been able to accept mutation rates much higher than those of the major haemoglobins whose physiological properties are more closely adapted to the mode of life and environment of the hagfish. If this is the case, we should therefore be justified in putting greater reliance on the major haemoglobins, when attempting to draw conclusions from comparisons with lampreys and other vertebrates.

13.1.2 Comparisons of cyclostome haemoglobin sequences with those of other vertebrates

Over the entire vertebrate series there are seven sites where no deviations have so far been recorded and a further nine positions where exceptions are limited to a single species. Of these invariant or near-invariant sites, the initial segment of the hagfish molecule shows identical residues at A3 and A12, where the sole exceptions are the human γ chain and the chick β-chain. In addition, the hagfish sequence, like that of the lamprey has a lysine at A5, where the only known substitutions are in myoglobin and the β-chain of the frog. On the limited evidence, it is probably safe to assume that most if not all, the other invariant residues will eventually turn out to be present in the hagfish. As a rough guide, we can estimate the probable percentage difference between the complete hagfish sequence and that of lampreys and other vertebrates by extrapolation from the known segment. Since this segment contains the unique agnathan sequence at its N terminus this will tend to exaggerate the differences in comparisons with other vertebrates. As a crude measure of divergence, the figures in Table 13.2 suggest that lamprey and hagfish sequences may differ to a rather similar extent from the fish α-chain, but a remarkable feature of the hagfish haemoglobin is that its first 30 residues shares no single amino acid in common with the homologous region of the bird α-chain. Consistent with the long period that has intervened since the divergence of the two cyclostome groups, the differences in the primary structure of their haemoglobins will almost certainly prove at least as great as those that separate the human and fish molecules, representative of the span of gnathostome evolution.

Table 13.2 Difference matrix for some vertebrate haemoglobins. Hagfish figures are based on the partial sequence. Other data from Dayhoff, 1972.

	Human	Carp	Lampreys	
			L. fluviatilis	*P. marinus*
Carp	50	0		
L. fluviatilis	73	75	0	
P. marinus	72	75	5	0
Hagfish	78	75	61	61

13.1.3 Cyclostomes and the functional evolution of haemoglobins

Based on the fact that the oxygenated lamprey monomer readily aggregates in the deoxygenated state to form dimers and temporary tetramers, it has been argued that the earliest vertebrate haemoglobins would have resembled those of lampreys, rather than the monomeric invertebrate globins. As in the lamprey, this primordial globin would have been capable of aggregating to form homotetramers, releasing oxygen more readily from its haem iron atoms. As seems to be the case in the lamprey, the first stages in the origins of co-operativity increased Bohr effects and decreased oxygen affinity may have been produced by the development of more constraining salt bridges in the deoxygenated state (leading to the aggregation of sub-units), but which break on oxygenation, with the liberation of protons and the separation of the sub-units (Goodman *et al.*, 1975).

Throughout these evolutionary developments, selective forces might be expected to operate most intensely on those sites in the molecule most intimately involved in haem functions, interchain co-operativity and the salt bridges related to Bohr effects. Haem contact sites, which would be indispensable in any globin molecule, must already have been well-established at a very early stage in its evolution. Among the invariant residues, the histidine at F8 is a constant feature in both vertebrates and invertebrates, but at E7 the histidine of the vertebrates is absent in invertebrates, where it is replaced by the non-polar leucine or isoleucine (Vuk-Pavlovic, 1975). This has led to the suggestion that the appearance of this polar histidine for the first time in the cyclostomes, may have been a very significant step in the evolution of haem interactions, reduced oxygen affinity and the oxygenation-dependent aggregation – disaggregation equilibrium of the monomer.

Examination of the rates of amino acid substitution have shown that those sites that now function as haem contacts in the α- and β-chains of the higher vertebrates have been the most slowly evolving positions and were probably already established at about the time of the agnathan–gnathostome divergence, at about 500 million years (Goodman et al., 1975). Some support for this view comes from comparisons of the residues listed as haem contact sites in human α- and β-chains with the corresponding positions in lamprey haemoglobin. This shows that nearly half these sites are occupied by the same residues, whereas the overall percentage of common residues for the molecule as a whole is only about 25 %. Another crucial function is that concerned with interchain co-operativity, facilitating oxygen delivery, and here again there is evidence for selective constraint in the agnathan stage of vertebrate evolution. In mammalian haemoglobins, each α-chain makes contact with two β-chains forming $\alpha^1\beta^1$ and $\alpha^1\beta^2$ contact sites which favour haem interactions by placing these groups in close proximity. By comparing the homologous sequences in lamprey and mammalian haemoglobins, it has been claimed that it may be possible to distinguish which of the cyclostome residues are likely to be involved in the aggregation of sub-units during deoxygenation (Riggs, 1972). A comparison of the $\alpha^1\beta^2$ contact sites shows that about half are identical in the mammal and lamprey; a correspondence that again is much higher than that of the whole molecule. The fact that such a high proportion of these residues have been conserved, also suggests that the changes that occur in the conformation of lamprey sub-units during oxygenation may be very similar to those of mammalian haemoglobins. Riggs has also pointed out that at one of these sites (C6), glutamic acid replaces the arginine of mammalian haemoglobin and that an identical substitution also occurs in a human mutant haemoglobin (J Cape Town), where it has the effect of reducing co-operativity.

The generally accepted interpretation of globin evolution postulates a series of gene duplications, leading first to the separation of myoglobins and α- and β-chains, followed by subsequent diversification within the mammals.

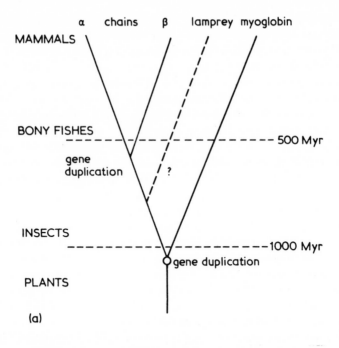

α   chains   β   lamprey myoglobin

MAMMALS

BONY FISHES — — — — — — — — — — — — — 500 Myr

gene
duplication   ?

INSECTS — — — — — — — — — — — — — 1000 Myr

⭘ gene duplication

PLANTS

(a)

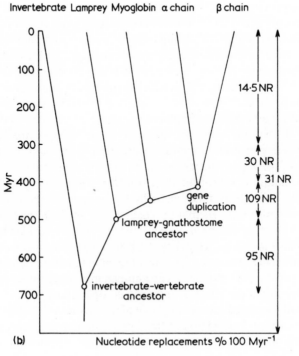

Invertebrate Lamprey Myoglobin α chain   β chain

0

100   14·5 NR

200

300
                                             30 NR
400   ⭘ gene   31 NR
           duplication   109 NR
500   ⭘ lamprey-gnathostome
            ancestor

600   95 NR

700 ⭘ invertebrate-vertebrate
          ancestor

(b)   Nucleotide replacements % 100 Myr^{-1}

Reconstructions of these events have been incorporated in phylogenetic trees, but these have differed widely in their time scales, evolutionary rates and in the order in which the significant events occur. For example, using a time scale derived from an average rate of globin evolution of 10.7 PAM 100 million years^{-1} calculated on the basis of sequence changes in frog β, carp α and mammalian chains, Dayhoff *et al.* (1972) put the myoglobin-haemoglobin divergence at 1070 million years followed by the β-α gene duplication at 500 million years (Fig. 13.2a). These authors believed that the lamprey sequence indicated a very ancient gene separation, which might have preceded the actual separation of the agnathan-gnathostome lineages. A different interpretation of these events has been advanced by Goodman *et al.* (1975). Their genealogical tree takes certain geological ages for important nodal points in globin evolution and, from these, rates of evolution have been estimated over defined periods in the history of the molecule (Fig. 13.2b). For example, the invertebrate–vertebrate dichotomy was placed at 680 million years and the agnathan–gnathostome (vertebrate–lamprey divergence) at 500 million years. Here, the gene duplications that produced the myoglobins and the vertebrate α- and β-chains are positioned after the separation of the lamprey haemo-globins at approximately 460 and 420 million years and would therefore be considerably more recent than the interpretation in Fig. 13.2a. This genealogi-cal tree indicates an average rate of evolution over the period from 680 million years to the present, of 31 NR $\%$ when expressed in terms of the numbers of nucleotide replacements (100 codons/10^8 years), although the rate at different periods in the evolution of the globins has varied widely with the intensity of selection for functional improvement. For example, the evolutionary rate was apparently accelerated sharply in the early vertebrates at the time when gene duplications were opening new fields for molecular diversity, with the introduction of the myoglobin and α- and β-chains. Thus, in the period from 680 million years–300 million years, the average rate of evolution was 46 NR $\%$ falling to only 15 NR $\%$ in the subsequent 300 million years. Moreover, during a more restricted period which would include the α/β gene duplication, that is from 500–400 million years, the estimated rate was 109 NR $\%$.

Where, as in the example in Fig. 13.2a, the sequence data may indicate that the age of a branch point in the phylogenetic tree is much older than the geological and palaeontological data would suggest, the solution may lie in a distinction between gene divergence and species divergence (Fig. 13.3). As mentioned in Section 13.1.1, it may be significant that statistical comparisons suggest that lamprey haemoglobin is closer to the mammalian α-chain in its amino acid composition, while that of the hagfish, in all species and virtually

Fig. 13.2 Views on haemoglobin evolution.
    (a) According to Dayhoff *et al.*, 1972.
    (b) Interpretation based on Goodman *et al.*, 1975.

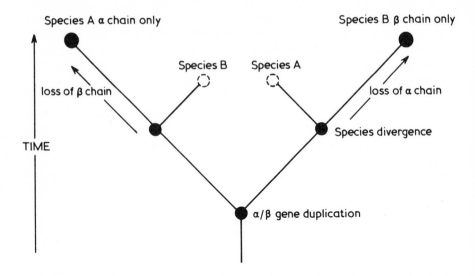

Fig. 13.3 Differences in the timing of species divergence and gene duplication resulting in
non-homologous protein chains and erroneous estimates of the age of the
species differentiation. Based on Dayhoff and Barker, 1972.

all components, appears to bear a closer resemblance to the β-chain. In view of
this, could it be that the gene duplication giving rise to α- and β-chains
occurred at a much earlier stage in vertebrate evolution than has generally been
supposed and that these genes were perhaps already present in the common
ancestors of agnathans and gnathostomes? Subsequently, with the divergence
of the lineages leading towards the myxinoids on the one hand, and the
lampreys on the other, there could have been a loss of either the α type or β type
genes in the two branches, resulting in divergent haemoglobins, which have
retained their respective affinities, however distant to the alpha and beta chains
of the higher vertebrates. Such a remote gene duplication, dating back to about
500 million years, has in fact been postulated by Dayhoff and Barker (1972),
basing their estimate on sequence data and a constant rate of haemoglobin
evolution.

Taken at their face value, the great variation in amino acid composition of
hagfish haemoglobins might be taken as evidence for a very ancient separation
of existing species. Moreover, this variability is in sharp contrast to the only
minor differences that separate the haemoglobins of the two lamprey genera,
Lampetra and *Petromyzon*. A much more likely explanation would be that this
difference between myxinoid and lamprey species has arisen from the
functional properties of their haemoglobins and that the greater degree of
interchain cooperativity, increased Bohr effects and lower oxygen affinities of
the lamprey molecules have involved much more intense selective pressures

than those that have operated during the evolution of the hagfish haemo-globins. Showing little if any evidence of cooperativity or Bohr effects and with much higher oxygen affinities, hagfish haemoglobins, might have been able to tolerate a far higher rate of random mutation than could be accommodated by the more adaptable haemoglobins of the lamprey.

13.2 Cyclostome insulins

Vertebrate insulin is a globular protein made up of two polypeptide chains α and β, linked by disulphide bonds. The biosynthesis of insulin involves the production of a large single-chained precursor molecule – proinsulin – containing about 86 amino acid residues. This ìs then enzymatically cleaved to produce the twin-chained hormonal molecule of insulin with 21 residues in the α-chain and 30 in the β-chain. The fragment remaining after cleavage of the proinsulin is known as the C-peptide. In the hagfish, the biosynthesis follows the same pattern, being preceded by proinsulin with a half life of about 12 hours for its conversion into insulin (Steiner *et al.*, 1973; Falkmer *et al.*, 1975).

13.2.1 Amino acid sequences of hagfish and other vertebrate insulins

Comparisons of the amino acid sequences of hagfish insulin, extracted from the islet organ of *Myxine*, show that most of the features common to other vertebrate species are also preserved in the myxinoid molecule (Fig. 13.4). For example, of the 24 residues that are invariant in other vertebrates all but one occur at the homologous sites in hagfish insulin. The exception is at position B18 where alanine is replaced for valine in the hagfish. In addition to these totally invariant sites, there are a further 5 positions where exceptions are so far known only from a single vertebrate species (A3, A5, B4, B22, B28) and at these sites also the hagfish insulin conforms to the normal vertebrate pattern. Conservative sites, that is positions where substitutions are usually restricted to amino acids with similar physico-chemical characteristics (acidic, aromatic, aliphatic, hydrophobic), have been identified at 16 positions in vertebrate insulin sequences. If hagfish insulin is compared with pig insulin, 9 of these positions are unchanged and a further three show amino acid replacements that retain the same characteristics. The remaining four positions represent radical changes in hagfish insulin. At 14 sites vertebrate insulins have been found to be highly variable. Of these, three are identical in pig and hagfish and three represent conservative substitutions. Residues which appear to be unique to the hagfish occur at no less than 16 sites, mainly in the β-chain and many of these represent radical substitutions (Fig. 13.4).

As in the haemoglobin molecule, selective constraints appear to have varied with the position of residues in the three-dimensional molecule and with the part that they play in maintaining its essential structural and functional characteristics. Thus, in the insulin molecule the relatively small number of

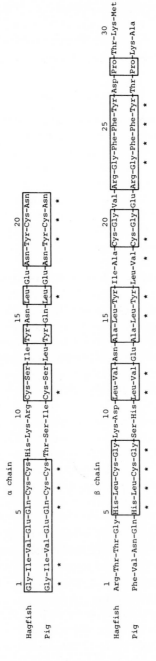

Fig. 13.4 Amino acid sequences of hagfish and pig insulins, with invariant vertebrate residues. Data from Peterson *et al.*, 1975; Falkmer *et al.*, 1975. * indicates invariant residues; boxes indicate residues common to pig and hagfish.

variable sites have been subjected to a rate of mutation about five times greater than that of the molecule as a whole. Because of this restriction, chance mutation and superimposed or back mutations will have been relatively more frequent. This will tend to obscure phylogenetic relationships and may account for the wide variations in sequence that have been observed even in species belonging to the same taxonomic group (Table 13.3).

Table 13.3 Partial difference matrix for vertebrate insulins (modified from Peterson *et al.*, 1975; Dayhoff, 1972). Values represent the range amongst various species.

	Mammals	Birds	Teleosts	Lamprey*
Birds	14–37	0		
Teleosts	22–34	16–27	0	
Hagfish	38–50	37	35–54	18*

*The value for the lamprey is based on the estimate by Peterson *et al.*, 1975.

From a study of the three-dimensional structure of insulin it has been concluded that the side-chains of the invariant or conservative amino acids are probably involved in stabilizing the structure of the monomer or favouring its aggregation to dimers or hexamers, while at variable sites, on the other hand, the residues mainly interact with the solvent at the surface (Peterson *et al.*, 1975). An interesting difference in hagfish insulin is that, although it may form dimers, it does not bind zinc ions and form tetramers. The explanation for this difference may lie in the replacement of histidine by aspartic acid at B10 and perhaps also to the substitution of phenylalanine and threonine for the hydrophobic residues normally present at B1 and B2. However, the fact that hagfish insulin retains most of the invariant or conservative residues suggests that its conformation may be similar to that of other vertebrate insulins and that the molecule is folded in a similar way. Furthermore, it has been pointed out that the 8 radical substitutions (compared to pig insulin), together with the 4 substitutions from the conservative group can probably occur without loss of biological activity. Three are in the α-chain disulphide loop on the surface of the molecule, two are at the beginning of the second helical region and three at each of the N and C terminals of the β-chain.

The biological effects of insulin result from its binding to the surface of the target cells and this binding region is thought to reside in relatively invariant sequences on the surface of the molecule. Because of this, the receptor binding affinities of the vertebrate insulins generally parallel their biological potency. In this respect, hagfish insulin is said to be unique. As measured by the lipogenic effect on rat fat cells its potency is about 5 % of pig insulin, whereas its relative receptor binding affinity is about 23 % (Emdin *et al.*, 1977). In contrast to these variable properties of the vertebrate insulins, the sites that are believed to be responsible for the negative co-operativity in insulin binding seem to have remained unchanged throughout the entire history of the vertebrates, since this same property has been observed in all the insulins tested, including that of the hagfish (de Meyts *et al.*, 1978).

13.2.2 The amino acid composition of lamprey insulin

Amino acid analyses of lamprey and hagfish insulins clearly show a general similarity, contrasting with the differences between the cyclostome molecule and that of various fish, bird and mammalian species. For 10 of the residues the proportions of individual amino acids are identical in both groups of cyclostomes and the SΔD value (40) indicates a relatively high degree of correspondence.

In the absence of sequence data for lamprey insulin it has been estimated that about 8 substitutions are likely to have occurred compared to the hagfish sequence (Falkmer *et al.*, 1975). This is based on the assumption that most of the changes in sequence would have been confined to the highly variable sites. Even if this estimate should eventually prove to be rather low, it is unlikely that the general picture would be seriously affected. Bearing in mind the degree of correspondence in the amino acid composition of lamprey and hagfish insulin, it is clear that the evolutionary distance between the two groups as measured by the divergence in their insulin molecule is much smaller than that suggested by the differences in their haemoglobins. This of course, is in line with the much slower rate of evolution of the insulin molecule which has been put at an average rate of 4 PAM 100 residues^{-1} 100 million years^{-1}, compared with an overall average of 14 PAM 100 residues^{-1} 100 million years^{-1} for the globins (Dickerson, 1972). Not only is the insulin molecule much smaller than the haemoglobin chain, but it also contains a very high proportion of invariant sites; nearly 50% compared to less than 5% in vertebrate haemoglobin. Quite apart from these considerations, the matrices for insulin and haemoglobin (Tables 13.2 and 13.3) suggest that other factors should be taken into account. For example, the probable percentage difference between lamprey and hagfish insulins (about 18%) is very much smaller than the difference between the hagfish and the mammal. On the other hand, for their haemoglobins the differences between the cyclostome and the higher vertebrates are not very much greater than the probable differences in the haemoglobins of hagfish and lamprey. The explanation for this disparity can probably be found in the different functions of the two molecules. Lampreys and hagfishes show many remarkable parallels in the morphology and cytology of their islet tissues and the physiological role of insulin must be very similar in both groups. This contrasts with the characteristics of their haemoglobins. That of the myxinoids still retains to a considerable degree the primitive features of a monomeric globin, with only slight haem interactions and Bohr effects. On the other hand, the haemoglobins of lampreys have become adapted to their much more active and migratory life and to the ability to move to and fro between varying marine and freshwater environments. This would have entailed more intense selection for functional improvements in the molecule, affecting co-operativity, haem interactions and Bohr effects. Since the metabolic demands on the insulin molecule are probably very similar in lampreys or hagfishes there would be less scope for this kind of divergent selection pressure.

13.3 Cytochrome *c*

As a basic element in the electron transport system of oxidative phosphorylation, cytochrome *c* occurs in all aerobic cells from the fungi to higher animals and plants. Like the globin molecule, it contains a haem group attached to the cysteine residues towards the amino terminal of the chain (Fig. 13.5). Throughout the vertebrates, the number of residues varies only from 103–104. Because of its universality, information on the primary structure of the molecule is more complete and extensive than for any other protein, but sequence data for the hagfish is at present restricted to a short segment of 12 residues around the haem group. The evolution of cytochrome has been characterized by a high degree of conservatism and a correspondingly low rate of mutation. This conservatism is reflected in the large number of invariant or conservative sites and by a long sequence of 14 residues (75–88) where there have been only two substitutions in the entire range of living organisms. Estimates of the average rate of mutation have been put at only 3 PAM 100 residues^{-1} 100 million years^{-1} (Dayhoff, 1972) or 5 PAM 100 residues^{-1} 100 million years^{-1} but, as in the case of haemoglobin, the rate of evolution is thought to have been accelerated in the earlier stages of vertebrate phylogeny (Moore *et al.*, 1976).

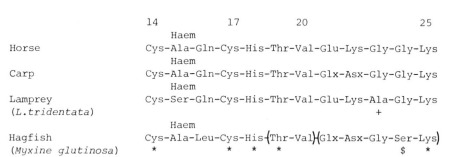

Fig. 13.5 Comparisons of partial sequences for hagfish, lamprey, carp and horse cytochrome *c*, consisting of 12 residues around the haem group. From Paleus and Braunitzer, 1972.

* Sites invariant throughout the vertebrates.

+ Sites invariant except in the lamprey.

$ Sites invariant except for the hagfish.

13.3.1 Comparisons of the primary structure and amino acid compositions of cyclostome and other vertebrate cytochromes

Within the short sequence of cyclostome cytochrome for which comparisons are possible (Fig. 13.5) there are the two completely invariant cysteines and the single histidine. Other invariant sites are the threonine at 18 and the lysine at 25. At positions 23 and 24 all gnathostomes have two glycines, but here the cyclostomes are unique; the lamprey alone having an alanine at 23, while at 24 the only organisms known to show substitutions are the hagfish and higher plants. On the rather slender evidence of this fragment, it has been suggested that hagfish cytochrome may show a closer resemblance to the invertebrates, whereas the lamprey is said to be closer to the higher vertebrates (Paleus and Braunitzer, 1972). Nevertheless, this point of view can be substantiated by evidence from the complete lamprey sequence as well as by statistical comparisons of amino acid analyses (Table 13.4). For example, all tetrapods have asparagine at position 62, while in lampreys and other fish species this is replaced by serine. Similarly, all fish except the dogfish have, like the lamprey, a valine at 66 in place of threonine, which is near invariant throughout the tetrapods. The lamprey also shares with teleosts the same exceptional substitution at 68, while at 73 and 74 all the lower vertebrates differ from the tetrapods in showing amino acid replacement for the methionine and glutamic acid that are invariant throughout the higher vertebrates.

Comparisons of the amino acid analyses of cyclostomes and gnathostomes based on the coefficient $S\Delta Q$, show that the values for the hagfish are consistently very much higher than for the lamprey when comparisons are made with species representative of other vertebrate groups (Table 13.4). In both cyclostome groups, the figures suggest a closer correspondence with the

Table 13.4 Values of the difference index S Δ Q for comparisons of the amino acid composition of cytochrome *c* from cyclostomes and other vertebrate species. Figures for the different groups are the mean values for the species listed.

| | S ΔD | |
Group and species	Hagfish*	Lamprey†
Mammals (man, horse, dog, rabbit)	124	45
Birds (hen, duck, penguin)	105	34
Reptiles (turtle, rattlesnake)	111	44
Amphibia (frog)	100	42
Teleosts (bonito, tuna)	53	32
Elasmobranch (dogfish)	68	14
Lamprey	84	–

Cyclostome data are based on amino acid analyses by Paleus and Braunitzer, 1972* for *Myxine glutinosa* and Uzzelli *et al.*, 1968 for *Lampetra tridentata*†. Other vertebrate data are from Dayhoff, 1972.

cytochromes of the fishes than with those of tetrapods and in the case of the lamprey, the agreement with the dogfish cytochrome is particularly striking. Significantly, the difference between the cytochrome of the hagfish and the lamprey is apparently greater than the disparities between lamprey cytochrome and those of other vertebrates.

In its amino acid composition, hagfish cytochrome shows a number of marked variations from the usual vertebrate patterns. The most extraordinary example of this is the small number of lysine residues. This hydrophilic amino acid is regarded as one of the more conservative residues, showing rates of substitution considerably lower than those of the majority of amino acids. This indicates the vital part that it plays in maintaining the structure of the molecule and its functional properties (Dickerson, 1972). No less than 13 lysine residues are absolutely invariant over the entire spectrum of living organisms and at a further two sites, substitutions for lysine are only known in some invertebrate species. The lamprey conforms to the normal vertebrate pattern, showing 18 lysine residues, whereas the hagfish analysis shows only 12, in place of the 16–19 recorded for other vertebrates. The nearest approach to the hagfish condition is found in some insects, but even here the number of lysine residues does not fall below 14 or 15. Other marked divergencies are the high number of alanine residues (10 in the hagfish compared to 5–7 in other vertebrates), a higher proportion of serine and lower numbers of threonine residues.

13.4 Cyclostome skin collagens

Collagen, the main component of the connective tissues is essentially a trimer, composed of three polypeptide chains forming left-handed helices and each containing about 1000 amino acid residues. These chains form in turn a three-stranded superhelix. An important feature in maintaining the structure of the molecule is its content of proline and hydroxyproline (imino acids) and throughout the vertebrate series the trend towards increased stability has been correlated with an increase in the proportion of imino acids and a corresponding decrease in the proportion of serine and threonine residues (Pikkarainen, 1968).

Judging by the temperature required for denaturation, the secondary structure of the hagfish collagen is the least stable of any of the vertebrates and, in this respect, is in marked contrast to the much greater structural stability of lamprey collagen. These differences are reflected in the higher content of imino acids in the lamprey (173 proline + hydroxyprolines 1000 residues^{-1}) compared with only 154 in the hagfish. In this respect, the lamprey more closely resembles the higher vertebrates. This is also true of the lower number of combined serine and threonine residues. In the lamprey, these total 81 residues 1000 residues^{-1} compared with 98 in the hagfish. Invertebrate collagens are also characterized by higher proportions of serine and threonine varying in different species from about 55–135 1000 residues^{-1} (Pikkarainen and Kulonen, 1969).

Differences in the amino acid constitution of hagfish and lamprey collagens are emphasized in the matrix prepared by Pikkarainen (1968) which compares them with species representing the major vertebrate classes. In this case, the figures represent the simple sum of the differences in each residue, expressed as a proportion of 1000 residues (Table 13.5). Except in comparisons with the ray or flounder, the differences between hagfish collagen and those of other vertebrate groups are very much larger than the corresponding comparisons with the lamprey. Moreover, the extent of the divergence between hagfish and lamprey collagens is similar to the difference between lamprey and dogfish and greater than the divergence between lamprey and frog.

Table 13.5 Differences in the amino acid composition of cyclostome and other vertebrate collagens, expressed as the sum of the differences in the proportions of individual residues per 1000 residues. From Pikkarainen, 1968.

	Pig	Chick	Frog	Flounder	Dogfish	Rayfish	Hagfish
Chick	30.5						
Frog	82.5	71					
Flounder	97.5	85	31				
Dogfish	63.5	57	58	67			
Rayfish	119	107	90	91	54		
Hagfish	102.5	113	51.5	36	85	85	
Lamprey	82.5	69	26	30	47	83	48

13.5 Cyclostome thyroglobulins

The biosynthesis of vertebrate thyroid hormones takes place within the protein thyroglobulin synthesized by the epithelial cells of the thyroid follicles. Characteristically, the thyroglobulin molecule has a sedimentation value of about 19 S and a molecular weight of about 680 000, although other components have been recognized in higher vertebrates, including one sedimenting at 12 S.

Amino acid analyses of the thyroglobulins of the lamprey, dogfish and mammals indicate that the evolution of this protein has been extremely conservative. This is shown by the very small values for the difference indices between human, lamprey and dogfish thyroglobulins (Fig. 13.6) as well as by the cross-reactions that occur between lamprey thyroglobulin and antisera raised against mammalian thyroglobulin (Wright *et al.*, 1978a, b). On the other hand, there are significant differences in the carbohydrate composition of the iodoproteins and the total carbohydrate composition of the lamprey protein is

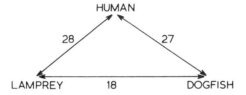

Fig. 13.6 Difference index S Δ Q for human, lamprey and dogfish thyroglobulins.

only about half that of the dogfish molecule and approximately one third of the mammalian values. In addition, cyclostome thyroglobulins are remarkable for their very low concentrations of T_4 (Table 11.1).

The ultracentrifugation pattern for the thyroglobulin of the hagfish, *E. stouti*, appears to be quite unusual in showing only a single peak at 3–8 S (Susuki *et al.*, 1975) with no trace of a 19 S component. The molecular weight is estimated to be only 155 000. In several species of lampreys a component sedimenting at 18–19S has been identified (Lachiver *et al.*, 1966; Salvatore *et al.* 1965; Aloj *et al.*, 1967; Susuki and Kondo, 1973; Monaco *et al.*, 1976). The presence of a 12S component was recognized in the iodoproteins of *L. fluviatilis* and a main 12 S fraction occurred in the purified iodoproteins of *L. tridentatus* (Susuki *et al.*, 1975). Investigations on the ammocoete of a non-parasitic lamprey, *L. reissneri*, suggested a parallel with conditions in the hagfish in so far as the main component sedimented at 3–5 S, with only a small proportion of 19 S iodoprotein whereas, after metamorphosis, the latter became the dominant form (Susuki *et al.*, 1975). This led to the suggestion that the presence of the lighter component in the larval endostyle and in the hagfish might be linked to the lack of any extracellular storage of iodoprotein and to the absence in the hagfish thyroid follicles of luminal colloid. This view assumes therefore that the 19 S fraction is the normal storage form of thyroglobulin and that the lighter components are intermediates associated with intracellular storage. The dominance of the lighter iodoproteins has not however been observed in the ammocoete of another non-parasitic species, *L. planeri*. Both in larvae and adults of this form the incorporation of [125]I- and (^{3}H)-labelled carbohydrates was followed over periods from 4–72 hours (Monaco *et al.*, 1978). Here the density gradient ultracentrifugation patterns consistently showed two mainpeaks at 18–19 S and 3–8 S with a shoulder representing the 12 S iodoprotein. With time, both these lighter components diminished, supporting the view that they are precursors of the 18–19 S thyroglobulin. The latter was synthesized by both ammocoete and adult, but in the endostyle its rate of elaboration proceeded more slowly, presumably in association with the purely intracellular storage of the thyroproteins.

13.6 Lamprey fibrinogen and fibrinopeptides

The process of blood clotting involves the thrombin-catalysed cleavage of the plasma glycoprotein, fibrinogen. This results in the formation of fibrin and the removal from the fibrinogen molecule of two fibrinopeptide fragments, A and B, whose main, if not sole function is to inhibit the polymerization of fibrinogen until the clotting mechanisms are initiated. For the initial stages of this polymerization, only the removal of the fibrinopeptide A is required and the release of the fibrinopeptide B follows more slowly. The fibrinogen molecule consists of three paired chains $\alpha_2\beta_2\gamma_2$ and in the process of clotting the fibrinopeptides A and B are released in sequence from the amino segments of the α- and β-chains; the cleavages taking place at arginyl-glycine bonds, leaving glycine at the amino terminus of both chains. During the formation of the clot, the fibrin is stabilized by the development of cross-linkages, involving the formation of γ–γ dimers and a slower cross linking of the α-chains.

Although the lamprey fibrinogen molecule seems to be significantly larger than that of other vertebrates so far investigated, it has the same $\alpha_2\beta_2\gamma_2$ sub-unit structure, the fibrinopeptides are cleaved at arginyl-glycine linkages and the fibrin molecule is stabilized by the formation of γ–γ dimers and multimers (Doolittle and Wooding, 1974). However, the lamprey molecule displays a number of quite distinctive characteristics. For example, the α-chains are unusually large and may be almost twice the size of the γ-chains (Murtaugh, 1974). On the other hand, the lamprey fibrinopeptide A, consisting of only six amino acid residues, is the smallest recorded in vertebrates and in the mammals the numbers range from a minimum of 13 in the guinea pig to a maximum of 19 in several species of artiodactyls (Fig. 13.7a). The β-chains are also comparatively large but, after the conversion to fibrin, they are greatly reduced by the splitting off of fibrinopeptide B. This is the longest fibrinopeptide chain on record, consisting of 36 residues with a covalently bound cluster of carbohydrates (Cottrell and Doolittle, 1976). Amongst the mammals, the smallest fibrinopeptide B chain is that of the rhesus monkey with only 9 residues and the largest which occur in the kangaroo and a number of species of artiodactyls have 21. The γ-chains, however, are of similar size in lampreys and mammals. There is still some uncertainty on the precise molecular weights of lamprey fibrinogen and its component chains, but Doolittle *et al.* (1976) have suggested that lamprey fibrinogen has a molecular weight of about 360000, with component α-, β-, and γ-chains of about 70000, 61000 and 47000.

In view of the evolutionary distance that separates the lampreys from the mammals, it is hardly surprising to find some species specificity in the operation of the thrombin–fibrinogen mechanisms from these two widely separated groups of animals. Thus, mammalian thombin is capable of removing only one of the two fibrinopeptides (B) from lamprey fibrinogen and lamprey thrombin releases only fibrinopeptide A from mammalian fibrinogen (Doolittle and

```
        1       3       5
A       Asp-Asp-Ile-Ser-Leu-Arg-                                        SO4
        1       3       5       7       9       11      13      15      17      19
B       Glu-Asp-Leu-Ser-Leu-Val-Gly-Gln-Pro-Glu-Asn-Asp-Tyr-Asp-Thr-Gly-Asp-Asp-Asx-Thr-
        21      23      25      27      29      31      33      35
        -Ala-Ala-Asp-Pro-Asp-Ser-Asn-Asn-Thr-Ala-Ala-Ala-Leu-Asp-Val-Arg
                                            |
                                            CHO
```

Fig. 13.7 (a) Amino acid sequences of lamprey fibrinopeptides A and B showing the considerable degree of homology of the amino terminal pentapeptide sequence. From Cottrell and Doolittle, 1976.

```
                                    site of thrombin
                                       cleavage
                    fibrinopeptide      ||       exposed fibrin sequence
                          A             ||
Human   α   ......  Gly-Val-Arg         ||    Gly-Pro-Arg-Val-Val-Glu-Arg  ......

Bovine  α   ......  Gly-Val-Arg         ||    Gly-Pro-Arg-Leu-Val-Glx-Arg  ......

Lamprey α   ......  Ser-Leu-Arg         ||    Gly-Pro-Arg-Leu- ? -Glx-Glx  ......

                    fibrinopeptide      ||       exposed fibrin sequence
                          B             ||
Human   β   ......  Ser-Ala-Arg-        ||    Gly-His-Arg-Pro-Leu-Asp-Lys  ......

Bovine  β   ......  Gly-Ala-Arg         ||    Gly-His-Arg-Pro-Tyr-Asx-Lys  ......

Lamprey β   ......  Asp-Val-Arg         ||    Gly-Val-Arg-Pro-Leu-Pro- ?   ......
```

Fig. 13.7 (b) Comparisons of the sequences cleaved by thrombins in Aα and Bβ chains of human, bovine and lamprey fibrinogens. From Cottrell and Doolittle, 1976.

Wooding, 1974). In its ability to form a clot after the application of mammalian thrombin with the removal of only the fibrinopeptide B, the lamprey is unique amongst vertebrates, although in the normal course of events, lamprey thrombin will of course release both fibrinopeptides from lamprey fibrinogen.

Because of the absence of a specific physiological function for the fibrinopeptide fragments once they have been cleaved from the fibrinogen molecule, it is not surprising that these chains have been able to accept a high rate of mutation; in fact the rate of evolution of the fibrinopeptides is the most rapid so far observed and is put at about 90 PAM 100 million years^{-1} (McLaughlin and Dayhoff, 1972). Like the haemoglobin or immunoglobulin chains, it has often been assumed that the α-, β- and γ-chains of fibrinogen have been evolved from a common ancestral chain by a process of gene duplication. On this interpretation it might be expected that in a primitive vertebrate like the lamprey, there would be some evidence for this in a closer homology between these chains, when compared with the resemblances between these chains in the higher vertebrates (Doolittle *et al.*, 1976). This has not turned out to be the case. Studies of the amino acid composition of lamprey fibrinogen have shown that the individual chains differ far more amongst themselves than do the mammalian chains. When the amino acid composition of the individual lamprey fibrinogen chains are compared with the corresponding human chains (Table. 13.6), it is clear that the differences are much more marked in the case of the α-chains and that of the lamprey is remarkable for its very high proportion of glycine, serine and threonine. These residues make up no less than 45 % of the total composition, whereas in the human molecule they contribute only 30 %. The composition of the lamprey β-chain is much less characteristic and to a somewhat lesser extent, this is true also of the γ-chain. In common with the immunological specificity of lower vertebrate fibrinogens, antibodies to mammalian fibrinogen do not cross-react with lamprey fibrinogen.

Table 13.6 Matrix for 'degrees of differentness'* among α-, β- and γ-chains of lamprey and human fibrinogen. From Doolittle *et al.*, 1976.

	Lamprey α	Lamprey β	Lamprey γ	Human α	Human β
Lamprey β	47.2				
Lamprey γ	54.1	25.1			
Human α	35.3	25.5	34.5		
Human β	56.8	16.3	25.3	32.1	
Human γ	53.6	21.9	21.3	33.8	18.8

*This was calculated by the authors as the average percentage difference between pairs of amino acids for pairs of chains. The boxed groups emphasize that the lamprey chains differ more among themselves than do the human chains.

Studies on the amino acid composition and sequences of the lamprey fibrinopeptides (Cottrell and Doolittle, 1976) have shown that although the sequences of the α- and β-chains on the fibrin side of the thrombin cleavage sites are almost the same as in the mammals, the fibrinopeptides of the lamprey are practically unrecognisable from those of the higher vertebrates (Fig. 13.7b). Comparing the two lamprey fibrinopeptides it may be seen that in spite of the vast disparity in their size, they nevertheless show some correspondence in the five N-terminal residues, although this homology is not a feature of any other vertebrate fibrinopeptides.

In view of the similarity in the terminal fibrin sequence of the α-chains of mammals and lampreys (Fig. 13.7b), it is surprising that mammalian thrombin should be unable to liberate the lamprey fibrinopeptide A. This may, as Cottrell and Doolittle (1976) suggest be due to the substitution in the lamprey fibrinopeptide of leucine for the valine in the penultimate position and this possibility is enhanced by the fact that out of 99 fibrinopeptides for which data is available, none has a leucine at this position. In fact, out of 52 fibrinopeptide A chains, valine occupies this site in all but three.

In keeping with the well-established interspecific variability of the fibrino-peptides, the lamprey chains show little resemblance to those of other vertebrates except that they exhibit a negative charge. However, fibrinopeptide B does share a tripeptide sequence Asp-Tyr-Asp with the B chain of mammalian species and, as in the latter, the tyrosine is sulphated.

The picture that emerges from a study of these lamprey peptides and the parent fibrinogen chains is one that bears convincing evidence of the common ancestry of the molecule in the shape of its basic structure and physiological function, but which nevertheless exhibits some features which as far as we know at the present time, are unique to the cyclostome. These include the ability of lamprey fibrinogen to form a clot after the removal of only the fibrinopeptide B, as well as the unusual sizes of the α- and β-chains. In the absence of a wider range of comparative data from other lower vertebrates, and especially from the myxinoids, it is quite impossible to say whether the special characteristics of lamprey fibrinogen or fibrinopeptides are to be regarded as primitive, or whether, as has been postulated by Doolittle *et al.* (1976), they may be attributed to large-scale deletions or duplications.

13.7 Immunoglobulins and immune responses

In the past there was a tendency to regard specific adaptive immunity as a vertebrate innovation, whose first manifestations were to be sought, albeit in rudimentary form, in the cyclostomes. The criteria normally used in identify-ing these defensive mechanisms include the ability to reject foreign tissue from an unrelated member of the same species (homograft rejection), the develop-ment of delayed hypersensitivity (allergy), immunological memory as exhi-bited by second set homograft rejection or secondary responses to antigens (anamnestic responses), the proliferation of immunocompetent cells after antigenic stimulation and the production of circulating antibodies (immuno-globulins). Although there is now considerable evidence for cellular immunity amongst various invertebrate groups, the cyclostomes are still important for comparative immunologists since they are the first organisms in the phyloge-netic series to show concurrent cellular and humoral immune responses and the first to exhibit circulating antibodies of defined specificity and of the immunoglobulin type (Thoenes, 1972; Marchalonis and Cone, 1973).

In general, the range and intensity of the immune responses are correlated with the degree of development of the lymphoid tissues. In cyclostomes, these tissues are relatively undifferentiated and respond weakly and only slowly to antigenic stimulation. Although lymphocyte-like cells occur in cyclostomes, plasma cells – the source of antibodies in higher vertebrates – do not appear to be present. Lympho-haemopoetic tissues comparable with the spleen and bone marrow are present in lampreys and in the pronephros and submucosa of the hagfish intestine, but it is doubtful whether either group possesses the equivalent of the vertebrate thymus.

13.7.1 Cell-mediated and humoral responses

(a) *Lampreys*

Although restricted in their scope, lampreys show the usual repertoire of vertebrate adaptive immunity; producing antibodies of the immunoglobulin type and showing cell-mediated immunity and skin graft rejection (Good *et al.*, 1972). The cells responsible for antibody production have not yet been identified, but proliferation in the lympho-haemopoetic tissues occurs after antigenic stimulation by killed bacterial cells (Linna *et al.*, 1975). Agglutinating antibodies have been produced to certain bacterial cells and to the H antigen of human red cells but, on the other hand, lampreys have been unresponsive to a wide range of soluble or particulate antigens. The response to killed *Brucella abortus* demonstrates the lamprey's capacity for immunological memory; antibody being produced after a sub-threshold priming dose, followed by a second injection of the same concentration. As in the anamnestic responses of higher vertebrates, repeated administration of antigen leads to higher titres of antibody. Delayed allergic reactions are also displayed after injections of old tuberculin, following previous sensitisation with Freunds adjuvant and *Mycobacterium tuberculosis*. Cellular responses in the form of skin allograft rejection began at three weeks and took about 6 weeks to complete. Second set rejection occurred more rapidly than the first accompanied by inflammatory infiltration and cellular destruction (Good *et al.*, 1972).

(b) *Hagfishes*

Earlier work had suggested that hagfishes were immunologically unresponsive, but with the application of improved handling and using temperatures above those to which the animals would normally be exposed in their normal environment, it has now been possible to demonstrate both cell-mediated and humoral immune mechanisms. Skin allografts are rejected (Hildemann and Thoenes, 1969) accompanied by the usual features of cell-mediated immunity – lymphocyte infiltration, haemorrhagic inflammation and the destruction of pigment cells. For first set allografts the median survival time at 18–19 °C was 72 days and 28 days for second set grafts (Fig. 13.8). Chronic allograft rejection, although more extreme in the cyclostomes, is nevertheless characteristic of elasmobranchs, chondrosteans, apodans and urodelans (Fig. 13.9) and is in sharp contrast to the acute and rapid rejection that generally occurs in teleosts and anuran amphibians (Cooper, 1976). The more primitive and chronic type of rejection has been related to a poorly developed system of histocompatability genes. These determine the degree of compatability between host and donor tissues and the development of general

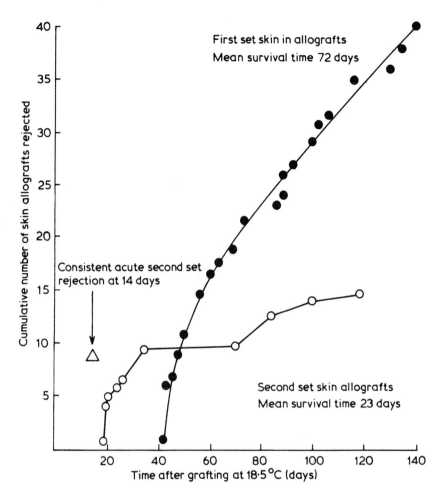

Fig. 13.8 Rejection of skin allografts by the hagfish, *E. stouti.*
Second set allografts were made while the first set were still viable, but had presumably sensitized their recipients. When second set grafts were placed within 0.56 days of the completion of the first set rejection these showed acute rejection within 14 days (indicated by arrow). Redrawn from Hildemann and Thoenes, 1969.

immunocompetence in the vertebrates may have been dependent on the increasing differentiation of tissue antigens (Thoenes, 1972).

Normal hagfish serum and mucus contains heat labile agglutinins to mammalian erythrocytes (Spitzer *et al.*, 1976; Tonder *et al.*, 1978), in addition to specific induced agglutinins that are produced by repeated injections of sheep red cells, and which result in increased titres. Unlike the naturally

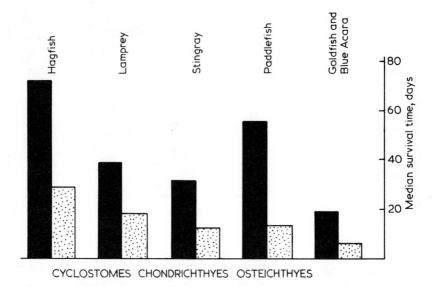

Fig. 13.9 Comparisons of skin allograft survival times in various species of fish and
cyclostomes. Data from Hildemann, 1970; Cooper, 1976; Good *et al*, 1972.
■ first set allografts; ▨ second set allografts.

occurring agglutinins, these induced antibodies are heat stable and highly
specific (Linthicum and Hildemann, 1970). Hagfish antibodies have also been
induced to keyhole limpet haemocyanin (Thoenes and Hildemann, 1970).
Bactericidal responses of the hagfish in response to injections of *Salmonella
typhi* and a gram-negative bacterium from the gut of the spiny lobster have
been studied by Acton *et al*., (1969, 1971).

13.7.2 Cyclostome immunoglobulins
Vertebrate antibodies are associated with a diverse group of serum proteins –
the immunoglobulins – normally composed of multiple polypeptide chains
and representing a wide range of molecular weights and physico-chemical
properties. Since the terminology of the these immunoglobulins was derived
from conditions in the mammals, there are some difficulties in applying it to
the antibodies of lower vertebrates. The various classes of mammalian
immunoglobulins are distinguished as IgA, IgD, IgE, IgG and IgM representing a range of molecular weights from 150 000–950 000, with characteristic
sedimentation coefficients and physicochemical properties. These molecules are
composed of light and heavy chains held together by disulphide bonds and
containing variable V regions and constant C regions (Fig. 13.10). The basic
unit of the constant region is an homologous sequence of some 110 amino
acids and the variable segments, responsible for immunological activity are
placed at the amino-terminal ends of the C regions. The various types of

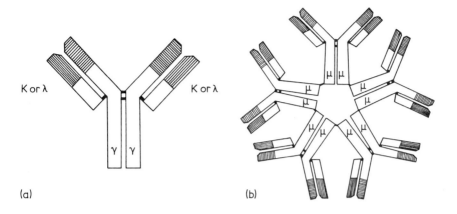

Fig. 13.10 Molecular structure of IgG (a) and IgM (b) immunoglobulins. From Marchalonis and Cone, 1971.
Shaded areas are variable (v) regions; unshaded, constant (c) regions γ, μ, heavy chains; k, λ, light chains.

immunoglobulin are distinguished by specific types of heavy chain (γ, α, μ, δ and ε), each of which may be combined with either of two types of light chain (k or λ). The IgG molecule, for example is composed of four chains, two light and two heavy, whereas the IgM molecule is a polymer consisting of five identical subunits each similar to the IgG molecule.

In the extent of its repertoire of immune response, the hagfish may be somewhat restricted and this has been linked to its intestinal peritrophic membranes, which may serve as a protection against pathogens (Thoenes and Hildemann, 1970). The inducible bactericidal antibodies to *Salmonella typhosa* 'H' antigen and to a bacterium isolated from the spiny lobster were heat-labile substances, showing no increase in titre with a second immunization, although the response was accelerated (Acton *et al.*, 1969, 1971). Associated with a macroglobulin fraction these antibodies were of only limited specificity, reacting to other gram-negative, but not to gram-positive organisms. The active serum fractions showed a major 20 S component after ultracentrifugation, with minor components sedimenting at 9 S and 13 S. After reduction with thiols there was no evidence of separation into light and heavy chains. The inducible antibodies to sheep erythrocytes or to the haemocyanin of the keyhole limpet (Linthicum and Hildemann 1970; Thoenes and Hildemann, 1970; Hildemann and Thoenes, 1971) were heat-stable and there was a marked increase in titres with repeated immunization. The anti-keyhole limpet haemocyanin antibody is described as IgM-like, occurring in the macroglobulin fraction of the serum and having a sedimentation coefficient of about 28 S.

Lampreys have produced antibodies to a number of antigens, including human 'O' cells (Boffa *et al.*, 1967; Drilhon, 1968; Pollara *et al.*, 1970), killed *Brucella abortus* cells, *Salmonella* antigens (Finstad and Good, 1964, 1966) and the F^2 bacteriophage (Marchalonis and Edelman, 1968b). In the estimation of molecular size by ultracentrifugation, mammalian antibodies show two main peaks at 19 S and 6.7 S, corresponding to IgM and IgG immunoglobulins (Fig. 13.11). Lamprey F^2 antiphage antibody activity however, differs in being spread over a wide range from 7–16 S, but purification yielded two antigenically identical antibodies with sedimentation coefficients of 6.6 S and 14 S; the latter regarded as an aggregate of the lighter subunits. From gel electrophoresis it was concluded that the 6.6 S component consisted in its turn of light and heavy chains similar to those of higher vertebrates.

Comparisons of the amino acid composition of the heavy chains of lampreys and other lower vertebrates with the human heavy chains (Table 13.7) show a close affinity with the human μ chain and the values for the $S\Delta Q$ index never exceed 48 units; comparable with the very conservative rate of evolution of cytochrome *c*. In these comparisons, the value for the lamprey-human chains (27) is distinctly larger than for representatives of the Osteichthyes or Chondrichthyes.

Rather different conclusions were reached from studies on lamprey antibodies to *Brucella* cells or to human 'O' erythrocytes (Litman *et al.*, 1970; Pollara *et al.*, 1966, 1970). Electrophoretically, these migrate in the fast β or α region and their charge density is said to be characteristic of an α-globulin. For the *Brucella* antibody, the molecular weight was considered to be intermediate between IgM and IgG and in the case of the antihuman 'O' cell antibody this was estimated at about 300 000. These immunoglobulins could not be dissociated into light or heavy chains, but a common and unique feature of these antibodies like the antiphage immunoglobulin, is their extreme lability; the molecule dissociating readily on storage or in weak solutions without the cleavage of disulphide bonds. Whether this is a primitive feature or alternatively a secondary loss, resulting from the loss or gain of one half of a cysteine residue is problematical (Marchalonis, 1977).

Whatever may be the explanation for the divergent interpretations of the structure of lamprey immunoglobulins, they lead to quite different views on their place in the phylogeny of antibody structure. On the one hand, the immunoglobulins isolated by Litman *et al.* (1970) are regarded as a unique 'ostracoderm' development, lacking the characteristic light and heavy chained structure of gnathostome immunoglobulins, with a distinctive tertiary structure as shown by their circular dichroic spectrum (Litman *et al.*, 1971) and consisting of a tetrameric molecule in which the four equivalent subunits are held together by non-covalent bonds (Good *et al.*, 1972). Marchalonis and Cone (1973) on the other hand believe that an IgM type of lamprey immunoglobulin is already in the main stream of vertebrate immunoglobulin evolution and that genetic mechanisms coding for light and heavy chains had

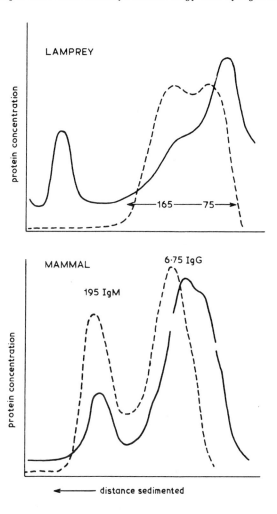

Fig. 13.11 Analysis of lamprey and mammalian antisera by ultracentrifugation on linear sucrose gradients. Redrawn from Marchalonis and Cone, 1973.
-----, Antibody activity; ————, total protein.

already been evolved by the agnathan stage of vertebrate evolution.

With the realization that immunoglobulins are universal throughout the vertebrates, it was natural that a search for clues to the origin of these molecules should be extended to the invertebrates, where naturally occurring bactericidins and agglutinins are widespread. Since similar bactericidins and haemagglutinins are known in the cyclostomes, could not these molecules be regarded as stepping stones in the phylogeny of the immunoglobulins? In unimmunized lampreys, natural haemagglutinins are large molecules with a

Table 13.7 Difference indices (SΔQ) for comparisons of amino acid composition of the heavy chains of lamprey, man and other lower vertebrates. From Marchalonis, 1977.

	Lamprey μ	Catfish μ	Gar μ	Dogfish μ	Human μ
Lamprey μ	0				
Catfish μ	31	0			
Gar μ	70	26	0		
Dogfish μ	46	12	20	0	
Human μ	27	14	18	15	0

sedimentation coefficient of 48S (Marchalonis and Edelman, 1968b) and their subunits can only be separated by reagents that cleave disulphide bonds. Like the agglutinins of invertebrates, they are potentiated by divalent ions and their amino acid composition is said to resemble that of the horseshoe crab (*Limulus*) and to an ever greater extent that of the starfish, *Asterias* (Table 13.8). This correspondence and the especially low S Δ Q values for starfish agglutinin when compared to the heavy chains of the lamprey or the eel, led Carton (1974) to suggest that the *Asterias* protein might well represent an immunoglobulin precursor, particularly in view of the supposed chordate affinities of the echinoderms. From what is known of the structure of the *Limulus* agglutinin, Marchalonis (1977), on the other hand, has argued that the invertebrate proteins are unlikely to have been directly ancestral to the immunoglobulins. Composed of 18 identical sub-units with molecular weights of 22 500 (Marchalonis and Edelman, 1968a), the molecule is described as a cylindrical toroid, bearing no resemblance to an immunoglobulin. However, as Marchalonis points out the question of molecular homology can only be resolved when complete sequence data are available for these invertebrate proteins.

The poorly developed lymphoid tissues of cyclostomes and what has been

Table 13.8 Difference indices (S Δ Q) for comparisons of amino acid composition of natural haemogglutinins and heavy chains of the immunoglobulins of lampreys and invertebrate species. Adapted from Carton, 1974.

	Agglutinins			
	Velesunio (Mollusca)	*Limulus* (Arthropoda)	*Asterias* (Echinodermata)	*Lamprey*
Immunoglobulin **μ chain**				
Lamprey	234	44	31	43
Agglutinins				
Lamprey	151	67	51	0
Asterias	229	45	0	
Limulus	241	0		
Velesunio	0			

regarded as their relatively primitive immunological systems, have led to a belief that this would be reflected in a low incidence or absence of neoplasms (Good and Finstad, 1969; Dawe, 1969; Mawdesley-Thomas, 1972). In fact, in an examination of some 25 000 *Myxine* from the West coast of Sweden, about 10% showed tumours in various organs, although three years later the incidence was less than 1%. For liver carcinoma, there was a clear correlation with length (and therefore age) (Falkmer *et al.*, 1977) and a similar relationship has been noted in the lamprey *L. fluviatilis*, where islet cell tumours tended to be more frequent towards the close of their life cycle (Hardisty, 1976).

13.8 Other biochemical characteristics of the cyclostomes

13.8.1 Acid mucopolysaccharides of skin, notochord and cartilage

The mucopolysaccharides of the connective tissues consist of carbohydrate chains in the form of linear polymers, made up of repetitive units of disaccharide, linked covalently to protein or polypeptide. In the skin, the usual vertebrate components include hyaluronic acid, dermatan sulphate and various dermatan polysulphates. The skin of the lamprey conforms to this pattern, in which the main components are hyaluronate and dermatan sulphate (Rahemtulla *et al.*, 1976). The skin of the hagfish, on the other hand, is unique in possessing a dermatan sulphate containing trisulphated disaccharide residues (Fig. 13.12) (Seno *et al.*, 1972). Lamprey cartilage contains a normal vertebrate 6-sulphate (chondroitin sulphate C), but the hagfish possesses a chondroitin 4-sulphate (chondroitin sulphate E), otherwise found only in the cartilage of invertebrates.

The mucopolysaccharide constitution of the cyclostome notochord is equally interesting. In the lamprey, this contains a chondroitin sulphate A, which also occurs in gnathostomes, but in the hagfish a novel component, chondroitin sulphate H (Fig. 13.12) has been identified (Anno *et al.*, 1971). In all these respects therefore, the myxinoids are separated from the rest of the vertebrates, whereas lamprey tissues contain a similar range of mucopolysaccharides to those of gnathostomes. Although the myxinoid pattern has been described as primitive (Mathews, 1967), it should be noted that the mucopolysaccharide content of the skin of *Amphioxus* is said to have a vertebrate rather than an invertebrate character and the constitution of its notochord is quite different from that of the cyclostomes (Anno and Kawai, 1975).

13.8.2 Bile salts

In his review of vertebrate bile salts, Haslewood (1968) remarked that their structure does not support the view that the cyclostomes are a 'monophyletic' group, or that the divergence of lampreys and hagfishes is of comparatively recent date. The unique bile salt of the hagfishes, *Myxine* and *Eptatretus*, myxinol disulphate (3α, 7α, 16α, 26(or 27) tetrahydroxy-5 cholestane, 3, 26 (or

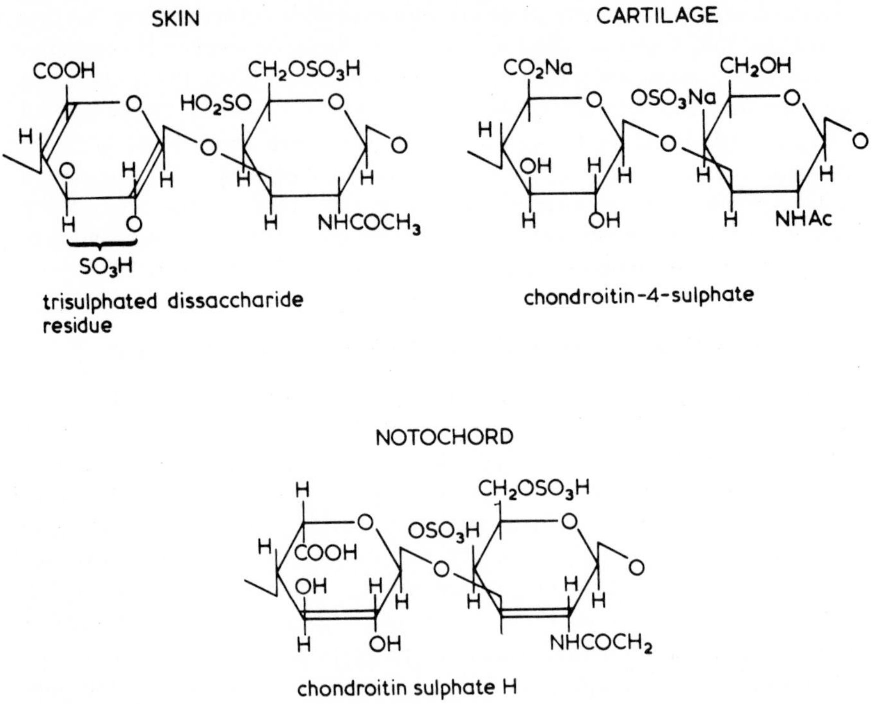

SKIN   CARTILAGE

trisulphated dissaccharide
residue

chondroitin-4-sulphate

NOTOCHORD

chondroitin sulphate H

Fig. 13.12 Mucopolysaccharides of hagfish connective tissues. From data of Mathews,
1967; Anno *et al.*, 1971; Seno *et al.*, 1972; Rahemtulla *et al.*, 1976.

27) disulphate) was described as the most primitive form known, in so far as it
still contains the complete carbon skeleton of cholesterol and contains 27
carbon atoms with only four OH groups (Fig. 13.13). All other bile alcohol C
27 sulphates have at least five OH groups, of which three are in the 3, 7 and 12
positions, and with only one sulphate ester.

Myxinol is said to be a rather ineffective emulsifying agent and would
presumably be required in rather large volumes during the digestion of lipids.
This has been advanced as a possible explanation for the relatively high levels
of plasma cholesterol in the hagfish, acting as a precursor in the synthesis of the
bile alcohols (Larssen and Fänge, 1977).

The main component in the lamprey, *P. marinus* is an entirely different form,
showing changes at five positions compared to myxinol. Petromyzonol, as it
has been named (3α, 7α, 13α, 24-tetrahydroxy-5α cholane) has 24 carbon
atoms, only one sulphate ester group and three of its OH groups are in the 3α,
7α, and 12α positions (Fig. 13.13).

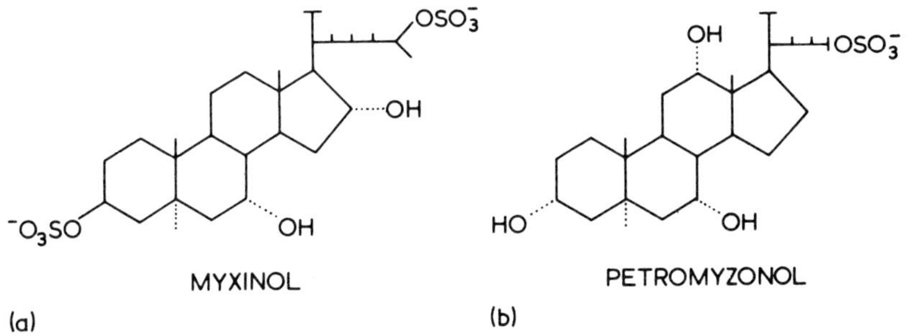

Fig. 13.13 Bile salts of the hagfish (a) and lamprey (b).

13.8.3 The composition and crystalline structure of cyclostome otoliths

The calcareous bodies (otoliths) of the vertebrate labyrinth occur either as single large statoliths or as masses of small particles varying from 1–50 μm–statoconia – which may be united in an organic gel. In the lamprey labyrinth (*L. fluviatilis*) the otoliths consist mainly of statoconia from 2–25 μm in diameter, but in some instances two or three may be fused to form larger bodies up to 250 μm. These are composed mainly of apatite – calcium phosphate – with small amounts of carbonate (Carlström, 1963). In the hagfish, *M. glutinosa*, only statoconia were present, also consisting of apatite and with apparently even less carbonate. In these respects the cyclostomes are said to differ from all other vertebrate groups, where the otoliths (statoconia and statoliths) are entirely composed of calcium carbonate in the form of either aragonite, calcite or vatarite.

13.8.4 Nitrogen metabolism

In many of the aquatic vertebrates, including the actinopterygian fishes, nitrogen is mainly excreted as ammonia and the small quantities of urea that they produce probably originate in the hydrolysis of arginine, or by the degradation of uric acid via the purine pathway. Such forms lack at least the complete series of enzymes that are involved in the ornithine-urea cycle, although arginase is usually retained.

Urea is present in only small amounts in the blood of lampreys and hagfishes and most of their nitrogen is excreted as ammonia. Neither in hagfish nor lamprey have the enzymes of the purine pathway been detected (Read, 1975; Florkin and Duchâteau, 1943). The hagfish liver is apparently unable to synthesize urea, since of the five enzymes involved in the ornithine-urea cycle only arginase was identified by Read (1975) and the absence of carbamoyl synthetase and ornithine transcarbamylase has been cited by Fänge *et al.* (1973). This cycle is also said to be inoperative in the lamprey and Read (1968) was able to find evidence for only arginase and carbamoyl phosphatase; the latter perhaps involved in pyrimidine synthesis.

The fact that these enzyme deficiencies are to be found in both groups of cyclostomes must throw some doubt on the widely held view that their absence in other aquatic vertebrates has been a secondary loss, attributable either to gene repression or deletion and it now seems quite possible that a complete series of enzymes of the purine pathway or the ornithine-urea cycle may not have been developed in the common ancestors of agnathans and gnathostomes.

13.8.5 Lens proteins

Soluble lens proteins are classified according to their electrophoretic mobilities and molecular weights, into α, β and γ crystallines. The highest concentrations of lens proteins transferred during evolution from one vertebrate class to another (organ-specific) are to be found in the more negatively charged β crystallines, while the proteins that tend to be restricted to one particular class (species specific), are most abundant in the less negatively charged γ crystallines. In the α crystallines, organ-specific proteins are more numerous than the species-specific forms.

In mammals the α component accounts for about 30% of the soluble lens protein and has a molecular weight of about 500 000 consisting of two main polypeptide chains. Since α- and β-chains are believed to be represented in lamprey crystallines, these are thought to have arisen as a result of gene duplication in the common ancestor of agnathans and gnathostomes. From the degree of divergence between the amino acid sequences of the α- and β-chain of the cow, it has been estimated that this duplication may have occurred at about 400 million years (Jong *et al.*, 1977).

The extremely low rate of evolutionary change in the lens proteins of vertebrates (estimated as about 2.5 PAM 100 million years^{-1}) is reflected in the fact that vertebrates from different classes share a number of lens antigens in common. This has been demonstrated in the elegant studies of Manski (1967a, b), in which antisera raised in the rabbit against several aquatic vertebrates (including the lamprey) were analysed by immuno-electrophoretic techniques against lens extracts of species representing the entire vertebrate series. As might be expected, these studies showed a complete absence of common lens antigens in cephalopod and lamprey lens proteins, but the lamprey pineal was found to possess some antigens in common with lamprey paired eyes, although these were not shared by the paired eyes of other vertebrates. Six agnathan antigens were identified in the lamprey, of which 5 components were recognized in the Dipnoi and six in members of the Actinopterygii. Elasmobranchs only retained four of the lamprey antigens and the same was found to be true of mammals.

13.8.6 Lactate dehydrogenase isozymes

The enzyme lactate hydrogenase concerned in the interconversion of lactate and pyruvate in the glycolytic pathway, exists in a number of forms or isozymes differing in their molecular structure. These have been intensively studied as a model for gene evolution or as a tool for the genetic analysis of animal populations. In the vertebrates these isozymes are tetramers with a molecular weight of about 140 000; the sub-units having masses of about 35 000–37 000. These sub-units, which are encoded by two structural genes A and B (believed to have arisen by gene duplication), may associate, either as homotetramers A_4 or B_4 or as heterotetramers A_3B_1, A_2B_2 and A_1B_3, thus producing the five isozymes characteristic of the birds and mammals. Within vertebrates there are characteristic differences in the distribution of the A and B sub-units; the A chain predominating in the white muscle, the B chain in heart muscle or brain. This suggests that the different isozymes vary in their metabolic roles. Where, as in the lower vertebrates, individual isozymes may be missing, this is usually explained by the failure to polymerize as heterotetrameric forms.

In an examination of the tissues of large numbers of the Pacific hagfish, *E. stouti* no fewer than 9 LDH phenotypes were identified electrophoretically by Ohno *et al.* (1967), suggesting the presence of two alleles at each of the two A and B gene loci. From the apparent absence of the heterotetrameric isozymes it was concluded that the hagfish was unique amongst vertebrates in having a monomeric LDH molecule, like that of its haemoglobin. However, this conclusion was not substantiated by Arnheim *et al.* (1967) who, in the same species, estimated the molecular weight as about 140 000, similar to that of other vertebrate LDH molecules.

In the tissues of the Atlantic hagfish, *M. glutinosa*, two LDH isozymes have been distinguished electrophoretically; a slower moving cathodal (LDH-1) type in the parietal and visceral lingual muscles and a faster moving anodal (LDH-5) type in the continuously active red fibres of the craniovelar and heart muscle (Dahl and Korneliussen, 1977).

In view of the evidence that the hagfish conforms to the usual vertebrate pattern, with two gene loci coding for A and B chains it is surprising to find that only one LDH isozyme has been identified in the lamprey although, like that of other vertebrates, this is a tetramer, showing homology with the A sub-units of other fishes when tested by immunological methods (Markert *et al.*, 1975). In this respect, therefore, we are driven to the conclusion either that the lamprey is more primitive than the hagfish and had not undergone the initial gene duplication that produced the A and B genes of other vertebrates including the myxinoids or, alternatively, that the B gene has been lost in the course of petromyzonid phylogeny.

13.9 Cyclostome chromosomes and the size of the genome

The chromosomes of the lampreys of the Northern Hemisphere (family Petromyzonidae) are characterized by their small size and exceptionally high diploid numbers (Fig. 13.14). For seven species representing four genera (*Lampetra, Petromyzon, Okkelbergia* and *Ichthyomyzon*) these numbers have varied only within a narrow range from 164–168 and are greater than in any other vertebrate species. Preliminary studies on the karyotype of the Southern Hemisphere genus, *Geotria*, suggest that this will turn out to have diploid numbers even higher than those of the Petromyzonidae, but this is in sharp contrast to the other Southern genus, *Mordacia*, where there are only 76 diploid chromosomes. A further distinction between the Mordaciidae and the Petromyzonidae is that, whereas in *Mordacia* species the centromere is usually located at or near the centre of the chromosome (metacentric), in the Northern lampreys it is usually positioned terminally or sub-terminally (acrocentric).

Although lamprey chromosomes are generally very small (in most cases about 0.5–1.5 μm in length) their total area is greater than in mammals and exceeds that of other amniotes (Robinson *et al.*, 1974). This is all the more surprising in view of the fact that the nuclear DNA content of lampreys is generally only 40 % of human values (Robinson *et al.*, 1975). A further interesting feature of the genome size is that, in spite of the similarity in chromosome numbers within the Petromyzonidae, one species, *P. marinus*, has a much higher DNA value (60 %) than the other Northern species. Furthermore, the members of the genus *Mordacia*, whose chromosome numbers and chromosomal areas are much smaller than those of the Petromyzonidae, nevertheless show very similar DNA values.

Counts on several hagfish species suggest that this group of cyclostomes is probably characterized by chromosome numbers of 46–52 combined with very high DNA values representing 78 % of human values. Most of these chromosomes are very large in comparison with those of lampreys and the chromosomal area also approaches mammalian figures. A majority of the chromosomes appear to be acrocentric. For a local Swedish population of *Myxine glutinosa* there has been a report of a remarkable range in chromosome numbers, varying from 17–116 (Nygten and Jahnke, 1972) and this has been related by Fernholm (1972) to their characteristic morphological and histological variability.

How are the differences between lamprey and hagfish karyotypes to be interpreted and what light do these differences shed on their relationships? Although it has been suggested that the high chromosome numbers of the Petromyzomidae may have arisen through the division of metacentric chromosomes (Kirpichnikov, 1974), a more widely favoured alternative is that they have been produced by polyploidy (Ohno *et al.*, 1968; Howell and

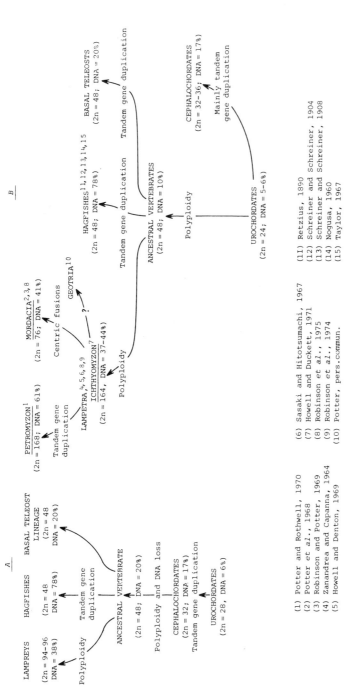

Fig. 13.14 Interpretations of the karyotype evolution of cyclostomes and lower vertebrates. (a) Based on Ohno, 1973; (b) Based on Potter and Robinson, 1971, 1973 and Robinson *et al.*, 1974.

(1) Potter and Rothwell, 1970
(2) Potter *et al.*, 1968
(3) Robinson and Potter, 1969
(4) Zanandrea and Capanna, 1964
(5) Howell and Denton, 1969
(6) Sasaki and Hitotsumachi, 1967
(7) Howell and Duckett, 1971
(8) Robinson *et al.*, 1975
(9) Robinson *et al.*, 1974
(10) Potter, pers.commun.
(11) Retzius, 1890
(12) Schreiner and Schreiner, 1904
(13) Schreiner and Schreiner, 1908
(14) Nogusa, 1960
(15) Taylor, 1967

Duckett, 1971; Robinson and Potter, 1969). However, if this view were to be accepted, how are we to explain the divergent karyotypes of the Mordaciidae and the Petromyzonidae? One possibility might be that these low numbers in *Mordacia* have been produced by centric fusions, in which the centromeres of acrocentric chromosomes fuse to form metacentrics. This mechanism would be consistent with the presence in *Mordacia* of a majority of metacentric chromosomes. A common ancestry of the Northern lampreys and the Mordaciidae at a remote geological period and the subsequent isolation of the latter forms within the Southern Hemisphere would be in harmony with the morphological evidence, which suggests that this group is the most specialized of all lampreys. However, by itself, this view of the origins of the *Mordacia* kayotype does not entirely explain the low numbers, which are less than half those of the Northern forms. Either a further loss of chromosomes (or a subsequent increase in those of the Petromyzonidae) must be a possibility.

Attempts to trace evolutionary trends in the karyotype cannot be divorced from a consideration of changes in genome size, as measured by nuclear DNA content. Throughout the vertebrate series there has been a general tendency towards increased amounts of DNA and the values in the mammals, for example are far greater than those of the protochordates and some fish species. This was interpreted by Ohno (1970, 1973) as indication that tetraploidisation may have occurred twice in the mammalian lineage; the first during the evolution of the ancestral vertebrate stock and the second at a later and possibly reptilian stage (Fig. 13.14). Of the two principal means of increasing the gene loci, polyploidy or tandem gene duplication, Ohno believes the first to have been a more important factor in the gene duplications that have occurred throughout the history of the vertebrates and of which many examples have been given in previous chapters. Because the functional gene loci form only a small fraction of the total genome, tandem gene duplication would be expected to result most frequently in the accumulation of non-functional or trivial DNA, rather than leading to a duplication of new structural genes. Tetraploidisation on the other hand, will by its very nature, duplicate every functional locus.

For Ohno, a chromosome number of 48 had a special significance and was regarded as the chromosomal complement characteristic of the ancestral vertebrate, combined with a DNA value of about 20 % of the mammalian genome. For example, about half of the mammalian species have chromosome numbers from 40–56, with a peak frequency at 48 (Ohno, 1973), and the same numbers occurs widely in many teleost orders. In his detailed interpretation of the evolution of the vertebrate karyotype, Ohno was influenced by an earlier report of a diploid chromosome number of 94–96 in the lamprey, *L. reissneri* (Nogusa, 1960), which suggested that tetraploidisation had occurred in the line from the ancestral vertebrate to the agnathans.

In this scheme, the hagfishes would retain the ancestral complement of 48, while vastly increasing their DNA content, presumably by tandem gene duplication (Atkin and Ohno, 1967).

With the more complete information that is now available on lamprey karyotypes, it is possible to modify Ohno's scheme, retaining an ancestral form with 48 chromosomes, but with a DNA value of only 10 % (Fig. 13.14). Here, tandem gene duplication is invoked to give a basal teleost DNA content of 20 % and polyploidy is held to be responsible for the higher chromosome numbers of the genera, *Lampetra, Geotria* and *Ichthyomyzon*. It would also explain satisfactorily their 40 % DNA values. In the scheme proposed by Ohno, the history of the vertebrate genome was traced back through the cephalochordates to the urochordates, where the chromosome number of 24 and the DNA value of 5–6 % could plausibly be related by tetraploidisation to the ancestral vertebrate genome. However, it now seems unlikely that *Amphioxus* can be regarded as on the main line of vertebrate evolution and the relationships of the protochordates as a whole to vertebrate ancestry remain obscure.

14

Conclusions and evolutionary perspectives

As formulated by Hennig (1966), Brundin (1968) and Løvtrup (1977), phylogenetic systematics starts from the assumption that every group in a phylogenetic system has a corresponding twin or sister group and that both are derived from a common ancestor (Fig. 14.1a). In this dichotomy, one of the sister groups retains to a greater extent the characters of the ancestral group (m_1) and is referred to as plesiomorphic (or in Løvtrup's terminology, plesiotypic) while the other, diverging more from the ancestral type, is described as apomorphic or apotypic. In making these distinctions it is clear that we must attempt to understand both the conservative character of the organism and its evolutionary innovations and be able to distinguish between the 'old' (primitive) and new (specialized) characters (Hansen, 1977). If, following Løvtrup, we apply this system to the relationship of the two cyclostome groups to the gnathostomes, the most likely interpretations are those represented by Figs. 14.1b, c and d. In Fig. 14.1b, the myxinoids and lampreys are portrayed as sister groups, here referred to as secondary twins, diverging from a common ancestor with the gnathostomes at t. The gnathostomes together with the myxinoids and lampreys are also sister groups constituting the primary twins. In the second and third alternatives, Fig. 14 c, d, each of the two cyclostome groups in turn is considered to be the secondary twin or sister group to the gnathostomes, sharing with the latter a common ancestor at t_2, while either the myxinoids or the lampreys and gnathostomes constitute the primary twins originating in the common vertebrate ancestor at t. A decision on the choice of these various alternatives rests on the degree of relationship that we can establish between the two cyclostome groups and the gnathostomes. Are the myxinoids and lampreys more closely related to one another as in Fig. 14.1b where they share a common ancestor than is either group to the gnathostomes? According to the principles of phylogenetic analysis the answer to this question rests on our ability to discriminate between plesiotypic and apotypic characters. Thus, those two groups that share the greatest number of apotypic characters are regarded as the

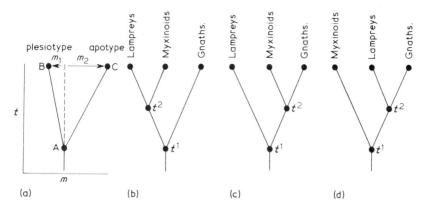

Fig. 14.1 Phylogenetic principles and the relations of cyclostomes to the gnathostomes. Based on Brundin, 1968 and Løvtrup, 1977.

(a) Variation from an ancestral form A of plesiotypic (b) and apotypic (c) sister species.

t. time; *m*. extent of variation from ancestral form. (b), (c) and (d) represent alternative interpretations of the position of the lampreys and myxinoids in relation to the gnathostomes (Gnath).

secondary twins and for this purpose, the common possession of plesiotypic characters is irrelevant. Since apotypic characters are inherited from the immediate common ancestor, these alone are an indication of close relationship.

In the selection of characters that should form the basis of phylogenetic classification, the traditionalist has usually regarded with suspicion the employment of physiological or biochemical criteria. On the other hand, without accepting completely the more extreme claim that priority must be given to non-morphological characters (Løvtrup, 1977), many students of evolutionary phylogeny would not disagree with Hansen's (1977) insistence that the total biology of the organism should be included in his field of study and that comparisons should be based on the exploitive, homeostatic and reproductive functions and/or their anatomical correlates. At the same time, it must be recognized that there are some difficulties in attempting to escape from the strait-jacket of comparative morphology. As Hansen has emphasized, much of our information on biosynthetic pathways or products is of little or no use in phylogenetic analysis, since it deals with ubiquitous processes that are 'unable to distinguish a virus from an elephant or a fungus from an octopus'. Moreover, unlike the comparative anatomist, the physiologist or biochemist is unable to appeal to the fossil record in deciding whether a given character is to be regarded as primitive or derived or to determine the direction of evolutionary change. In addition, a great deal of our biochemical and physiological information cannot readily be expressed in absolute terms, or categorized as the presence or absence of this or that process or product, and is amenable only to relative and quantitative statements. Hennig (1966) has insisted on the rule that, when distinguishing between plesiomorphous and apomorphous conditions of a given character, no attention should be paid to the magnitude of the differences between them. For example, if a is a plesiomorphic character and a^1 and a^2 its apomorphic forms, the bearer of a^1 must be regarded as more closely related to the carrier of a^2 even if, quantitatively, a^1 should differ less from a than from a^2. This procedure would be difficult to reconcile with the methods used in molecular evolutionary studies, where the branches of the phylogenetic tree based on the numbers of mutational steps, are regarded as quantitative representations of phyletic distances.

From a detailed analysis of both morphological and non-morphological characters of the two cyclostome groups *vis à vis* the gnathostomes, Løvtrup has concluded that the lampreys should be regarded as the secondary twin to the gnathostomes (Fig. 13.11). In his list of diagnostic vertebrate characters, Løvtrup found that they shared 10 apotypic morphological characters with the higher vertebrates – the presence of rudimentary vertebral elements, radial fin muscles, spiral fold in the intestine, absence of a persistent pronephros, 'morphology' of the mesonephros, the presence of collecting tubules, absence of accessory hearts, more than one semicircular canal, synaptic ribbons in the retinal receptors and the histology of the adenohypophysis. On the other hand, only one morphological character united the myxinoids and gnathostomes – the union of the dorsal and ventral nerve roots. This argument based on anatomical criteria was re-inforced by a further list of 9 physiological or biochemical characters common to the lampreys and the gnathostomes – a blood volume of less than 10%, hyperosmotic regulation, the 'properties of insulin,' perfected immune responses, nervous regulation of the heart, chondroitin-6-sulphate in cartilage, absence of the unique dermatan sulphate found in the hagfish, the amino acid composition of collagen and the peptide pattern and amino acid composition of haemoglobin. In contrast, Løvtrup was unable to find any non-morphological characters that were synapotypic to the myxinoids and gnathostomes, although as we have seen the condition of the lactate dehydrogenase isozymes provides one such example (Section 13.8.6).

Løvtrup's conclusion that the lampreys are, in practically all respects, closer to the higher vertebrates than the myxinoids is in agreement with the evidence examined in this volume. A similar point of view has also been adopted by Janvier (1978) who has argued that the absence of paired fins in lampreys is a secondary regression and has proposed that together with the cephalaspidomorphs and gnathostomes they should be grouped in a monophyletic taxon – the Myopterygii, regarded as a sister group to the myxinoids and heterostracans. At the same time, it must be recognized that there are some difficulties when we attempt to fit some of Løvtrup's non-morphological criteria into the rigid strait-jacket of phylogenetic analysis. For example, there is no evidence of absolute differences in the physiological properties of lamprey or hagfish insulins, neither is it known whether the primary structure of lamprey insulin is any closer to the very variable insulin molecules of higher vertebrates than is that of the hagfish. Similarly, it is far from clear that immunologically the hagfish is 'inferior' or more primitive than the lamprey and too little is known of the constitution of cyclostome immunoglobulins to make use of them in phylogenetic arguments. In the case of the haemoglobins, it has been shown in Chapter 13 that, as judged by the admittedly inadequate standards of amino acid difference indices, the lamprey molecule may be rather closer to the mammalian α-chain than is the hagfish haemoglobin to the β-chain, but the variability of the individual hagfish components creates a large area of uncertainty. On the other hand, in their physiological properties lamprey haemoglobins are certainly closer to the heterotetrameric haemoglobins of the higher vertebrates, showing lower oxygen affinities, more marked haem interactions and greater Bohr effects. Yet, in spite of this, it must be admitted that, on the limited evidence of the partial sequence of hagfish haemoglobin, the primary structure of the myxinoid molecule may eventually prove to differ from gnathostome haemoglobins to a similar extent to that of the lamprey. A much more convincing picture has emerged from the comparisons of cyclostome and gnathostome cytochromes. Here, the difference indices for lamprey and gnathostome molecules are consistently smaller than for comparisons between the cytochrome of the hagfish and those of higher vertebrates. Similar trends have also been disclosed in the composition and physical properties of cyclostome and gnathostome collagens.

In the past, discussions of the interrelationships of the cyclostomes have tended to centre round the question of whether they should be regarded as a 'diphyletic' or monophyletic group although, as Løvtrup (1977) has emphasized, these terms have no strict relevance in phylogenetic classification, where all taxa of whatever rank are by definition monophyletic. At the same time, the widespread use of these terms by palaeozoologists is a reflection of their preoccupation with the problem of the origins of the hagfishes and lampreys from this or that group of fossil agnathans. As Jollie (1968) has remarked, this problem has tended to resolve itself into an acceptance or rejection of Stensio's views on the essential 'diphylety' of the group; the myxinoids derived from the heterostracans and the lampreys from the cephalaspidomorphs, whereas a number of other alternatives are possible which, while fully recognizing the basic agnathan character of the cyclostomes, at the same time acknowledge the 'great gulf of dissimilarity' that separates the two groups. Jarvik (1968), for example, has rejected a derivation of the myxinoids from the heterostracans but, at the same time, believes that both may have stemmed from a common pteraspidomorph ancestor, which in its turn would be the sister group of the cephalaspidomorph ancestor of the petromyzonids (Chapter 3 and Fig. 3.13). In its essentials, therefore, this interpretation would place the hagfishes and lampreys as secondary twins to the gnathostomes as in Fig. 14.1b and would fail to take account of the weight of biochemical and physiological evidence showing on the one hand, tendencies for the petromyzonids to approach more closely to the gnathostome conditions and on the other, the more 'primitive' or even 'invertebrate' character of so many aspects of hagfish biology.

Amino acid sequences for lamprey haemoglobin and cytochrome *c* have been widely used in the construction of phylogenetic trees representing the evolutionary history of these molecules. Here the constitution of the lamprey protein has been used as one of the key branch points in the evolutionary sequence, marking the agnathan—gnathostome divergence and permitting the hypothetical sequence for the common invertebrate—vertebrate ancestor to be computed. From the survey of the comparable hagfish protein data in Chapter 13, it is clear that the choice of the lamprey as the sole representative of the agnathan stage in molecular evolution may be rather unfortunate and will almost certainly have given rise to misleading conclusions. Had hagfish data been available for haemoglobin, and more especially for cytochrome *c*, there seems little doubt that the inclusion of this information in the construction of computerised phylogenetic trees would alter significantly the picture of the earlier phases of vertebrate evolution. If the interpretation represented in Fig. 14.1(d) is valid, the hagfish would, of course, be closer to the common vertebrate ancestor than the lamprey and nearer to the invertebrate—vertebrate branch point. However, the greatest area of uncertainty in these procedures would still remain — the geological dating of branch points — and here, the palaeozoological evidence is probably on more solid ground in the case of the lamprey, whose relationship to the cephalaspidomorphs is more securely based on morphological grounds than is the much more debatable origins of the myxinoids. What seems certain is that the use of relevant hagfish protein sequences would materially alter the branch lengths in the phylogenetic trees and probably change significantly the calculations of mutation rates in the early history of the cytochrome or haemoglobin molecules.

Apart from the large number of morphological divergencies between the lampreys and myxinoids summarized in Table 1.2, many other biochemical and physiological differences have been noted in subsequent chapters and these are reproduced in Table 14.1. In many of these features the condition in the hagfish can be regarded as primitive (or plesiotypic) and that of the lamprey as synapotypic with the higher vertebrates. Among these may be counted the absence in the hagfish of differentiated Mauthner and Muller neurones or Hinterzelle, the insensitivity of the myxinoid heart to acetylcholine or catecholamines, the greater cytodifferentiation of the lamprey adenohypophysis and the absence in the hagfish of a differentiated meta-adenohypophysis producing an MSH type hormone, the absence in hagfish

Table 14.1 Some histological, physiological and biochemical comparisons between the hagfishes and lampreys.

Hagfishes	Lampreys
Slow muscle fibres on ventral and lateral surfaces of myotomes	Slow fibres on all except medial surfaces
Fast central fibres with common basement membrane and individual innervation	Fast fibres without common basement membrane and electrically coupled
Labyrinth shows no response to vibrational stimuli	Responds to vibrational stimuli
No differentiated Muller and Mauthner neurones	Present
Dorsal or 'Hinterzellen' absent	Present
Heart insensitive to acetylcholine	Acteylcholine exerts positive chronotropic and negative inotropic effects
Heart does not respond to catecholamines	Positive inotropic and chronotropic effects with large doses of noradrenalin
Low metabolic rates	Adults with high metabolic rates
High oxygen affinity of haemoglobin	Adult with low oxygen affinity haemoglobins
Haem interactions slight or absent	Distinct haem interaction
Bohr effects slight or absent	Marked Bohr effects
Amino acid composition of haemoglobins more closely related to mammalian β-chain	Greater resemblance to mammalian α-chains
No water re-absorption from glomerular filtrate	Constant rate of water re-absorption
Simple undifferentiated kidney tubule	Highly differentiated with collecting ducts
Isosmotic and marine	Hyperosmotic and hyposmotic regulation
Absence of differentiated meta-adenohypophysis and of MSH-type hormones	Present
Presence of unique prehypophysial plexus	Absent
Exogenous adrenalin does not result in hyperglycaemia	Transient hyperglycaemia
Gonadal steroidogenesis probably not dependent on gonadotrophins	Steroid output of gonads controlled by adenohypophysis
Thyroglobulin with 3–8 S but no 19 S component	19S component present
Skin collagen unique among vertebrates with low content of imino acids and high proportions of serine and threonine	Amino acid composition closer to that of higher vertebrates
Cytochrome *c* exceptional in the low proportion of lysine residues	Composition closer to that of higher vertebrates
Bile alcohol with 3β hydroxylation regarded as primitive (myxinol)	Bile alcohol with 3α hydroxylation (petromyzonol)
Lactate dehydrogenase isozymes controlled by two gene loci	Only one gene locus
Only one type of granulocyte with heterophilic granules	Eosinophilic, basophilic and heterophilic granulocytes
Karyotype of 48 believed to be similar to common vertebrate ancestor	Except in *Mordacia*, about 160 chromosomes
Small batches of large yolky eggs shed at intervals throughout a probably indefinite life span	Single spawning of large numbers of small eggs, followed by death

Table 14.2 Common physiological and biochemical characteristics of cyclostomes.

Poorly vascularized endocrine tissues and secretion of hormones *via* the connective tissues
Frequency of cysts in the adenohypophysis and islet tissues
Islet tissues producing only insulin
Iodoproteins with very low levels of thyroxine
Immunoreactivity to gastrin, glucagon and CCK in gut endocrine cells
No evidence for a growth hormone-like effect on blood sugar concentrations
No evidence for prolactin-like hormone sharing common antigenic sites with mammalian prolactin
Only a single neurohypophysial octapeptide hormone – arginine vasotocin
Slow clearance of glucose loads
Prolonged insulin hypoglycaemia
Ability to maintain blood sugar levels in prolonged starvation
High concentrations of catecholamines and of chromaffin cells in the heart
No enzymes of the purine pathway
Deficiency in the enzymes of the ornithine-urea cycle
Otoliths consist mainly of calcium phosphate (apatite)
Monomeric haemoglobins with unique C-terminal segment
Haemoglobin functions not modified by organic phosphates
No antibodies in gamma globulin fractions of serum
Chronic rejection of skin allografts
Absence of plasma cells

thyroglobulins of a heavy component sedimenting at 19 S and the apparent independence of gonadal steroidogenesis on pituitary control. Yet, in spite of these further indications of the gulf that separates the lamprey from the hagfish they nevertheless share in common a number of non-morphological characteristics that separate the agnathans from the gnathostomes (Table 14.2). Such features as the condition of the endocrine pancreas, the monomeric haemoglobins, restricted hormonal repertoire and the character of their immunological responses can confidently be regarded as primitive traits that must have been present in the common vertebrate ancestors.

If we accept that the phylogenetic scheme of Fig. 13.1d is the one that best fits the evidence discussed in this volume, we must now regard the hagfish as an earlier offshoot from the early vertebrate stock and in this respect as more 'primitive' than the lamprey, although the extent of the evolutionary regression that it has suffered, faces us on every hand with the problem of deciding how far the apparent simplicity of this or that character is plesiotypic or a result of reduction. Here, embryology can sometimes provide us with useful clues. An example of this is the simple ring-like labyrinth of the myxinoids, which it is tempting to regard as yet another aspect of their sensory degeneration. On the other hand, as Thornhill (1972) has shown, the lamprey labyrinth also passes through an embryonic stage in which it resembles the toroidal hagfish organ and only later acquires the ability to respond to mechanical stimuli that distinguishes it from the myxinoid labyrinth. We cannot therefore, exclude the possibility that the simplicity of the myxinoid labyrinth is plesiotypic, reflecting an early stage in the evolutionary history of the vertebrate organ. Similar doubts arise in connection with the absence in hagfishes of extrinsic eye muscles derived from the second and third somites (together with their motor nerves). Interpretations of heterostracan head structure by Whiting and Halstead suggest that these muscles had not yet been developed by this group of Palaeozoic agnathans and similar conclusions have also been reached by Janvier (1975) in his comparative study of pteraspid and myxinoid eyes.

The designation of the lamprey as secondary twin to the gnathostomes has a number of implications for the way we interpret some of its more 'advanced' features. If this group is now regarded as sharing a common ancestor with the higher vertebrates there is no reason why (as has often been the case in the past) we should explain the common possession of such characters by lampreys and gnathostomes as being a result of convergent evolution. For example, the highly differentiated kidney tubule, osmoregulatory functions that parallel those of teleosts, similarities in gill structure, the presence of a 'precursor' of the exocrine pancreas or of haemoglobins foreshadowing in their functional properties the tetrameric molecules of higher vertebrates, can now be seen as consequences of shared common ancestry without recourse to the improbabilities of repeated convergence.

Irrespective of the view that is taken of the origins of the cyclostomes, we are inevitably confronted by the problem of the survival of these relict forms in the face of the mass extinction that overtook the ostracoderms towards the end of the Devonian. Is it only a coincidence that both surviving agnathans are naked forms with smooth, slimy skins and an undulatory method of locomotion? As we have seen (Chapter 6), there is much in favour of the view that the cyclostomes have suffered a secondary loss of calcified tissues and of the exoskeleton characteristic of the ostracoderms, although the evidence for this regression is somewhat more convincing in the case of the lamprey. Here, the fossil *Mayomyzon* is of crucial importance. This creature demonstrates that if skeletal reduction has in fact occurred in the lamprey lineage, it must already have been complete by the late Carboniferous, nearly 300 million years ago. As the end result of what may have been a long and continuous evolutionary trend we should be forced to the conclusion that this process had begun very much earlier, presumably at least as far back as the Devonian, a period when Orvig (1968) believes that there is evidence for skeletal regression in the cephalaspidomorphs. In the case of the myxinoids we have no such guide lines and it is quite possible that their ancestors were primarily naked forms (Speldnaes, 1967). On the other hand, the assumption that skeletal regression has occurred independently in the lines leading to present-day lampreys and hagfishes does not necessarily involve an improbable degree of convergence, since similar trends are known to have occurred in a number of other vertebrate groups. By losing their exoskeleton and assuming more active habits, these relatively mobile agnathan ancestors might have been better able to compete with the rapidly radiating gnathostomes, where perhaps the more heavily armoured ostracoderms would have been at some disadvantage. This hypothesis cannot however, be completely satisfying since it fails to explain the extinction of the more mobile and fish-like anaspids.

Yet a further clue to their survival may lie in the direction of their concealed burrowing habit and nocturnal activity, which though more characteristic of the hagfish and larval lamprey, are nevertheless still retained to a lesser degree by adult lampreys. An early commitment to this kind of life might well have accompanied skeletal regression, replacing mechanical protection by evasion. The burrowing habit of the cyclostomes is linked with light-avoidance responses and nocturnal activity, associated, at least in the lamprey, with innate diurnal metabolic rhythms, probably regulated by the light-sensitive pineal. Since this structure had already been evolved by the ostracoderms, similar deep-rooted physiological characteristics may have been developed at an early period in the phylogeny of the cyclostomes, although in the myxinoids the regression of the pineal would have led to the loss of this form of metabolic control.

Despite the many behavioural parallels between the lampreys and hag-fishes, the two groups have adopted quite different evolutionary strategies. The myxinoids presumably escaped the fate of their Palaeozoic relatives by evasion − seeking out a niche on the sea bed, where much of their sensory equipment (apart from chemosensory and tactile functions) became re-dundant and relying for their food on the supply of decomposing organic material descending from the more densely populated surface waters. In this restricted habitat it was well able to compete with other vertebrate and invertebrate scavengers, using its ability to escape predation by swift evasive action, the production of copious slime and the ability to conceal itself under the mud, from which as the paired eyes regressed, it emerged to feed only during the night. In an environment where the availability of food is low and unpredictable, opportunism and speed in feeding would be at a premium and here the jaws of the hagfish have proved to be a highly efficient device for rapidly demolishing the occasional morsels of organic material that come within range of its sensitive olfactory receptors. Low metabolic rates are characteristic of deep sea fishes and these, combined with extensive energy stores in the form of neutral lipids, are physiological adaptations that enable them, like the hagfish, to withstand long intervals of starvation, punctuated by sporadic bursts of voracious feeding (Smith, 1978).

By developing morphological and physiological specializations similar to those of cave dwelling or deep sea animals, hagfishes have been able to establish a biological equilibrium with the occupants of surrounding niches, although at the price of confining themselves to an evolutionary cul-de-sac.

Yet, within a benthic habit, present day species show a considerable range of adaptation to a deep water environment and in this respect the members of the genus *Myxine* with their highly degenerate eyes, loss of skin pigmentation and single exhalant branchial opening may represent the most extreme and apotypic forms. Little is known of the physiological characteristics of different species of hagfish, but there are some indications of differences in metabolic rates and in the properties of their haemoglobins, which may reflect divergent habits and ecological requirements. Like *E. burgeri* today, it is possible that the ancestral myxinoids were more active animals, living in shallower water and more responsive to diurnal or seasonal changes. The existence within a group of animals of varying degrees of specialization is predictable on cladistic principles and as Brundin (1968) has remarked 'in sister groups of every rank there is a dual trend towards conservatism (plesiotypy) on the one hand and specialization (apotypy) on the other. The relative primitiveness of one of the two sister groups ensuring the survival of the evolutionary potential of the common ancestral group should the experiment represented by the apotypic and specialized sister group prove unsuccessful'.

The lamprey route to survival has been altogether more adventurous than that of the myxinoids. Although retaining much of the common basic features of the agnathans, they developed (perhaps in common with some other cephalaspidomorphs) the kind of differentiated kidney tubule and ion transport mechanisms in the gills and the gut that enabled them first, to colonise brackish and freshwater environments and later, as their body size increased, to re-invade the marine environment. Yet, in spite of these developments, they were never irrevocably committed to any one habitat and have shown great flexibility in adjusting their life cycle to meet changing environmental conditions, either as landlocked or as non-parasitic species. During their initial penetration into freshwater, they probably developed for the first time, an extended larval phase, taking advantage of the rich organic deposits accumulating in the river systems as the land flora developed. Because of its protected burrowing habits, mortality during this larval phase would have been relatively low and the slow growth rates associated with microphagous feeding might have favoured its extension until finally, it came to represent the major part of the life cycle. So conditioned have generations of zoologists become to treating the ammocoete as an 'archetype' of the ancestral craniate (Gaskell, 1908; Leach, 1944; Carter, 1967) that to suggest that this larval stage has been a secondary introduction into the life history of the petromyzonids may almost be regarded as heretical. However, this is by no means a novel concept and, as far back as 1944, Leach had remarked that the most striking feature of the ammocoete is 'its prolonged retention of many structural characters which are present only during the embryonic differentiation of vertebrates. The transformation of ammocoetes therefore, involves what might be looked upon as a long delayed completion of an interrupted embryonic development'; a phrase that clearly implying the involvement of heterochrony in the evolution of the larval phase.

It has been argued that selection for locomotor efficiency in animals can be related to its effects on fecundity (Alexander, 1967), since any saving in energy could lead to a higher proportion of the total energy budget being diverted to the development of the gonads and the production of larger numbers of progeny. Although there may be some doubt whether the energy costs of locomotion in fish are sufficiently high (Priede, 1977), to confer a very great selective advantage on 'economical swimmers' it could be argued that the reproductive patterns of the hagfish might be related to their sedentary life and low metabolic rates, permitting a relatively high proportion of their energy input to be devoted to the production of a few, large and yolk-laden eggs. In an analogous way, it may be noted that the entire oocyte stock of the lamprey gonad is laid down during the relatively sedentary larval stage, when its metabolic rates are low and the assured and apparently unlimited food supply (brought in by the respiratory current) requires little further energy expenditure for its collection. In the context, the isosmotic condition of the hagfish offers a significant advantage in reducing the energy costs of osmotic and ionic regulation, compared with those that would be incurred by lampreys in either a marine or freshwater environment. For example, in the case of the teleost, *Tilapia* or the rainbow trout it has been estimated that the additional energy expenditure involved in osmotic regulation in 30 % sea water represents no less than 27–29 % of its total oxygen consumption (Rao, 1968; Farmer and Beamish, 1969). Thus, although the hagfish has reaped significant physiological advantages from its confinement to a marine environment, this has been at the expense of the evolutionary adaptability conferred on the lampreys by their well-developed osmoregulatory mechanisms.

Referring to the many primitive chordate characteristics that have been retained by the cyclostomes, Young (1971) has questioned whether or not this could in some way be related to their 'jawless' conditions and modes of feeding. It has been claimed that the extinction of the ostracoderms may have owed something to their inability to compete with the more active gnathostomes, whose hinged jaws would have made them more flexible in their choice of diet. Although we can only speculate on the methods of feeding of the fossil agnathans, the development of the unique biting and rasping mechanisms of the myxinoids and lampreys would have helped them to escape from the limitations of planktonic or mud-sifting habits and might have enabled them to hold their own in the face of competition from the jawed fishes. Evidence from the fossil *Mayomyzon* indicates that, as far back as the Carboniferous, these lamprey ancestors already possessed at least a rudimentary tongue mechanism, which although probably not used in parasitic feeding, could have been used in rasping and suctorial feeding on a wide variety of dead or decomposing organic materials; a habit that is said to be retained even today, by the Danube river lamprey, *E. danfordi*. If, like present-day myxinoids, the ancestral cyclostomes were indiscriminate scavengers or carrion feeders, this could have given them advantages over other ostracoderms, where food gathering may have been dependent on feeding currents set up by ciliary or velar mechanisms.

In both groups of cyclostomes, the development of the massive system of protractor and retractor muscles that operate the feeding apparatus must have been associated with changes in the respiratory system, although these took an entirely different course in the ancestors of the lampreys and hagfishes; in the former involving the separation of the alimentary and respiratory functions of the pharynx through the development of a separate water tube; in the latter, the backward shift of the gill pouches and the use of the nasohypophysial canal to provide a passage for the respiratory water current. By making the feeding mechanisms independent of the respiratory current, the lampreys were able to develop their unique suction mechanism and the tidal pumping system of gill ventilation made it possible to maintain higher metabolic rates than those that could be sustained by the essentially velar methods of the hagfish. Furthermore, the suction mechanism must have made an important contribution to their ability to colonise freshwater habitats where, with their modest swimming abilities and limited staying power, the ability to reduce energy expenditure by anchoring to the substrate, would have been almost a precondition of life in fast-running waters. Although it is not yet possible to make comparisons between the respiratory efficiency of lampreys and myxinoids, there is no doubt that these developments in the pharynx and branchial system of the petromyzonids have resulted in very effective mechanisms for gas exchange, which as judged by any of the available parameters, compare favourably with those of the more active fish species, coping equally well with changes in oxygen concentrations or increased locomotory activity. The high metabolic rates of the adult lamprey, linked to a cardiovascular system capable of maintaining relatively high pressures, is in sharp contrast to conditions in the hagfish, with its low rates of oxygen consumption, open blood system and low blood pressures, requiring accessory hearts to maintain the circulation. Such conditions, suitable for a relatively sedentary animal, indulging in only occasional bursts of vigorous activity, may represent physiological characteristics that the myxinoids have retained from their benthic agnathan ancestors and which have been partly responsible for their ecological limitations. On the other hand, the presence in *Mayomyzon* of a closed pericardium (suggested by the interpretation in Fig. 3.8 of a pericardial fenestra) together with indications from the size of the pericardial concavity on the anterior face of the liver, that it probably possessed a relatively large heart, could mean that even at that time the ancestral petromyzonids had already developed the more efficient circulation that distinguishes them from hagfishes.

The contrast between the two groups of cyclostomes is nowhere more striking than in their reproduction and development and in the pattern of their life cycles. Discussing the application of the concepts of *K* and *r* selection to the life cycles of animals, Gould (1977) stresses the importance of heterochrony as a source of evolutionary change. In so far as their deep sea habitat is stable and predictable, hagfish populations may be considered as subject to *K* selection under density-dependent conditions with low food availability. Such conditions would be expected to favour the kind of reproductive cycle that we find in the myxinoids, involving the production of only a few, relatively well-developed young. As Gould points out, in reference to amphibian life cycles, direct development like that of the hagfish, tends to be associated with stable or aseasonal environments, while more complex life cycles are linked with short-term and seasonal environmental changes. At least in their adult life, lampreys are subjected to much more severe and unpredictable conditions, but the carrying capacity of their environment would only rarely be a limiting factor. Such environments could be regarded as *r* selective, putting a premium on the attainment of maximum fecundity.

The operation of these two forms of selection in the life cycles of amphibians has been illustrated by Gould by reference to populations of facultative paedomorphic salamanders, where metamorphosis becomes rarer as the aquatic larval habitat becomes more predictable and the terrestrial habitat more hostile and unproductive; a situation that is analogous with the origin of the non-parasitic lampreys (Chapter 2). By retarding larval development and extending the age at which the ammocoete enters metamorphosis, heterochrony must have played a major role in the speciation of lampreys, as witnessed by the almost universal occurrence of these dwarf species among a majority of the recognized genera. By a strict definition of these terms, it may not be possible to apply the concepts of *K* and *r* selection to the larval and adult phases of the lamprey life history, but there can be little doubt that selection has operated in different ways on ammocoete and adult lamprey populations. The environment of the ammocoete is certainly stable and mortality is probably low, yet it is unlikely that the availability of microscopic food is a limiting factor. For these reasons, where the environment of the adult may become more difficult (either through increased mortality or shortage of host fishes), the balance of advantage would swing against parasitic feeding and selection might be expected to favour an extension of the protected, stable larval phase and a postponement of metamorphosis, even though this must involve a decrease in the body size of the adult and a corresponding reduction in its fecundity. At the same time, for reasons discussed earlier (Section 12.3) in lampreys (as also in frogs), heterochrony has never been carried to its ultimate conclusion by the complete abandonment of adult life.

Heterochrony in lampreys has been primarily concerned in changes in somatic developmental rates rather than with gonadal development and therefore has had little or no effect on the duration of the life span. To say that the post-spawning death of these animals is a genetically programmed event is merely to state that sexual maturation is an autonomous process, based on internal rhythms that are independent of somatic development. In emphasizing this dissociability of maturation from growth and development, Gould has drawn on evidence from the amphibians suggesting that, where populations consist of both paedomorphic and normally transforming animals, both appear to reach sexual maturity at the same age. If indeed, the gonadal maturation of the lamprey proceeds according to its own inherent rhythm and has been genetically fixed so as to ensure the co-ordination of activity in seasonally breeding populations made up predominantly of a single age class, this might help to explain the failure of endocrinologists to obtain more decisive evidence for the hormonal control of the processes of sexual maturation (Chapter 10). As Dodd (1972) has pointed out, pituitary influence on gametogenesis in the lamprey appears to be only rate-limiting and, although in its absence, the ovaries might eventually reach maturity, its timing would be out of step with the seasonal cycle. Mandatory pituitary control of the gonad presumably only developed after the separation of the petromyzonid and gnathostome lineages, since it is first found in the elasmobranchs. The results of experimental work, involving the ablation of endocrine tissue or the administration of exogenous hormones, tend to indicate that the activation of the many processes involved in sexual maturation are dependent on the differentiation of the gonads themselves (Larsen, 1969b). Thus the laboratory physiologist, in attempting to influence the timing of these even by experimental manipulation, may be flying in the face of deep-seated rhythms and control systems, developed over vast periods of evolutionary time to ensure the synchrony between the reproductive cycle and the particular set of environmental conditions that operate during the normal breeding season. Despite extensive studies on the pituitary and other endocrine tissues of the cyclostomes, it must be admitted that we have made little progress in our understanding of the role of hormones and neuroendocrine regulatory mechanisms in the normal life of these animals. What is more, what little information we have is in some cases paradoxical. For example, in spite of the apparent inactivity of the adenohypophysial tissue of the myxinoids and the scarcity of granular cells, we have much stronger evidence for pituitary control of thyroidal biosynthesis than we have for the much more active lamprey pituitary. In a similar way, we have some indication that the production of adrenocorticosteroids in the hagfish can be stimulated by ACTH, even though in these animals the site of the interrenal tissue has not been established with certainty, whereas in the case of the lamprey the biosynthetic capacities of this tissue remain in doubt. Ultrastructural studies have made it clear that the neurohypophysis contains hypothalamic nerve endings with a variety of secretory granules and yet it has not been possible to obtain convincing evidence of hypothalamic regulation of the adenohypophysis. However, at least in the lampreys, it is possible that these neuroendocrine controls are primarily directed towards the meta-adenohypophysis, whose secretions might well have wider metabolic effects than the purely pigmentary control with which it is generally credited.

If throughout this volume, we have no longer looked upon the agnathans as directly ancestral to the higher vertebrates, this does not detract from their phylogenetic significance or their interest for comparative studies. Representing one of the more important watersheds in the history of animal evolution, the cyclostomes may, as we have seen, owe some of their unique character to their failure to evolve hinged jaws of the gnathostome type. This limitation may have been at least partly responsible for their isolation in specialized niches; an isolation that removed them from direct competition with other populations that could have brought about their extinction and that has ensured their survival from the dawn of vertebrate history. The striking morphological parallels between the lampreys and the cephalaspidomorphs leave us in little doubt that they have been an extremely conservative group and that, in spite of specialized features related to their particular mode of life, they have nevertheless retained a basic plan of organization throughout the 600 million years that spans the entire history of vertebrate evolution. This evolutionary conservatism is especially emphasized by the fossil *Mayomyzon*, which shows that the petromyzonids have undergone only minor changes in their structure in the 280 million years since the late Carboniferous. If so much of the structural plan of the early vertebrates has been retained by the surviving agnathans, can it be seriously doubted that a similar degree of conservatism will be reflected in their functional characteristics and that, by studying these animals, we may hope to gain some insight into the conditions that may have existed in their Palaeozoic ancestors?

Stimulated to some extent, by the need to control the depredations of the North American landlocked sea lamprey, recent years have seen considerable advances in our knowledge of the ecology and physiology of these animals. This is now sufficient to allow us at least a glimpse into some of their unique adaptations to a way of life that has remained virtually unchanged through vast periods of geological time. Because of their inaccessible environment, we know far less of the myxinoids and there are wide areas of their physiology that remain totally unexplored. Until we are able to learn more about the life of these animals in their natural habitat, laboratory studies are deprived of much of their significance and an extension of our understanding can only come from the increased use of the techniques, now being developed, for underwater observation. In confronting the problems that the cyclostomes present in the wider context of vertebrate phylogeny, it seems likely that the contribution that can be made to these questions by comparative morphologists has now been largely exploited and that for further illumination we must look more and more to the efforts of the comparative physiologists and biochemists. This volume will have served its purpose if it has drawn attention to the more obvious gaps in our knowledge of cyclostome biology and to the many fascinating problems posed by this group of vertebrates over the entire spectrum of the biological sciences.

References

Abakumov, V. A. (1956), [The mode of life of the Baltic river lamprey] *Vop. Ikhtiol.*, **16**, 128–133. [In Russian].

Acton, R. T., Weinheimer, P. F., Hildemann, W. H. and Evans, E. E. (1969), Induced bactericidal antibody response in the hagfish. *J. Bact.*, **99**, 626–628.

Acton, R. T., Weinheimer, P. F., Hildemann, W. H. and Evans, E. E. (1971), Bactericidal antibody response in the Pacific hagfish, *Eptatretus stouti*. *Infection and Immunity*, **4**, 160–166.

Adam, H. (1963a), The pituitary gland. In: *The Biology of Myxine*, (Brodal, A. and Fänge, R., eds.) pp. 459–476, Universitetsforlaget, Oslo.

Adam, H. (1963b), Structure and histochemistry of the alimentary canal. In: *The Biology of Myxine*, (Brodal, A. and Fänge, R., eds.) pp. 256–288, Universitetsforlaget, Oslo.

Adam, H. and Strahan, R. (1963), Systematics and distribution. In: *The Biology of Myxine*, (Brodal, A. and Fänge, R., eds.) pp. 1–8, Universitetsforlaget, Oslo.

Adinolfi, M., Chieffi, G. and Siniscalco, M. (1959), Haemoglobin pattern of the cyclostome, *Petromyzon planeri* during the course of development. *Nature*, **184**, 1325–1326.

Aisen, P., Leibmann, A. and Sia, C-L. (1972), Molecular weight and subunit structure of hagfish transferrin. *Biochemistry*, **11**, 3461–3464.

Aler, G., Båge, G. and Fernholm, B. (1971), On the existence of prolactin in cyclostomes. *Gen. comp. Endocrinol.*, **16**, 498–503.

Alexander, R. McN. (1967), *Functional design in fishes*, Hutchinson, London.

Allis, E. P. (1924), On the homologies of the skull of the Cyclostomata. *J. Anat.*, **58**, 256–265.

Allis, E. P. (1931), Concerning the mouth opening and certain features of the visceral endoskeleton of *Cephalaspis*. *J. Anat.*, **65**, 509–527.

Allison, A. C., Cecil, R., Charlwood, P. A., Gratzer, W. B., Jacobs, S. and

Snow, N. S. (1960), Haemoglobin of the river lamprey, *Lampetra fluviatilis. Biochim. biophys. Acta*, **42**, 43–48.

Alnaes, E., Jansen, J. K. S. and Rudjord, T. (1964), Spontaneous junctional activity of fast and slow parietal fibres of the hagfish. *Acta physiol. scand.*, **60**, 240–255.

Aloj, S., Salvatore, G. and Roche, J. (1967), Isolation and properties of a native subunit of lamprey thyrogiobulin. *J. biol. Chem.*, **242**, 3810–3814.

Andersen, M. E. and Gibson, Q. H. (1971), A kinetic analysis of the binding of oxygen and carbon monoxide to lamprey haemoglobin. *J. biol. Chem.*, **246**, 4790–4799.

Andersen, P., Jansen, J. K. S. and Loyning, Y. (1963), Slow and fast muscle fibres in the Atlantic hagfish (*Myxine glutinosa*). *Acta physiol. scand.*, **57**, 167–179.

Andres, K. H. (1975), Neue morphologische Grundlagen zur Physiologie des Riechens und Schmeckens. *Arch. Oto-Rhino-Laryng.*, **210**, 1–41.

Anno, K. and Kawai, Y. (1975), Mucopolysaccharides from the connective tissues of the *Amphioxus, Branchiostoma belcheri. Comp. Biochem. Physiol.*, **52B**, 547–549.

Anno, K., Seno, N., Mathews, M. B., Yagamoto, T. and Susuki, S. (1971), A new dermatan polysulphate, chondroitin sulphate H, from hagfish notochord. *Biochim. biophys. Acta*, **237**, 173–177.

Antonini, E., Wyman, J., Bellelle, D., Rumen, N. and Siniscalco, M. (1964), The oxygen equilibrium of some lamprey haemoglobins. *Archs. Biochem. Biophys.*, **105**, 404–408.

Applegate, V. C. (1950), Natural history of the sea lamprey (*Petromyzon marinus*) in Michigan. *Spec. Sci. Rep. U.S. Fish. Wild. Serv.*, **55**, 1–237.

Arlock, P. (1975), Electrical activity and mechanical response in the systemic and portal vein heart of *Myxine glutinosa. Comp. Biochem. Physiol.*, **51A**, 521–522.

Arnheim, N., Cocks, G. T. and Wilson, A. C. (1967), Molecular size of hagfish muscle lactate dehydrogenase. *Science*, **157**, 568–569.

Atkin, N. B. and Ohno, S. (1967), DNA values of four primitive chordates. *Chromosoma*, **21**, 181–188.

Autuori, F. and Bertolini, B. (1965), A study of some acid hydrolases in the liver of the larval and adult lamprey. *Z. Zellforsch. mikrosk. Anat.*, **68**, 818–829.

Båge, G. (1969), The thyrotropic cells of the adenohypophysis of *Lampetra fluviatilis. Gen. comp. Endocrinol.*, **13**, 491.

Båge, G. and Fernholm, B. (1975), Ultrastructure of the pro-adeno-hypophysis of the river lamprey, *Lampetra fluviatilis* during gonad maturation. *Acta Zool.* (Stockh.), **56**, 95–118.

Bainbridge, R. (1963), Caudal fin and body movement in the propulsion of some fish. *J. exp. Biol.*, **40**, 25–56.

Bannai, S., Sugita, Y. and Yoneyama, Y. (1972), Studies on haemoglobin from the hagfish, *Eptatretus burgeri. J. biol. Chem.*, **247**, 505–510.

Bardack, D. and Zangerl, R. (1971), Lampreys in the fossil record. In: *The Biology of Lampreys*. (Hardisty, M. W. and Potter, I. C., eds.) Vol. 1, pp. 67–84, Academic Press, London.

Barnes, K. and Hardisty, M. W. (1972), Ultrastructural and histochemical studies on the testis of the river lamprey, *Lampetra fluviatilis* (L.). *J. Endocrinol.*, **53**, 56–69.

Barrington, E. J. W. (1942), Blood sugar and the follicles of Langerhans in the ammocoete larva. *J. exp. Biol.*, **19**, 49–55.

Barrington, E. J. W. (1963), *An Introduction to General and Comparative Endocrinology*. Clarendon Press, London.

Barrington, E. J. W. (1969), Unity and diversity in comparative endocrinology. *Gen. comp. Endocrinol.*, **13**, 282–488.

Barrington, E. J. W. (1972), The pancreas and intestine. In: *The Biology of Lampreys*. (Hardisty, M. W. and Potter, I. C., eds.) Vol. 2, pp. 135–169, Academic Press, London.

Barrington, E. J. W. (1974), Biochemistry of primitive Deuterostomians. In: *Chemical Zoology*. (Florkin, M. and Scheer, B. Y., eds.) Vol. 8, pp. 61–96, Academic Press, New York and London.

Barrington, E. J. W. and Dockray, G. J. (1970), The effect of intestinal extracts of lampreys (*Lampetra fluviatilis* and *Petromyzon marinus*) on pancreatic secretion in the rat. *Gen. comp. Endocrinol.*, **14**, 170–177.

Barrington, E. J. W. and Franchi, L. L. (1956), Some cytological characteristics of thyroidal function in the endostyle of the ammocoete larva. *Q. J. microsc. Sci.*, **97**, 393–409.

Barrington, E. J. W. and Sage, M. (1963a), On the response of the glandular tracts and associated regions of the endostyle of the larval lamprey to goitrogens and thyroxine. *Gen. comp. Endocrinol.*, **3**, 153–165.

Barrington, E. J. W. and Sage, M. (1963b), On the responses of iodine binding regions of the endostyle of larval lampreys to goitrogens and thyroxine. *Gen. comp. Endocrinol.*, **3**, 669–679.

Barrington, E. J. W. and Sage, M. (1966), On the response of the endostyle of the hypophysectomised larval lamprey to thiourea. *Gen. comp. Endocrinol.*, **7**, 463–474.

Barrington, E. J. W. and Sage, M. (1972), The Endostyle and thyroid. In: *The Biology of Lampreys* (Hardisty, M. W. and Potter, I. C., eds.) Vol. 2, pp. 105–134, Academic Press, London.

Battle, H. I. and Hayashida, K. (1965), Comparative study of the intraperitoneal alimentary tract of parasitic and non-parasitic lampreys from the Great Lakes region. *J. Fish. Res. Bd. Can.*, **22**, 289–306.

Batueva, I. V. and Shapovalov, A. I. (1974), Synaptic effects evoked by supraspinal and intraspinal stimulation in lamprey motoneurones.

Neurophysiology, **6**, 629–635.

Batueva, I. V. and Shapovalov, A. I. (1977a), Synaptic effects evoked in motoneurons by direct stimulation of single presynaptic fibres in the lamprey. *Neirofiziologiya*, **9**, 390–396.

Batueva, I. V. and Shapovalov, A. I. (1977b), Electrotonic and chemical EPSPs evoked in lamprey motoneurons by descending tract and dorsal root afferent stimulation. *Neirofiziologiya*, **9**, 512–517.

Bauer, C., Engels, U. and Paleus, S. (1975), Oxygen binding to haemoglobins of the primitive vertebrate, *Myxine glutinosa* L. *Nature*, **256**, 66–68.

Beamish, F. W. H. (1973), Oxygen consumption of adult *Petromyzon marinus* in relation to body weight and temperature. *J. Fish. Res. Bd. Can.*, **30**, 1367–1370.

Beamish, F. W. H. (1974), Swimming performance of adult sea lamprey in relation to weight and temperature. *Trans. Am. Fish. Soc.*, **103**, 355–358.

Beamish, F. W. H. (1979), Migration and spawning energetics of the anadromous sea lamprey, *Petromyzon marinus* L. *Env. Biol. Fish* (in press).

Beamish, F. W. H. and Potter, I. C. (1972), Timing of changes in the blood, morphology and behaviour of *Petromyzon marinus* during metamorphosis. *J. Fish. Res. Bd. Canada*, **29**, 1277–1282.

Beamish, F. W. H., Potter, I. C. and Thomas, L. (1979), Proximate composition of the adult anadromous sea lamprey, *Petromyzon marinus* in relation to feeding, migration and reproduction. *J. Anim. Ecol.* (in press).

Beamish, F. W. H., Strachan, P. D. and Thomas, L. (1978b), Osmotic performance of the anadromous sea lamprey, *Petromyzon marinus*. *Comp. Biochem. Physiol.*, **60A**, 435–443.

Beamish, F. W. H. and Williams, N. E. (1976), A preliminary report on the effects of river lamprey (*Lampetra ayresii*) predation on salmon and herring stocks. *Fish. Mar. Serv. Res. Dev. Tech. Rep.*, **611**, 1–26.

de Beer, G. R. (1952), Embryology and taxonomy. In: *The New Systematics*, (Huxley, J., ed.), pp. 365–393, Oxford University Press, London.

Behlke, J. and Scheler, W. (1970), Der Einflüss von Liganden auf den Assoziationsgrad des Desoxy-Hämoglobins der Flüssneunaugen (*Lampetra fluviatilis* L.). *FEBS Letters*, **7**, 177–179.

Belenkii, M. A. (1975), Ultrastructure of the neurohypophysis of the river lamprey, *Lampetra fluviatilis*. *Zh. Evol. Biokhim. Fiziol.*, **11**, 605–611.

Bentley, P. J. and Follett, B. K. (1963), Kidney function in a primitive vertebrate, the cyclostome, *Lampetra fluviatilis*. *J. Physiol.*, **169**, 902–918.

Bentley, P. J. and Follett, B. K. (1965), The effects of hormones on the carbohydrate metabolism of the lamprey, *Lampetra fluviatilis. J. Endocrinol.*, **31**, 127–137.

Beringer, T. and Hadek, R. (1972), Cardiac innervation in the lamprey, *Petromyzon marinus. Anat. Rec.*, **172**, 269–270.

Bertolini, B. (1965), The structure of the liver cells during the life cycle of a brook lamprey (*Lampetra zanandreai*), *Z. Zellforsch. mikrosk. Anat*, **67**, 297–318.

Bird, D. J., Lutz, P. L. and Potter, I. C. (1976), Oxygen dissociation curves of the blood of larval and adult lampreys, *Lampetra fluviatilis* (L.). *J. exp. Biol.*, **65**, 449–458.

Birnberger, K. L. and Rovainen, G. M. (1971), Behavioural and intracellular studies of a habituating fin reflex in the sea lamprey. *J. Neurophysiol.*, **34**, 983–989.

Bjerring, H. C. (1968), The second somite with special reference to the evolution of its myotomic derivatives. In: *Current Problems in Lower Vertebrate Phylogeny.* (Orvig, T., ed.) pp. 341–357, Nobel Symposium No. 4. Interscience, New York.

Bjerring, H. (1977). A contribution to structural analysis of the head of craniate animals. *Zool. Scripta* **6**, 127–183.

Blight, A. B. (1977), The muscular control of vertebrate swimming movements. *Biol. Rev.*, **52**, 181–218.

Bloom, G., Östlund, E. and Fänge, R. (1963), Functional aspects of the cyclostome heart in relation to recent structural findings. In *The Biology of Myxine* (Brodal, A. and Fänge, R. eds.) pp. 317–339, Universitetsforlaget, Oslo.

Boddeke, R., Slijper, E. J. and van der Stelt, A. (1959), Histological characteristics of the body musculature of fish in connection with their mode of life. *Proc. Koninki Neder. Akad. Wet.*, **62**, 576–588.

Boenig, H. (1927), Studien zur Morphologie und Entwicklungsgeschichte des Pankreas beim Bachneunauge (*Lampetra (Petromyzon) planeri*). I. *Z. mikrosk. anat. Forsch.*, **8**, 489–511.

Boenig, H. (1928), Studien zur Morphologie und Entwicklungsgeschichte des Pankreas beim Bachneunauge. II. *Z. mikrosk. anat. Forsch.*, **12**, 537–594.

Boenig, H. (1929), Studien zur Morphologie und Entwicklungsgeschichte des Pankreas beim Bachneunauge. III. *Z. mikrosk. anat. Forsch.*, **17**, 125–184.

Boffa, G. A. and Drilhon, A. (1970), Les transferrines de la lamproie marine; relation entre les unités moléculaires et leur valeur fonctionelle. *C. R. Séanc. Soc. Biol.*, **164**, ·25–32.

Boffa, G. A., Fine, J. M., Drilhon, A. and Amouch, P. (1967), Immunoglobulins and transferrin in marine lamprey sera. *Nature*, **214**, 700–702.

Bond, C. E. and Kan, T. T. (1973), *Lampetra (Entosphenus) minima* n. sp.; a dwarfed parasitic lamprey from Oregon. *Copeia*, 568–576.

Bone, Q. (1963), The central nervous system. In: *The Biology of Myxine.* (Brodal, A. and Fänge, R., eds.) pp. 50–91, Universitetsforlaget, Oslo.

Bone, Q. (1966), On the function of two types of muscle fibre in elasmobranch fish. *J. Mar. Biol. Ass. U.K.*, **46**, 321–349.

Bone, Q. (1975), Muscular and energetic aspects of fish swimming. In: *Swimming and Flying in Nature.* (Wu, T. Y. T., Brokaw, C. J. and Brenner, C., eds.) pp. 493–528, Plenum Press, New York and London.

Botticelli, C. R., Hisaw, F. L. and Roth, W. D. (1963), Estradiol-17β, estrone and progesterone in the ovaries of lamprey (*Petromyzon marinus*). *Proc. Soc. exp. Biol. Med.*, **114**, 255–257.

Braunitzer, G. and Fujiki, H. (1969), Zur Evolution der Vertebraten. Die Konstitution und Tertiarstruktur des Hämoglobins des Flüssneunauges. *Naturwissenschaften*, **56**, 322–323.

Brinn, J. E. and Epple, A. (1976), New types of islet cells in a cyclostome, *Petromyzon marinus* L. *Cell Tiss. Res.*, **171**, 317–330.

Brodal, A. and Fänge, R. (eds.) (1963), *The Biology of Myxine.* Universitetsforlaget, Oslo.

Bruckmoser, P. (1971), Elektrische Antworten im Vorderhirn von *Lampetra fluviatilis* (L.) bei Reizung des Nervus olfactorius. *Z. vergl. Physiol.*, **75**, 69–85.

Bruckmoser, P. and Dobrylko, A. K. (1972), Evoked potentials in the telencephalon of the lamprey, *Lampetra fluviatilis* from stimulation of olfactory nerve. *Zh. Evol. Biokhim. Fiziol.*, **8**, 558–560.

Brundin, L. (1968), Application of phylogenetic principles in systematics and evolutionary theory. In: *Current problems in Lower Vertebrate Phylogeny.* (Orvig, T., ed.) pp. 473–495, Nobel Symposium *4*. Interscience, New York.

Burian, R. (1910), Funktion der Nierenglomeruli und Ultrafiltration. *Arch. Ges. Physiol. Menschen Tiere*, **136**, 741–760.

Busson-Mabillot, S. (1967), Structure ovarienne de la lamproie adulte (*Lampetra planeri* Bloch). II. Les enveloppes de l'ovocyte; cellules folliculaires et stroma ovarienne. *J. Microscopie*, **6**, 807–838.

Buus, O. and Larsen, L. O. (1975), Absence of known corticosteroids in blood of river lamprey (*Lampetra fluviatilis*) after treatment with mammalian corticotrophin. *Gen. comp. Endocrinol.*, **26**, 96–99.

Bytinski-Salz, H. (1956), The survival of senile tissues. I. The transplantation of whole eyes from adult to larval stages in the European brook lamprey, (*Lampetra planeri*). *Riv. Biol.*, **48**, 3–36.

Campbell, B. (1940), Integration of locomotor behaviour patterns of the hagfish. *J. Neurophysiol.* **3**, 323–328.

Carlisle, D. B. and Olssen, R. (1965), An effect of extracts of the pituitary gland of fish and cyclostomes and of the neural complex of ascidians on the melanophores of crabs. *Gen. comp. Endocrinol.*, **5**, 662.

Carlström, D. (1963), A crystallographic study of vertebrate otoliths. *Biol. Bull.*, **125**, 441–463.

Carter, G. S. (1967), *Structure and Habit in Vertebrate Evolution*. Sidgewick Jackson, London.

Carton, Y. (1974), Parentées entre les hémagglutinines naturelles d'échinodermes et les chaînes des immunoglobulines des Vertebrés. *Ann. Immunol.*, **125**, 731–745.

Casley-Smith, J. R. and Casley-Smith, J. R. (1975), The fine structure of the blood capillaries of some endocrine glands of the hagfish, *Eptatretus stouti*; implications for the evolution of blood and lymph vessels. *Rev. Suisse Zool.*, **82**, 35–40.

Chapman, C. B., Jensen, D. and Wildenthal, K. (1963), On circulatory control mechanisms in the Pacific hagfish. *Circ. Res.*, **12**, 427–440.

Chappuis, P. A. (1939), Über die Lebensweise von *Eudontomyzon danfordi* Regan. *Arch. Hydrobiol.*, **34**, 645–658.

Charnov, E. L., Maynard-Smith, J. and Bull, J. J. (1976), Why be an hermaphrodite? *Nature*, **263**, 125–126.

Chester-Jones, I. (1963), Adrenocorticosteroids. In: *The Biology of Myxine*. (Brodal, A. and Fänge, R., eds.) pp. 488–502, Universitetsforlaget, Oslo.

Chester-Jones, I., Phillips, J. G. and Bellamy, (1962), Studies on water and electrolytes in cyclostomes and teleosts with special reference to *Myxine glutinosa* L. (the hagfish) and *Anguilla anguilla* (the Atlantic eel). *Gen. comp. Endocrinol.* Suppl. *I.* 36–47.

Chieffi, C. and Botte, V. (1962), Il tessuto interstitiale del testiculo dei Ciclostomi. Ricerche istologiche ed istochimiche in *Lampetra fluviatilis* e *Lampetra planeri*. *Rend. hist. Scient. Univ. Camerino*, **3**, 90–95.

Cholette, C., Gagnon, A. and Germain, P. (1970), Isosmotic adaptations in *Myxine glutinosa* L. I. Variations of some parameters and the role of the amino acid pool of the muscle cells. *Comp. Biochem. Physiol.*, **33**, 333–346.

Christensen, B. N. (1976), Morphological correlates of synaptic transmission in the lamprey spinal cord. *J. Neurophysiol.*, **39**, 197–212.

Ciaccio, G. and Dell'Agata, M. (1973), Research on the histology and histochemistry of myotube striated muscle fibres in *Lampetra planeri* (Bloch). *Riv. Biol.*, **65**, 341–345.

Claridge, P. N. and Potter, I. C. (1974), Heart ratios at different stages in the life cycle of lampreys. *Acta zool. (Stockh.)*, **55**, 61–69.

Claridge, P. N. and Potter, I. C. (1975), Oxygen consumption, ventilatory frequency and heart rate of lampreys (*Lampetra fluviatilis*) during their spawning run. *J. exp. Biol.*, **63**, 193–206.

Claridge, P. N., Potter, I. C. and Hughes, G. M. (1973), Circadian rhythms of activity, ventilatory frequency and heart rate in the adult river lamprey,

Lampetra fluviatilis. J. Zool. **171**, 239–250.

Claydon, G. J. (1938), The premandibular region of *Petromyzon planeri. Proc. Zool. Soc.* Lond., **108**, 1–16.

Coates, M. L. (1975), Haemoglobin function in the vertebrates. An evolutionary model. *J. Mol. Evol.*, **6**, 285–307.

Cole, F. J. (1905), A monograph on the general morphology of the myxinoid fishes based on a study of *Myxine. I.* The anatomy of the skeleton. *Trans. R. Soc. Edinb.*, **31**, 749–788.

Cole, F. J. (1913), A monograph on the general morphology of the myxinoid fishes based on a study of *Myxine. V.* The anatomy of the gut and its appendages. *Trans. R. Soc. Edinb.*, **49**, 293–344.

Conel, J. L-R. (1931), The genital system of the myxinoids. A study based on notes and drawings of these organs in *Bdellostoma* made by Bashford Dean. The Bashford Dean Memorial Volume. *Archaic Fishes.* I. (1930–1933). 64–102.

Cooper, E. L. (1976), *Comparative Immunity.* Prentice-Hall, New Jersey.

Cottrell, B. A. and Doolittle, R. F. (1976), Amino acid sequences of lamprey fibrinopeptides A and B and characterisation of the junctions split by lamprey and mammalian thrombins. *Biochim. biophys. Acta*, **453**, 426–438.

Csaba, G. and Nagy, S. O. (1975), Iodine incorporation by embryonic entoderm. *Z. mikrosk. anat. Forsch.*, **89**, 599–605.

Czopek, J. C. (1970), Surviving ability in the river lamprey under conditions of partial or complete elimination of gill respiration. *Bull. Acad. pol. Sci (Biol)*, **18**, 237–240.

Dahl, H. A. and Korneliussen, H. (1977), Lactate dehydrogenase isoenzymes in different types of muscle fibres in the Atlantic hagfish (*Myxine glutinosa* L.). *Comp. biochem. Physiol.*, **55B**, 381–385.

Damas, H. (1944). Recherches sur le développement de *Lampetra fluviatilis Archs. Biol.*, Paris, **55**, 1–284.

Damas, H. (1954), La branchie préspiraculaire des Cephalaspides. *Anns. Soc. r. Zool. Belg.*, **85**, 89–102.

Damas, H. (1958), Crâne des Agnathes. In: *Traité de Zoologie.* (Grasśe, P. P. ed.). Vol. 13, pp. 22–39, Masson, Paris.

D'Ancona, U. (1943), Nuova ricerche sulla determinazione sessuale dell'anguilla. *Arch. Océanogr. Limnol.*, **3**, 159–271.

D'Ancona, U. (1950), Détermination et différentiation du sexe chez les poissons. *Archs. Anat. microsc. Morph. exp.*, **38**, 274–294.

Dawe, C. J. (1969), Phylogeny and oncogeny. *Nat. Cancer Inst. Monogr.*, **31**, 1–39.

Dawson, J. A. (1963), The oral cavity, the 'jaws' and the horny teeth of *Myxine glutinosa.* In: *The Biology of Myxine.* (Brodal, A. and Fänge, R., eds.) pp. 231–235, Universitetsforlaget, Oslo.

Dayhoff, M. O. (ed.) (1972), *Atlas of Protein Sequence and Structure*. Nat. Biomed. Res. Foundation, Washington, D. C.

Dayhoff, M. O. and Barker, W. C. (1972), Mechanisms in molecular evolution; examples. In: *Atlas of Protein Sequence and Structure*. (Dayhoff, M. O., ed.) pp. 41–45, Nat. Biomed. Res. Foundation, Washington, D. C.

Dayhoff, M. O., Hunt, L. T., McLaughlin, P. J. and Jones, D. D. (1972), Gene duplication in evolution; the globins. In: *Atlas of Protein Sequence and Structure*. (Dayhoff, M. O., ed.) pp. 17–30, Nat. Biomed. Res. Foundation, Washington, D. C.

Dayton, P. K. and Hessler, R. R. (1972), Role of biological disturbance in maintaining diversity in the deep sea. *Deep Sea Res.*, **19**, 199–208.

Dean, B. (1899), On the embryology of *Bdellostoma stouti*. A general account of myxinoid development from the egg and segmentation to hatching. *Festschr. Carl von Kupfer*, Jena, 221–276.

Dean, B. and Sumner, F. B. (1898), Notes on the spawning habits of the brook lamprey (*Petromyzon wilderi*). *Trans. N. Y. Acad. Sci.*, **16**, 321–324.

Dedman, J. R., Gracy, R. W. and Harris, B. G. (1974), A method for estimating sequence homology from amino acid composition. The evolution of *Ascaris* employing aldolase and glyceraldehyde-3-phosphate dehydrogenase. *Comp. Biochem. Physiol.*, **49B**, 715–731.

Denison, R. H. (1961), Feeding mechanisms of agnathans and early gnathostomes. *Am. Zool.*, **1**, 177–181.

Dickerson, R. E. (1972), The structure and history of an ancient protein. In: *The Chemical Basis of Life*. pp. 82–95, Butterworths, San Fransisco.

Dickhoff, W. W., Crim, J. W. and Gorbman, A. (1978), Lack of effect of thyrotropin releasing hormone on Pacific hagfish (*Eptatretus stouti*) pituitary-thyroid tissues *in vitro*. *Gen. comp. Endocrinol.*, **35**, 96–98.

Dickhoff, W. W. and Gorbman, A. (1977), *In vitro* thyrotropic effect of the pituitary of the Pacific hagfish, *Eptatretus stouti*. *Gen. comp. Endocrinol.*, **31**, 75–79.

Dodd, J. M. (1972), Ovarian control in cyclostomes and elasmobranchs. *Am. Zool.*, **12**, 325–359.

Dodd, J. M., Evennett, P. J. and Goddard, C. K. (1960), Reproductive endocrinology in cyclostomes and elasmobranchs. *Symp. Zool. Soc, Lond.* **1**, 77–103.

Dohi, Y., Sugita, Y. and Theyama, Y. (1973), The self-association and oxygen equilibrium of haemoglobin from the lamprey, *Entosphenus japonicus*. *J. biol. Chem.*, **248**, 2354–2363.

Doolittle, R. F., Cottrell, B. A. and Riley, M. (1976). Amino acid composition of the subunit chains of lamprey fibrinogen. Evolutionary significance of some structural anomalies. *Biochim. biophys. Acta*, **453**, 439–452.

Doolittle, R. F. and Wooding, G. L. (1974), The subunit structure of lamprey fibrinogen and fibrin. *Biochim. biophys. Acta.*, **271**, 277–282.

Doving, K. B. and Holmberg, K. (1974), A note on the function of the olfactory organ of the hagfish, *Myxine glutinosa*. *Acta physiol. scand.*, **91**, 430–432.

Drilhon, A. (1968), Étude des protéines sériques des poissons. *Ann. Biol.*, **7**, 652–682.

Drilhon, A., Fine, J. M. and Boffa, G. A. (1968), Étude comparée des protéines sériques de deux cyclostomes, *Petromyzon marinus* et *Lampetra fluviatilis*. *C. R. Séanc. Soc. Biol.*, **162**, 862–869.

Eaton, R. C., Bombadieri, R. A. and Meyer, D. L. (1977), The Mauthner initiated startle response in teleost fish. *J. exp. Biol.*, **66**, 65–81.

Eddy, J. M. P. (1969), Metamorphosis and the pineal complex in the brook lamprey, *Lampetra planeri*. *J. Endocrinol.*, **44**, 451–452.

Eddy, J. M. P. (1972), The pineal complex. In: *The Biology of Lampreys*. (Hardisty, M. W. and Potter, I. C., eds.) Vol. 2, pp. 91–103, Academic Press, London.

Eddy, J. M. P. and Strahan, R. (1970), The structure of the epiphyseal complex of *Mordacia mordax* and *Geotria australis* (Petromyzonidae). *Acta zool. (Stockh).*, **51**, 67–84.

Eisenbach, G. M., Weise, M., Weise, R., Hanke, K., Stollte, H. and Baylan, J. W. (1971), Renal handling of protein in the hagfish, *Myxine glutinosa*: *Bull. Mt. Desert Isl. Biol. Lab.*, **11**, 11–15.

Emdin, S. O., Gammeltoff, S. and Gliemann, J. (1977), Degradation, receptor binding affinity and potency of insulin from the Atlantic hagfish (*Myxine glutinosa*) determined in isolated rat fat cells. *J. biol. Chem.*,**252**, 602–608.

Enequist, P. (1937), Das Bachneunauge als ökologische Modifikation des Flüssneunauges. Uber die Flüss und Bachneunaugen Schwedens. *Ark. Zool.*, **29**, 1–22.

Epple, A. and Lewis, T. L. (1973), Comparative histophysiology of the pancreatic islets. *Am. Zool.*, **13**, 567–590.

Ericsson, J. L. E. and Seljelid, R. (1968), Endocytosis in the ureteric duct epithelium of the hagfish (*Myxine glutinosa* L.) *Z. Zellforsch*, **90**, 263–272.

Ermisch, A. (1966), Beiträge zur Histologie und Topochemie des Inselsystems der Neunaugen unter natürlichen und experimentellen Bedingungen. *Zool. Jb. (Anat.).*, **83**, 52–106.

Etkin, W. and Goa, A. G. (1974), Evolution of thyroid function in poikilothermous vertebrates. In: *Handbook of Physiology*. Endocrinology, (Greer, M. A. and Solomon, S. J., eds.) Vol. 3, pp. 5–20, Am. Physiol. Soc., Washington.

Evenett, P. J. and Dodd, J. M. (1963), Endocrinology of reproduction in the river lamprey. *Nature*, **197**, 715–716.

Falkmer, S. (1972), Insulin production in vertebrates and invertebrates. *Gen. comp. Endocrinol.*, Suppl. 3, 184–191.

Falkmer, S., Cutfield, J. F., Cutfield, S. M., Dodson, G. G., Glieman, J., Gammeltoft, S., Marques, M., Peterson, J. D., Steiner, D. F., Emdin, S. O., Havu, N., Östberg, Y. and Winbladh, L. (1975). Comparative endocrinology of insulin and glucagon production. *Am. Zool.*, **15**, Suppl. 1, 255–270.

Falkmer, S., Emden, S., Havu, N., Lundgren, G., Marques, M., Östberg, Y., Steiner, D. F. and Thomas, N. W. (1973), Insulin in invertebrates and cyclostomes. *Am. Zool.*, **13**, 625–638.

Falkmer, S., Emdin, S. O., Östberg, Y., Mattison, A., Johanssen, M-L., Sjöbeck, and Fänge, R. (1977), Tumour pathology of the hagfish, *Myxine glutinosa* and the river lamprey, *Lampetra fluviatilis*. *Prog. exp. Tumour Res.*, **20**, 217–250.

Falkmer, S. and Matty, A. J. (1966), Blood sugar regulation in the hagfish, *Myxine glutinosa*. *Gen. comp. Endocrinol.*, **6**, 334–346.

Falkmer, S. and Patent, G. J. (1972), Comparative and embryological aspects of the pancreatic islets. In: *Handbook of Physiology*. Endocrinology, (Steiner, D. F. and Fraenkel, N., eds.) Vol. 1, pp. 1–28, Am. Physiol. Soc., Washington.

Falkmer, S., Thomas, N. W. and Boquist, L. (1974), Endocrinology of the Cyclostomata. In: *Chemical Zoology*. (Florkin, M. and Scheer, B. Y., eds.) Vol. 8, pp. 195–257, Academic Press, London and New York.

Falkmer, S. and Wilson, S. (1967), Comparative aspects of the immunology and biology of insulin. *Diabetologia*, **3**, 519–528.

Falkmer, S. and Winbladh, L. (1964), An investigation of the pancreatic islet tissue of the hagfish (*Myxine glutinosa*) by light and electron microscopy. In: *The Structure and Metabolism of the Pancreatic Islets*. (Brolin, S. E. Hellman, B. and Knutson, H. eds.). pp. 17–22, Pergamon Press, Oxford.

Fänge, R. (1972), The circulatory system. In: *The Biology of Lampreys* (Hardisty, M. W. and Potter, I. C., eds.) Vol. 2, pp. 241–259, Academic Press, London.

Fänge, R. (1973), The lymphatic system of *Myxine*. In: *Myxine glutinosa*. (Fänge, R., ed.) pp. 57–64, Acta Reg. Soc. Sci. Litt., Gothenburg.

Fänge, R. and Edström, A. (1973), Incorporation of thymidine, uridine and leucine into blood cells of *Myxine glutinosa*. In: *Myxine glutinosa*. (Fänge, R., ed.) pp. 49–52, Acta Reg. Soc. Sci. Litt., Gothenburg.

Fänge, R., Larsson, A. and Lidman, U. (1973), Liver function and bile composition in *Myxine*. In: *Myxine glutinosa*. (Fänge, R., ed.) pp. 88–92, Acta Reg. Soc. Scient. Litt., Gothenburg.

Farmer, G. J. and Beamish, F. W. H. (1969), Oxygen consumption of *Tilapia nilotica* in relation to swimming speed and salinity. *J. Fish. Res. Bd. Can.*, **26**, 2807–2821.

Farmer, G. J., Beamish, F. W. H. and Robinson, G. A. (1975), Food consumption of the adult landlocked sea lamprey, *Petromyzon marinus* L. *Comp. Biochem. Physiol.*, **50A**, 753–757.

Fernholm, B. (1969), A third embryo of *Myxine*. Considerations on hypophysial ontogeny and phylogeny. *Acta. zool. (Stockh.).*, **50**, 169–177.

Fernholm, B. (1972a), Ultrastructure of the adenohypophysis of *Myxine glutinosa*. *Z. Zellforsch. mikrosk. Anat.*, **132**, 451–472.

Fernholm, B. (1972b), Is there any steroid formation in the ovary of the hagfish, *Myxine glutinosa*? *Acta zool. (Stockh.)*, **53**, 235–242.

Fernholm, B. (1974), Diurnal variations in the behaviour of the hagfish, *Eptatretus burgeri*. *Mar. Biol.*, **27**, 351–366.

Fernholm, B. (1975), Ovulation and eggs of the hagfish, *Eptatretus burgeri*. *Acta zool. (Stockh)*, **56**, 199–204.

Fernholm, B. and Holmberg, K. (1974), The eyes in three genera of hagfish (*Eptatretus, Paramyxine* and *Myxine* – a case of degenerative evolution. *Vision Res.*, **15**, 253–259.

Fernholm, B. and Olsson, R. (1969), A cytopharmacological study of the *Myxine* adenohypophysis. *Gen. comp. Endocrinol.*, **13**, 336–356.

Finstad, J., Fänge, R. and Good, R. A. (1969), The development of lymphoid systems, immune response and radiation sensitivity in lower vertebrates. In: *Lymphatic Tissue and Germinal Centers in Immune Response*. 21–31. (Fiore-Donati, L. and Hanna, M. G., eds.) pp. 21–31, Plenum Press, New York.

Finstad, J. and Good, R. A. (1964), Evolution of the immune response. III. Immunologic responses of the lamprey. *J. exp. Med.*, **120**, 1151–1168.

Finstad, J. and Good, R. A. (1966), Phylogenetic studies of adaptive immune responses in the lower vertebrates. In: *Phylogeny of Immunity*. (Smith, R. T., Miescher, P. A. and Good, R. A., eds.) p. 173, Univ. Florida Press, Gainsville.

Fischer, G. and Albert, W. (1971), Ein biologisch wirksames Peptid aus der Haut von Neunaugen. *Naturwissenschaften*, **58**, 363.

Flood, P. R. (1969), Fine structure of notochord in *Myxine glutinosa*. *J. Ultrastruct. Res.*, **29**, 573–574.

Flood, P. R. (1973), The skeletal muscle fibre types of *Myxine glutinosa* L. related to those of other chordates. *Acta Reg. Soc. Sci. Litt. Gothenburg* (Zool)., **8**, 17–20.

Florkin, M. (1966), *A Molecular Approach to Phylogeny*. Elsevier, Amsterdam.

Florkin, M. and Duchâteau, G. (1943), Les formes du système enzymatique de l'uricolyse et l'évolution du catabolisme purique chez les animaux.

Arch. int. Physiol., **52**, 267–307.

Follenius, E. (1964), Structure fine des cellules interstitielles du cyclostome, *Lampetra planeri*. *C. R. Séanc. Acad. Sci. Paris*, **259**, 450–452.

Follett, B. K. and Heller, H. (1964), The neurohypophysial hormones of bony fishes and cyclostomes. *J. Physiol (Lond.)*, **172**, 74–91.

Fominykh, M. A. (1970), Morphological differentiation of the neuromuscular system of *Lampetra fluviatilis*. *Neuroscience Transl.*, **16**, 92–98.

Fontaine, M. (1930), Recherches sur le milieu interieur de la lamproie marine, (*Petromyzon marinus*). Ses variations en fonction de celles du milieu extérieur. *C. R. Séanc. Acad. Sci. Paris*, **191**, 736–737.

Fontaine, M. (1938), La lamproie marine. Sa pêche et son importance économique. *Bull. Soc. Océanogr. Fr.*, **17**, 1681–1687.

Fontaine, M. and Leloup, J. (1950), L'iodémie d'un cyclostome marin (*Petromyzon marinus*) au moment de sa migration réproductrice. *C. R. hebd. Séanc. Acad. Sci. Paris*, **230**, 1538–1539.

Fontaine, J., Le Lievre, C. and Le Douarin, N. M. (1977), What is the developmental fate of the neural crest cells which migrate into the pancreas in the avian embryo? *Gen. comp. Endocrinol.*, **33**, 394–404.

Foss, G. (1963), *Myxine* in its natural surroundings. In: *The Biology of Myxine*. (Brodal, A. and Fänge, R. eds) pp. 42–49, Universitetsforlaget, Oslo.

Foss, G. (1968), Behaviour of *Myxine glutinosa* L. in natural habitat. Investigation of the mud biotope by a suction technique. *Sarsia*, **31**, 1–13.

Fujita, H. (1975), X-ray microanalysis on the thyroid follicle of the hagfish, *Eptatretus burgeri* and the lamprey, *Lampetra japonica*. *Histochemistry*, **43**, 283–290.

Fujita, H. and Honma, Y. (1966), Electron microscopical studies on the thyroid of a cyclostome, *Lampetra japonica* during its upstream migration. *Z. Zellforsch. mikrosk. Anat.*, **73**, 559–575.

Fujita, H. and Honma, Y. (1968), Some observations on the fine structure of the endostyle of larval lampreys, ammocoetes of *Lampetra japonica*. *Gen. comp. Endocrinol.* **11**, 111–131.

Fujita, H. and Honma, Y. (1969), Iodine metabolism of the endostyle of larval lampreys, ammocoetes of *Lampetra japonica*. *Z. Zellforsch. mikrosk. Anat.*, **98**, 527–537.

Fujita, T. and Kobayashi, S. (1974), The cells and hormones of the GEP endocrine system. The current studies. In: *Gastro-entero-pancreatic endocrine system. A cell-biological approach.* (Fujita, T., ed.) pp. 1–16, G. Thieme, Stuttgart.

Fujita, H. and Shinkawa, Y. (1975), Electron microscopic studies on the thyroid gland of the hagfish, *Eptatretus burgeri*. *Arch. histol. jap.*, **37**, 277–289.

Gaskell, W. H. (1908), *Origin of the Vertebrates*. Longmans Green, London.

Gauthier, F. G., Lowey, S. and Hobbs, A. W. (1978), Fast and slow myosins in developing muscle fibres. *Nature*, **274**, 25–29.

George, J. C. and Beamish, F. W. H. (1974), Haemocytology of the supraneural myeloid body in the sea lamprey during several phases of the life cycle. *Can. J. Zool.*, **52**, 1585–1589.

Goncharevskaya, O. A. (1975), Microdissection study of the nephron of the lamprey. *Zh. Evol. Biokhim. Fiziol.*, **11**, 88–90.

Good, R. A., Finstad, J. and Litman, G. W. (1972), Immunology. In: *The Biology of Lampreys* (Hardisty, M. W. and Potter, I. C., eds.) Vol. 2, pp. 405–432, Academic Press, London.

Good, R. A. and Finstad, J. (1969), Essential relationship between the lymphoid system immunity and malignancy. *Natn. Cancer Inst. Monogr.*, **31**, 41–58.

Goodman, M., Moore, G. W. and Matsuda, G. (1975), Darwinian evolution in the genealogy of haemoglobin. *Nature*, **253**, 603–608.

Gorbman, A. (1959), *Comparative Endocrinology*. Wiley, New York.

Gorbman, A. (1963), The myxinoid thyroid gland. In: *The Biology of Myxine*. (Brodal, A. and Fänge, R., eds.) pp. 477–480, Universitetsforlaget, Oslo.

Gorbman, A. (1965), Vascular relation between the neurohypophysis and adenohypophysis of cyclostomes and the problem of hypothalamic neuroendocrine control. *Archs. Anat. microsc. Morph. exp.*, **54**, 163–194.

Gorbman, A. and Bern, H. A. (1962), *A Textbook of Comparative Endocrinology*. Wiley, New York.

Gorbman, A. and Tsuneki, K. (1975), A technique for hypophysectomy of the Pacific hagfish; first observations. *Gen. comp. Endocrinol.*, **26**, 420–422.

Gorbunova, M. P. (1975), [Study of Endostyle of lamprey larvae associated with the problem of thyroid gland evolution] *Zh. Obshch. Biol.*, **36**, 173–188. [In Russian with English summary]

Gould, S. J. (1977), *Ontogeny and Phylogeny*. Harvard Univ. Press, Cambridge, Mass. and London.

Gradwell, N. (1972), Hydrostatic pressures and movements of the lamprey, *Petromyzon marinus* during suction, olfaction and gill ventilation. *Can. J. Zool.*, **50**, 1215–1223.

Grassle, J. F. and Sanders, H. L. (1973), Life histories and the role of disturbance. *Deep Sea Res.*, **20**, 643–659.

Grigg, G. C. (1974), Respiratory function of blood in fishes. In: *Chemical Zoology*. (Florkin, M. and Scheer, B. Y., eds.) Vol. 8, pp. 331–368, Academic Press, London and New York.

Grossu, A., Homei, V., Barbu, P. and Popescu, A. (1962), Contribution à l'étude des Petromyzonides de la République Populaire Roumaine. *Trav. Mus. Hist. nat. Gr. Antipa*, **3**, 253–279.

Hallam, A. (1977), Secular changes in marine inundation of USSR and North America through the Phanerozoic. *Nature*, **269**, 769–772.

Halstead, L. B. (1971), The presence of a spiracle in the Heterostraci (Agnatha). *J. Linn. Soc.*, **50**, 195–197

Halstead, L. B. (1973a), Affinities of the Heterostraci (Agnatha). *J. Linn. Soc.*, **5**, 339–349.

Halstead, L. B. T. (1973b), The heterostracan fishes. *Biol. Rev.*, **48**, 279–332.

Hansen, E. D. (1977), *The Origin and Early Evolution of Animals*. Wesleyan Univ. Press, Connecticut.

Hansen, S. J. and Youson, J. H. (1978), Morphology of the epithelium in the alimentary tract of the larval lamprey. *J. Morph.*, 155, 193–218.

Hardisty, M. W. (1956), Some aspects of osmotic regulation in lampreys. *J. exp. Biol.*, **33**, 431–447.

Hardisty, M. W. (1957), Osmotic conditions during the embryonic and early larval life of the brook lamprey (*Lampetra planeri*). *J. exp. Biol.*, **34**, 237–252.

Hardisty, M. W. (1961), Studies on an isolated spawning population of the brook lamprey (*Lampetra planeri*) *J. Anim. Ecol.*, **30**, 339–355.

Hardisty, M. W. (1963), Fecundity and speciation in lampreys. *Evolution*, Lancester, Pa., **17**, 17–22.

Hardisty, M. W. (1964), The fecundity of lampreys. *Arch. Hydrobiol.*, **60**, 340–357.

Hardisty, M. W. (1965a), Sex differentiation and gonadogenesis in lampreys. 1. The ammocoete gonads of the brook lamprey, *Lampetra planeri*. *J. Zool. (Lond.)*, **146**, 305–345.

Hardisty, M. W. (1965b), Sex differentiation and gonadogenesis in lampreys. 11. The ammocoete gonads of the landlocked sea lamprey, *Petromyzon marinus*. *J. Zool. (Lond.)*, **146**, 346–387.

Hardisty, M. W. (1971), Gonadogenesis, sex differentiation and gametogenesis. In: *The Biology of Lampreys*. (Hardisty, M. W. and Potter, I. C., eds.) Vol. 1, pp. 295–359, Academic Press, London.

Hardisty, M. W. (1972a), Quantitative and experimental studies on the interrenal tissues of the upstream migrant stage of the river lamprey, *Lampetra fluviatilis* L. *Gen. comp. Endocrinol.*, **18**, 501–514.

Hardisty, M. W. (1972b), The Interrenals. In: *The Biology of Lampreys*. (Hardisty, M. W. and Potter, I. C., eds.) Vol. 2, pp. 171–192, Academic Press, London.

Hardisty, M. W. (1976), Cysts and neoplastic lesions in the endocrine pancreas of the lamprey. *J. Zool. (Lond.)*, **178**, 305–317.

Hardisty, M. W. and Baines, M. E. (1971), The ultrastructure of the interrenal tissue of the lamprey. *Experientia*, **27**, 1072–1075.

Hardisty, M. W. and Barnes, K. (1968), Steroid 3β-dehydrogenase activity in the cyclostome gonad. *Nature*, **218**, 880.

Hardisty, M. W. and Huggins, R. W. (1970), Larval growth in the river lamprey, *Lampetra fluviatilis*. *J. Zool. (Lond.)*, **161**, 549–559.

Hardisty, M. W. and Potter, I. C. (1971a), The behaviour, ecology and growth of larval lampreys. In: *The Biology of Lampreys*. (Hardisty, M. W. and Potter, I. C., eds.) Vol. 1, pp. 85–125, Academic Press, London.

Hardisty, M. W. and Potter, I. C. (eds.) (1971b), The general biology of adult lampreys. In: *The Biology of Lampreys*. Vol. 1, pp. 127–206, Academic Press, London.

Hardisty, M. W. and Potter, I. C. (eds.) (1971c). Paired species. In: *The Biology of Lampreys*. Vol. 1, pp. 249–277, Academic Press, London.

Hardisty, M. W. and Potter, I. C. (1977), Diets and culture of lampreys. In: *Handbook of Nutrition and Food*. (Rechcigl, M., ed.) Vol. 2, pp. 261–270, CRC Press, Cleveland.

Hardisty, M. W., Rothwell, B. R. and Steele, K. (1967), The interstitial tissue of the testis of the river lamprey, *Lampetra fluviatilis*. *J. Zool. (Lond.)*, **152**, 9–18.

Hardisty, M. W., Zelnik, P. R. and Moore, I. A. (1975), The effects of subtotal and total isletectomy in the river lamprey, *Lampetra fluviatilis*. *Gen. comp. Endocrinol.*, **27**, 179–192.

Hardisty, M. W., Zelnik, P. R. and Wright, V. C. (1976), The effects of hypoxia on blood sugar levels and on the endocrine pancreas, interrenal and chromaffin tissues of the lamprey, *Lampetra fluviatilis* (L.). *Gen. comp. Endocrinol.*, **28**, 184–204.

Harris, C. E. and Teller, D. D. (1973), Estimation of primary sequence homology from amino acid composition of evolutionary related proteins. *J. Theor. Biol.* **38**, 347–362.

Haslewood, G. A. D. (1968), Bile salt differences in relation to taxonomy and systematics. In: *Chemotaxonomy and Serotaxonomy*. (Hawkes, J. G., ed.) pp. 159–172, Academic Press, London.

Healey, J. G. (1972), The central nervous system. In: *The Biology of Lampreys*. (Hardisty, M. W. and Potter, I. C., eds.) Vol. 2, pp. 307–372, Academic Press, London.

Heath-Eves, M. J. and McMillan, D. B. (1974), The morphology of the kidney of the Atlantic hagfish, *Myxine glutinosa* (L.). *Am. J. Anat.*, **139**, 309–334.

Heintz, A. (1958), The head of the anaspid *Birkenia elegans*, Traq. In: *Studies on Fossil Vertebrates*. (Westoll, S. T., ed.) pp. 71–85, Athlone Press, London.

Heintz, A. (1963), Phylogenetic aspects of myxinoids. In: *The Biology of Myxine*. (Brodal, A. and Fänge, R., eds.) pp. 9–21, Universitetsforlaget, Oslo.

Hesse, R. (1921), Das Herzgewicht der Wirbeltiere. *Zool. Jb. Abt. Allg. Zool. Physiol.*, **38**, 243–364.

Henderson, N. E. (1972), Ultrastructure of the neurohypophysial lobe of the hagfish, *Eptatretus stouti* (Cyclostomata). *Acta zool. (Stockh.)*, **53**, 243–266.

Henderson, N. E. (1975), Thyroxine concentrations in plasma of normal and hypophysectomised hagfish, *Myxine glutinosa* (Cyclostomata). *Can. J. Zool.*, **54**, 180–184.

Henderson, N. E. and Gorbman, A. (1971), Fine structure of the thyroid follicle of the Pacfic hagfish, *Eptatretus stouti*. *Gen. comp. Endocrinol.*, **16**, 409–429.

Henderson, N. E. and Lorschneider, F. L. (1975), Thyroxine and protein-bound iodine concentrations in plasma of the Pacific hagfish, *Eptatretus stouti* (Cyclostomata). *Comp. Biochem. Physiol.* **51A**, 723–726.

Hendrickson, W. A. and Love, W. E. (1971), Structure of lamprey haemo-globin. *Nature New Biol.*, **232**, 197–203.

Hennig, W. (1966), *Phylogenetic Systematics*. Univ. Illinois Press, Urbana.

Higashi, H., Hirao, S., Yamada, J. and Kikuchi, R. (1958), Vitamin content in the lamprey, *Entosphenus japonicus* Martens. *J. Vitamin.* **4**, 88–99.

Hildemann, W. H. (1970), Transplantation immunity in fishes, agnatha, chondrichthyes, and osteichthyes. *Transplantation Proc.*, **2**, 253–259.

Hildemann, W. H. and Thoenes, G. H. (1969), Immunological responses of the Pacific hagfish. 1. Transplantation immunity. *Transplantation*, **7**, 506–521.

Hirose, K., Tamaski, B., Fernholm, B. and Kobayashi, H. (1975), *In vitro* bioconversions of steroids in the mature ovary of the hagfish, *Eptatretus burgeri*. *Comp. Biochem. Physiol.*, **51B**, 403–408.

Hoheisel, G. (1969), Untersuchungen zur funktionellen Morphologie des Endostyls und der Thyroidea von Bachneunauge (*Lampetra planeri* Bloch). I. Untersuchungen am Endostyl. *Gegenbaurs Morph. Jb.*, **114**, 204–307.

Hoheisel, G. (1970), Untersuchungen zur funktionellen Morphologie des Endostyls und der Thyroidea von Bachneunauge (*Lampetra planeri* Bloch). II. Untersuchungen an der Thyroidea. *Gegenbaurs Morph. Jb.*, **114**, 337–358.

Hoheisel, G. and Sterba, G. (1963), Über die Wirkung von Kalium perchlorate ($KClO_4$) auf Ammocoeten von *Lampetra planeri* (Bloch). *Z. mikrosk. anat. Forsch.*, **70**, 490–516.

Holmberg, K. (1968), Ultrastructure and response to background illumi-nation of the melanophores of the hagfish, *Myxine glutinosa* L. *Gen. comp. Endocrinol.* **10**, 421–428.

Holmberg, K. (1971), On the light sensitivity of the Atlantic hagfish, *Myxine glutinosa*, maintained against a white background under continuous illumination. *Gen. comp. Endocrinol.* **17**, 232–239.

Holmberg, K. (1972), On the possibility of melanophore-dispersing activity in the adenohypophysis of the Atlantic hagfish, *Myxine glutinosa*. *Acta zool. (Stockh.)*, **53**, 173–178.

Holmberg, K. and Ohman, P. (1976), Fine structure of retinal synaptic organelles in lamprey and hagfish photoreceptors. *Vision Res.*, **16**, 237–239.

Holmes, R. L. and Ball, J. N. (1974), *The Pituitary Gland: a Comparative Account.* Cambridge Univ. Press, Cambridge.

Holmgren, N. (1946), On two embryos of *Myxine*. *Acta zool. (Stockh.)*, **27**, 1–90.

Homma, S. (1975), Velar motoneurones of lamprey larvae. *J. comp. Physiol.* **104**, 175–183.

Hornsey, D. J. (1977), Triiodothyronine and thyroxine levels in the thyroid and serum of the sea lamprey, *Petromyzon marinus* L. *Gen. comp. Endocrinol.*, **31**, 381–383.

Howell, W. M. and Denton, T. E. (1969), Chromosomes of ammocoetes of the Ohio brook lamprey, *Lampetra aepyptera*. *Copeia*, **1969**, 393–395.

Howell, W. M. and Duckett, C. R. (1971), Somatic chromosomes of the lamprey, *Ichthyomyzon gagei*: Agnatha, Petromyzonidae. *Experientia*, **27**, 22–23.

Hoyt, J. W. (1975), Hydrodynamic drag reduction due to fish slimes. In: *Swimming and Flying in Nature.* (Wu, T. Y. M., Brokaw, C. J. and Brennan, C., eds.) Vol. 2, pp. 653–672, Plenum Press, London and New York.

Hubbs, C. L. and Potter, I. C. (1971), Distribution, phylogeny and taxonomy. In: *The Biology of Lampreys.* (Hardisty, M. W. and Potter, I. C., eds.) Vol. 1, pp. 1–65, Academic Press, London.

Hubbs, C. L. and Trautman, M. B. (1937), A revision of the lamprey genus *Ichthyomyzon*. *Misc. Publs. Univ. Mich. Mus. Zool.*, **35**, 1–109.

Hughes, G. M. (1972), Morphometrics of fish gills. *Resp. Physiol.*, **14**, 1–25.

Hughes, G. M. and Morgan, M. (1973), The structure of fish gills in relation to their respiratory function. *Biol. Rev.*, **48**, 419–475.

Idler, D. R. and Burton, M. D. M. (1976), The pronephroi as the site of presumptive interrenal cells in the hagfish, *Myxine glutinosa* L. *Comp. Biochem. Physiol.*, **53A**, 73–77.

Idler, D. R., Sangalong, G. B. and Weisbart, M. (1971), Are corticosteroids present in the blood of all fish? In: *Proc. 3rd. Int. Congr. Horm. Steroid.* Hamburg, 1970. (James, V. H. T. and Martin, L., eds.) Excerpta Medica Found, Amsterdam.

Idler, D. R. and Truscott, B. (1972), In: *Steroids in non-mammalian vertebrates.* (Idler, D. R., ed.) pp. 127–252, Academic Press, New York.

Inui, Y. and Gorbman, A. (1977), Sensitivity of the Pacific hagfish, *Eptatretus stouti* to mammalian insulin. *Gen. comp. Endocrinol.*, **33**, 423–427.

Inui, Y. and Gorbman, A. (1978), Role of the liver in regulation of carbohydrate metabolism in hagfish, *Eptatretus stouti. Comp. Biochem. Physiol.*, **60A**, 181–183.

Inui, Y, Yu and Gorbman, A. (1978). Effect of bovine insulin on the incorporation of (^{14}C) Glycine into protein and carbohydrates in liver and muscle of hagfish, *Eptatretus stouti. Gen. Comp. Endocrinol.* **36**, 133–141.

Jackson, I. M. D. and Reichlin, S. (1974), Thyrotropin-releasing hormone (TRH): distribution in hypothalamic and extrahypothalamic brain tissue of mammalian and submammalian Chordates. *Endocrinol.*, **65**, 854–862.

Jacoby, O. H. and Hickman, C. P. (1966), A study of circulating iodocompounds of rainbow trout, *Salmo gairdneri* by the method of isotopic equilibrium. *Gen. comp. Endocrinol.*, **7**, 245–254.

Jansen, J. K. S. (1930), The brain of *Myxine glutinosa. J. comp. Neur.* **22**, 359–507.

Jansen, J. K. S. and Andersen, P. (1963), Anatomy and physiology of the skeletal muscles. In: *The Biology of Myxine*. (Brodal, A. and Fänge, R., eds.) pp. 111–194, Universitetsforlaget, Oslo.

Janvier, P. (1971), La position et la forme du sac nasal chez les *Osteostraci. C. R. Acad. Sci. Paris*, **272**, 2434–2436.

Janvier, P. (1974), The structure of the naso-hypophysial complex and the mouth in fossil and extant Cyclostomes with remarks on Amphiaspiformes. *Zool. Scripta*, **3**, 193–200.

Janvier, P. (1975), Les yeux des Cyclostomes fossiles et le problème de l'origine des Myxinoides. *Acta zool. (Stockh.)*, **56**, 1–9.

Janvier, P. (1976), Rémarques sur l'orifice naso-hypophysaire des *Cephalaspidomorphes. Annls. Palaéontol. vert.*, **61**, 3–16.

Janvier, P. (1977). Contribution à la connaissance de la systématique et de l'anatomie du genre *Boreaspis* Stensiö (Agnatha, Céphalaspidomorphi, Ostéostraci) du Dévonien inférieur du Spitsberg. *Annl. Palaéont.* **63**, 1–32.

Janvier, P. (1978). Les nageoires paires des Ostéostracés et la position systématique des Céphalaspidomorphes. *Annl. Palaéont.* **64**, 113–142.

Jarvik, E. (1964), Specialisations in early vertebrates. *Annals. Soc. r. Belg.*, **94**, 11–95:

Jarvik, E. (1965), Die Raspelzunge der Cyclostomen und die pentadactyl Extremität der Tetrapoden als Beweis fur monophyletische Herkunft. *Zool. Anz.*, **175**, 8–143.

Jarvik, E. (1968)., Vertebrate phylogeny. In: *Current Problems in Lower Vertebrate Phylogeny* (Orvig, T., ed.) Nobel Symposium **4**, pp. 497–527, Interscience, New York.

Jasper, D. (1967), Some structural features of the T-system and sarcolemma. *J. Cell Biol.*, **32**, 219–226.

Jensen, D. (1963), Eptatretin: a potent cardioactive agent from the branchial heart of the Pacific hagfish, *Eptatretus stouti*. *Comp. Biochem. Physiol.*, **10**, 129–151.

Jensen, D. (1965), The aneural heart of the hagfish. *Ann. N. Y. Acad. Sci.*, **127**, 443–458.

Jensen, D. (1966), The hagfish. *Sci. Am.*, **214**, 82–90.

Jensen, D. (1969), Intrinsic cardiac rate regulation in the sea lamprey, *Petromyzon marinus* and rainbow trout, *Salmo gairdneri*. *Comp. Biochem. Physiol.*, **30**, 685–690.

Jesperson, A. (1975), Fine structure of spermiogenesis in Eastern Pacific species of hagfish (*Myxinidae*). *Acta zool. (Stockh.)*, **56**, 189–198.

Johansen, K. (1960), Circulation in the hagfish, *Myxine glutinosa L*. *Biol. Bull.*, **118**, 289–295.

Johansen, K. (1963), The cardiovascular system of *Myxine glutinosa L*. In: *The Biology of Myxine* (Brodal, A. and Fänge, R., eds.) pp. 289–316, Universitetsforlaget, Oslo.

Johansen, K., Lenfant, C. and Hansen, D. (1973), Gas exchange in the lamprey, *Entosphenus tridentata*. *Comp. Biochem. Physiol.*, **44A**, 107–119.

Johansen, K. and Strahan, R. (1963), The respiratory system of *Myxine glutinosa*. In: *The Biology of Myxine*. (Brodal, A. and Fänge, R., eds.) pp. 352–371, Universitetsforlaget, Oslo.

John, T. M., Thomas, E., George, J. C. & Beamish, F. W. H. (1977). Effect of vasotocin on plasma free fatty acid level in the migrating anadromous sea lamprey. *Arch. Int. Physiol. Biochim.* **85**, 865–870.

Johnels, A. G. (1949), On the development and morphology of the skeleton of the head of *Petromyzon*. *Acta zool. (Stockh.)*., **29**, 139–279.

Jollie, M. (1962), *Chordate Morphology*. Reinholt, New York.

Jollie, M. (1968), Some implications of the acceptance of a delamination principle. In: *Current problems of lower vertebrate phylogeny*. (Orvig, T., ed.) Nobel Symp. 4, pp. 89–107, Interscience, New York.

de Jong, W. W., Terwindt, E. C. and Groenewoud, G. (1977), Subunit composition of vertebraté α crystallins. *Comp. Biochem. Physiol.*, **55B**, 49–56.

Jørgensen, C. B. and Larsen, L. O. (1967), Neuroendocrine mechanisms in lower vertebrates. In: *Neuroendocrinology* (Martini, L. and Ganong, W. F., eds.) Vol. 2, pp. 485–528, Academic Press, New York.

Joss, J. M. P. (1973), The pineal complex, melatonin and color change in the lamprey, *Lampetra*. *Gen. comp. Endocrinol.*, **21**, 188–195.

Joss, J. M. P. (1977), Hydroxyindole-O-methyltransferase (HIOMT) activity and the uptake of ^{3}H- melatonin in the lamprey, *Geotria australis* Gray. *Gen. comp. Endocrinol.*, **31**, 270–275.

Kamer, J. C. van de and Schreurs, A. F. (1959)., The pituitary of the brook lamprey (*Lampetra planeri*) before during and after metamorphosis. *Z. Zellforsch*, **49**, 605–630.

Karamyan, A. I. (1972), Development of the structural and functional organisation of the palaeo-, archi- and neocortex in the phylogenesis of submammalian vertebrates. *J. Evol. Biochem. Physiol.*, **8**, 324–332.

Karamyan, A. I. (1975), The views of A. A. Ukhtomskii on the hierarchic organisation of the central nervous system and the modern achievements of evolutionary neurophysiology. *J. Evol. Biochem. Physiol.*, **11**, 218–224.

Karamyan, A. I., Veselkin, N. P., Belekhova, M. G. and Zagorulko, T. K. (1966), Electrophysiological characteristics of tectal and thalamo-cortical divisions of the visual system in lower vertebrates. *J. comp. Neurol.*, **127**, 559–576.

Karamyan, A. I., Zagorul'ko, T. M., Belekhova, M. G., Veselkin, N. P. and Kosareva, A. A. (1975), Corticalisation of two divisions of the visual system during vertebrate evolution. *J. Evol. Biochem. Physiol.*, **7**, 12–20.

Keibel, F. (1928), Beiträge zur Anatomie zur Entwicklungsgeschichte und zur Stammesgeschichte der Sehorgane der Cyklostomen. *Z. mikrosk-anat. Forsch.*, **12**, 391–456.

Kennedy, M. C. and Rubinson, K. (1977), Retinal projections in larval, transforming and adult sea lamprey, *Petromyzon marinus*. *J. comp. Neurol.*, **171**, 465–480.

Kerkoff, P. R., Boschwitz, D. and Gorman, A. (1973), The response of hagfish thyroid tissue to thyroid inhibitors and to mammalian thyroid stimulating hormone. *Gen. comp. Endocrinol.*, **23**, 231–240.

Kiaer, J. (1924), The Downtonian fauna of Norway. I. Anaspida. *Skr. norske Vidensk Akad. I. Mat. nat. Kl.*, **6**, 1–139.

Kille, R. A. (1960), Fertilisation of the lamprey egg. *Exp. Cell Res.*, **20**, 12–27.

Kirpichnikov, V. S. (1974), [Genetic mechanisms and the evolution of heterosis]. *Genetika.*, **10**, 165–179. [In Russian, English Summary].

Kleerekoper, H. (1972), The sense organs. In: *The Biology of Lampreys*. (Hardisty, M. W. and Potter, I. C., eds.). Vol. 2, pp. 373–404, Academic Press, London.

Knowles, F. G. W. (1939), The influence of anterior pituitary and testicular hormones on the sexual maturation of lampreys. *J. exp. Biol.*, **16**, 435–547.

Kobayashi, H. and Uemura, H. (1972), The neurohypophysis of the hagfish, *Eptatretus burgeri* (Girard). *Gen. comp. Endocrinol. Suppl.* **3**, 114–124.

Korneliussen, H. (1973), Monoaminergic innervation of slow non-twitch muscle fibres in the Atlantic hagfish (*Myxine glutinosa*). *Z. Zellforsch*, **140**, 425–432.

Korneliussen, H. and Nicolaysen, K. (1973), Ultrastructure of four types of striated muscle fibres in the Atlantic hagfish (*Myxine glutinosa*) *Z. Zellforsch.*, **143**, 273–290.

Korolewa, N. W. (1964)., [Water respiration of lamprey and survival in a moist atmosphere]. *Isv. vses. nauchno-issled. Inst. ozern. rechn. ryb. Khoz.*, **58**, 186–190. [In Russian.]

Kosareva, A. A., Veselkin, N. P. and Ermakova, T. V. (1977). Retinal projections in the lamprey, *Lampetra fluviatilis* as revealed by optic nerve transport of horseradish peroxidase. *Zh. Evol. Biokhim. Fiziol.* **13**, 405–406.

Kott, E. (1973), Epidermal pigmentation in the sea lamprey, *Petromyzon marinus* L. *Can. J. Zool.*, **51**, 101–104.

Krogh, A. (1919), The rate of diffusion of gases through animal tissues with some remarks on the coefficient of diffusion. *J. Physiol.*, **52**, 391–403.

Kuhlenbeck, H. (1968), *The Central Nervous System*. Karger, Basle.

Kuhn, K., Stollte, H. and Reale, E. (1975), The find structure of the kidney of the hagfish (*Myxine glutinosa* L.). *Cell Tiss. Res.*, **164**, 201–213.

Kux, Z. (1965), Beiträge zur Verbreitung der Neunaugen (Petromyzonidae) in der Tschechoslowakei. *Act. Mus. Mor.*, **54**, 203–222.

Kux, Z. and Steiner, H. M. (1971/72). *Lampetra lanceolata*, eine neue Neunaugenart aus dem Einzugebiet des schwarzen Meeres in der nordöstlichen Turkei. *Act. Mus. Mor.*, **36/37**, 375–384.

Lachiver, F., Fontaine, Y. and Martin, A. (1966), The iodination *in vivo* of thyroid iodoproteins in various vertebrates. In: *Current topics in Thyroid Research*. (Cassano, C. and Andreoli, M, eds.) pp. 182–192, Academic Press, New York.

Lanzing, W. J. R. (1954), The occurrence of a water balance, a melanophore expanding and an oxytocic principle in the pituitary gland of the river lamprey, (*Lampetra fluviatilis* L.). *Acta Endocrinol.* **16**, 277–284.

Lanzing, W. J. R. (1958), Structure and function of the suction apparatus of the lamprey. *Proc. K. ne. Akad. Wet.*, **61**, 300–307.

Lanzing, W. J. R. (1959), Studies on the river lamprey, *Lampetra fluviatilis* during the anadromous migration. *Uitgeversmaatschappij Neerlandia*, Utrecht. (Thesis).

Larsen, L. O. (1965), Effects of hypophysectomy in the cyclostome, *Lampetra fluviatilis* (L.) Gray. *Gen. comp. Endocrinol.*, **5**, 16–30.

Larsen, L. O. (1969a), Effect of hypophysectomy before and during sexual maturation in the cyclostome, *Lampetra fluviatilis* (L.) Gray. *Gen. comp. Endocrinol.*, **12**, 200–208.

Larsen, L. O. (1969b), Hypophysial function in river lampreys. *Gen. comp. Endocrinol.*, Suppl. 2, 522–527.

Larsen, L. O. (1970), The lamprey egg at ovulation (*Lampetra fluviatilis* L. Gray.) *Biol. Rep.*, **2**, 37–47.

Larsen, L. O. (1972), Endocrine control of intestinal atrophy in normal lampreys and of intestinal hypertrophy in gonadectomised lampreys, *Lampetra fluviatilis. Gen. comp. Endocrinol.*, **18**, 602 (Abstract).

Larsen, L. O. (1973), *Development in Adult Freshwater River Lampreys and its Hormonal Control*. Copenhagen: Trykteknik A/S.

Larsen, L. O. (1974), Effects of testosterone and oestradiol on gonadectomised and intact male and female river lampreys (*Lampetra fluviatilis* (L.) Gray.) *Gen. comp. Endocrinol.*, **24**, 305–313.

Larsen, L. O. (1976a), Blood glucose levels in intact and hypophysectomised river lampreys (*Lampetra fluviatilis* L.) treated with insulin, 'stress' or glucose before and after the period of sexual maturation. *Gen. comp. Endocrinol.*, **29**, 1–13.

Larsen, L. O. (1976b), Regulation of blood glucose in the river lamprey; the probable physiological role of insulin and hyperglycaemic hormone. In: *The Evolution of the Pancreatic Islets*. (Grillo, T. A., Leibson, L. and Epple, A., eds.) pp. 285–290, Pergamon Press, Oxford.

Larsen, L. O. (1978), Sub-total hepatectomy in intact or hypophysectomised river lampreys (*Lampetra fluviatilis* L.). Effects on regeneration, blood glucose regulation and vitellogenesis. *Gen. comp. Endocrinol.*, **35**, 197–204.

Larsen, L. O. and Rosenkilde, P. (1971), Iodine metabolism in normal, hypophysectomised and thyrotropin treated river lampreys, *Lampetra fluviatilis* L. Gray (Cyclostomata). *Gen. comp. Endocrinol.*, **17**, 94–104.

Larsen, L. O. and Rothwell, B. (1972), The adenohypophysis. In: *The Biology of Lampreys* (Hardisty, M. W. and Potter, I. C., eds.) Vol. 2, pp. 1–67, Academic Press, London.

Larssen, A. and Fänge, R. (1977), Cholesterol and free fatty acids (FFA) in the blood of marine fish. *Comp. Biochem. Physiol.*, **57B**, 191–196.

Larssen, A., Johansson-Sjöbeck, Maj-Lis and Fänge, R. (1976), Comparative study of some haematological and biochemical blood parameters in fishes from the Skagerrak. *J. Fish Biol.*, **9**, 425–440.

Larssen, L. I. and Rehfeld, J. F. (1977), Evidence for a common evolutionary origin of gastrin and cholecystokinin. *Nature*, **269**, 335–358.

Leach, W. J. (1944), The archetypal position of *Amphioxus* and ammocoetes and the role of endocrines in chordate evolution. *Am. Nat.*, **58**, 341–357.

Leach, W. J. (1946), Oxygen consumption of lampreys with special reference to metamorphosis and phylogenetic position. *Physiol. Zool.*, **19**, 365–374.

Leach, W. J. (1951), The hypophysis of lampreys in relation to the nasal apparatus. *J. Morph.*, **89**, 217–246.

Leatherland, J. F. (1975). Structure and fine structure in the pars distalis in Cyclostome, Holostean and Teleostean representatives. *Gen. comp. Endocrinol.*, **26**, 2–15.

Lee, R. F. and Puppione, D. P. (1974), Serum lipoproteins of hagfish (*Eptatretus deani*). *J. Am. Oil Chem. Soc.*, **51**, 174.

Leibson, L. C. and Plisetskaya, E. M. (1969), Hormonal control of blood sugar levels in Cyclostomes. *Gen. comp. Endocrinol.*, Suppl. 2, 528–534.

Leibson, L., Plisetskaya, E. and Leibush, B. (1976), The comparative study of mechanisms of insulin action on muscle carbohydrate metabolism. In: *The Evolution of Pancreatic Islets.* (Grillo, T. A. I., Leibson, L. and Epple, A., eds.) pp. 345–362, Pergamon Press, Oxford.

Leloup, J. and Hardy, A. (1976), Hormones thyroidiennes circulantes chez Cyclostome et des Poissons. *Gen. comp. Endocrinol.*, **29**, 258. (Abstr.).

Lessertisseur, J. and Robineau, D. (1970), Le mode d'alimentation des premiers vertébrés et l'origine des machoires. II. Les corrélations et les conséquences. *Bull. Mus. Hist. nat.*, **42**, 102–121.

Lethbridge, R. C. and Potter, I. C. (in press). The oral fimbriae of the lamprey, *Geotria australis*. *J. Zool. (Lond.)*.

Lewis, S. V. (1976), Respiration and gill morphology of the paired species of lampreys, *Lampetra fluviatilis* (L.) and *Lampetra planeri* (Bloch). Ph. D. Thesis, Univ. Bath.

Lewis, S. V. and Potter, I. C. (1976), Gill morphometrics of the lampreys, *Lampetra fluviatilis* (L.) and *Lampetra planeri* (Bloch). *Acta zool. (Stockh.)*, **57**, 103–112

Lewis, S. V. and Potter, I. C. (1977), Oxygen consumption during the metamorphosis of the parasitic lamprey, *Lampetra fluviatilis* (L.) and its non-parasitic derivative, *Lampetra planeri* (Bloch). *J. exp. Biol.*, **69**, 187–198.

Li, S. L. and Riggs, A. (1970), The amino acid sequence of haemoglobin V from the lamprey, *Petromyzon marinus*. *J. biol. Chem.*, **245**, 6149–6169.

Li, S. L. and Riggs, A. (1972), The partial sequence of the first 30 residues from the amino terminus of haemoglobin B in the hagfish, *Eptatretus stouti*; homology with lamprey haemoglobin. *J. Mol. Evol.*, **1**, 208–210.

Li, S. L., Tomita, S. and Riggs, A. (1972), The haemoglobin of the Pacific hagfish, *Eptatretus stouti*. 1. Isolation, characterisation and oxygen equilibria. *Biochim. biophys. Acta*, **278**, 344–354.

Lie, H. K. (1973), An intermediate muscle type in the river lamprey, *Lampetra fluviatilis* L. In: *Acta Reg. Soc. Sci. Litt. Gothenburg (Zool)* (Fänge, R., ed.), Vol. 8, pp. 21–22.

Lignon, J. (1975), Influence de la colchicine sur les éffets de l'acetylcholine et de pilocarpine sur le ventricule isolée de l'ammocoete de *Lampetra planeri* (Bloch). *C. R. Acad. Sci. Paris*, **280**, 181–184.

Lindström, Th. (1949), On the cranial nerves of cyclostomes with special reference to N. trigeminus. *Acta zool. (Stockh.)*, **30**, 315–458.

Linna, J. J., Finstad, J. and Good, R. A. (1975), Cell proliferation in epithelial and lympho-haemopoetic tissues of cyclostomes. *Am. Zool.*, **15**, 29–38.

Linthicum, D. S. and Hildemann, W. H. (1970), Immunologic responses of Pacific hagfish. III. Serum antibodies to cellular antigens. *J. Immunol.*, **105**, 912–918.

Litman, G. W., Frommel, D., Finstad, J. Howell, J., Pollara, B. W. and Good, R. A. (1970), The evolution of the immune response. VIII. Structural studies of the lamprey immunoglobulin. *J. Immunol.*, **105**, 1278–1285.

Litman, G. W., Rosenberg, A., Frommel, D., Pollara, B., Finstad, J. and Good, R. A. (1971), Biophysical studies of the immunoglobulins. The circular dichroic spectra of the immunoglobulins; a phylogenetic comparison. *Int. Archs. Allergy appl. Immun.*, **40**, 551–575.

Lofts, B. and Bern, H. A., (1972), The functional morphology of steroidogenic tissues. In: *Steroids in non-mammalian vertebrates*. (Idler, D. R., ed.) pp. 37–125, Academic Press, New York.

Løvtrup, S. (1977), *The Phylogeny of Vertebrata*. Wiley and Sons, New York.

Lowenstein, O. (1970), The electrophysiological study of the responses of the isolated laybrinth of the lamprey, (*Lampetra fluviatilis*) to angular acceleration, tilting and mechanical vibration. *Proc. R. Soc. Lond. Ser. B*, **174**, 419–434.

Lowenstein, O. (1973), The labyrinth in the making. *Adv. Oto-Rhino-Laryng.*, **19**, 31–34.

Lowenstein, O. and Thornhill, R. A. (1970), The labyrinth of *Myxine*. Anatomy, ultrastructure and electrophysiology. *Proc. R. Soc. Lond. Ser. B*, **176**, 21–42.

Lowe, D. R. and Beamish, F. W. H. (1973), Changes in the proximate body composition of the landlocked sea lamprey, *Petromyzon marinus* (L.) during larval life and metamorphosis. *J. Fish Biol.*, **5**, 673–682.

Lukomskaya, N. J. and Michelson, M. J. (1972), Pharmacology of the isolated heart of the lamprey, *Lampetra fluviatilis. Comp. gén. Pharmac.*, **3**, 213–225.

Luppa, H. (1964), Histologiche und topochemische Untersuchungen am Epithel des Oesophagus und Mitteldarmes von *Lampetra planeri*. *Z. mikrosk. anat. Forsch.*, **71**, 85–113.

Luppa, H. and Ermisch, A. (1967), Untersuch zur Struktur und Funktion des exokrinen Pankreas von Neunaugen. *Morph. Jb.*, **110**, 245–269.

Lutz, P. L. (1975a), Adaptive and evolutionary aspects of the ionic content of fishes. *Copeia*, **1975**, 369–373.

Lutz, P. L. (1975b), Osmotic and ionic composition of the blood plasma and muscle of the polypteroid, *Erpetoichthys calabaris* and a comparison with teleosts. *Copeia*, **1975**, 119–123.

Macallum, A. H. (1910), The inorganic composition of the blood in vertebrates and its origin, *Proc. R. Soc. Lond. Ser. B.*, **83**, 602–604.

Mangia, F. and Palladini, G. (1970), Recherches histochimiques sur le mucocartilage de la lamproie pendant son ontogenesis larvaire. *Arch. anat. micros. Morphol. exp.* **59**, 283–288.

Manski, W., Halbert, S. P., Auerbach-Pascal, T. and Javier, P. (1967a), On the use of antigenic relationships among species for the study of molecular evolution. *1.* The lens proteins of the Agnatha and Chondrichthyes. *Int. Arch. Allergy*, **31**, 38–56.

Manski, W., Halbert, S. P., Auerbach-Pascal, T. and Javier, P. (1967b), *3.* The lens proteins of the late Actinopterygii. *Int. Arch. Allergy*, **31**, 529–545.

Manwell, C. (1963), The blood proteins of Cyclostomes. A study in phylogenetic and ontogenetic biochemistry. In *The Biology of Myxine*. (Brodal, A. and Fänge, R., eds.) pp. 372–455, Universitetsforlaget, Oslo.

Marchalonis, J. J. (1977), *Immunity in Evolution*. Arnold, London.

Marchalonis, J. J. and Cone, R. E. (1973), The phylogenetic emergence of vertebrate immunity. *Aust. J. exp. Biol. Med. Sci.*, **51**, 461–488.

Marchalonis, J. J. and Edelman, G. M. (1968a), Isolation and characterisation of a natural haemagglutinin from *Limulus polyphemus*. *J. molec. Biol.* **32**, 453.

Marchalonis, J. J. and Edelman, G. M. (1968b), Phylogenetic origins of antibody structure. III. Antibodies in the primary immune response of the sea lamprey, *Petromyzon marinus. J. exp. Med.* **127**, 891–914.

Marchalonis, J. J. and Wellman, J. K. (1971), Relatedness among proteins. A new method of estimation and its application to immunoglobulins. *Comp. Biochem. Physiol.* **38B**, 609–625.

Marinelli, W. and Strenger, A. (1954), *Vergleichende Anatomie und Morphologie der Wirbeltiere*. Vol. 1. *Petromyzon fluviatilis*. Vol. 2. *Myxine glutinosa*. Deuticke, Wien.

Markert, C. L., Shaklee, J. B. and White, G. S. (1975), Evolution of a gene. *Science*, **189**, 102–114.

Martin, A. R., Wickelgren, W. O. and Beranek, R. (1970), Effects of iontophoretically applied drugs on spinal interneurones of the lamprey. *J. Physiol. (Lond.)*, **207**, 653–665.

Martin, A. R. and Wickelgren, W. O. (1971), Sensory cells in the spinal cord of the sea lamprey. *J. Physiol. (Lond.)* **212**, 65–83.

Martin, R. J. and Bowsher, D. (1977), An electrophysiological investigation of the projections of intramedullary primary afferent cells of the lamprey ammocoete. *Neurosci. Letters*, **5**, 39–44.

Mathers, J. S. and Beamish, F. W. H. (1974). Changes in serum osmotic and ionic concentrations in landlocked *Petromyzon marinus. Comp. Biochem. Physiol.*, **49A**, 677–688.

Mathews, M. B. (1967), Macromolecular evolution of connective tissue. *Biol. Rev.* **42**, 499–551.

Mathews, G. and Wickelgren, W. O. (1978). Trigeminal sensory neurons of the sea lamprey *J. comp. Physiol.* **123**, 329–333.

Matty, A. J. (1966), Endocrine glands in lower vertebrates. *Int. Rev. gen. exp. Zool.*, **2**, 43–137.

Matty, A. J. and Falkmer, S. (1965a), Hormonal control of carbohydrate metabolism in *Myxine glutinosa. Gen. comp. Endocrinol.* **5**, 701–

Matty, A. J. and Falkmer, S. (1965b), Hypophysectomy and blood sugar regulation in a cyclostome, *Myxine glutinosa. Nature*, **207**, 533–534.

Matty, S. and Gorbman, A. (1978), The effects of isletectomy and hypophysectomy on some blood plasma constituents of the hagfish, *E. stouti*. Abs. IXth. Conf. Eur. Comp. Endocrin. *Gen. comp. Endocrinol.*, **34**, 94.

Matty, A. J., Tsuneki, K., Dickhoff, W. W. and Gorbman, A. (1976), Thyroid and gonadal function in hypophysectomised hagfish, *Eptatretus stouti. Gen. comp. Endocrinol.* **30**, 500–516.

Mawdesley-Thomas, L. E. (1972), Some tumours of fish. *Symp. Zool. Soc. London*, **30**, 191–283.

Mazeaud, M. M. (1969), Adrenalinémie et noradrenalinémie chez la lamproie marine (*Petromyzon marinus* L.). *C. R. Séanc. Soc. Biol.* **163**, 349–352.

Mazeaud, M. M. (1972), Epinephrine biosynthesis in *Petromyzon marinus* (Cyclostomata) and *Salmo gairdneri* (Teleost). *Comp. gén. Pharmac.*, **3**, 457–468.

McFarland, W. N. and Munz, F. W. (1965), Regulation of body weight and serum composition of hagfish in various media. *Comp. Biochem. Physiol.*, **14**, 383–398.

McInerney, J. E. (1974), Renal sodium reabsorption in hagfish, *Eptatretus stouti. Comp. Biochem. Physiol.* **49A**, 273–280.

McInerney, J. E. and Evans, D. O. (1970), Habitat characteristics of the Pacific hagfish, *Polistotrema stouti. J. Fish. Res. Bd. Can.* **27**, 966–968.

McKeown, B. A. and Hazlett, C. A. (1975), The effect of salinity on pituitary, thyroid and interrenal cells in immature adults of the landlocked sea lamprey, *Petromyzon marinus. Comp. Biochem. Physiol.* **50A**, 379–381.

McLaughlin, P. J. and Dayhoff, M. O. (1972), Evolution of species and proteins: a time scale. In: *Atlas of Protein Sequence and Structure* (Dayhoff, M. O., ed.). pp. 42–52, Nat. Biomed. Res. Foundation, Washington, D. C.

Meiniel, A. (1971), Étude cytophysiologique de l'organe parapineal de *Lampetra planeri. J. neuro-visc. relationships.*, **32**, 157–199.

Meiniel, A. and Collins, J. P. (1971), Le complex pineal de l'ammocoete (*Lampetra planeri* Bl.) *Z. Zellforsch.* **117**, 354–380.

Metzger, H., Shapiro, M. B., Mosimann, J. E. and Vinton, J. E. (1968),

Assessment of compositional relatedness between proteins. *Nature*, **219**, 1166–1168.

Meyts de, P., Obberghen, E. V., Roth, J., Wollmer, A. and Brandenburg, D. (1978), Mapping of the residues responsible for the negative co-operativity of the receptor binding region of insulin. *Nature*, **273**, 504–509.

Molnár, B. and Szabo, S. (1968), Histological study of the hypophysis of the Transylvanian lamprey, (*Eudontomyzon danfordi* Regan). *Acta biol. hung.*, **19**, 373–379.

Molnár, B. and Szabo, S. (1975), Actiunea clorpromazinei asupra sistemuliu neurosecretor hipotalamo-hipofizar la *Eudontomyzon danfordi* Regan. *Acad. Rep. Pop. Rom. Fil. Cluj. Stud. Cerc. Biol.* **27**, 101–104.

Monaco, F. Andreoli, M., Cataudella, S. and Roche, J. (1976), Sur la biosynthèse de la thyroglobuline chez une lamproie adulte, *Lampetra planeri* (Bloch). *C. R. Séanc. Soc. Biol. Fil.*, **170**, 59–64.

Monaco, F., Andreoli, M., La Posta, A. and Roche, J. (1978), Thyroglobulin biosynthesis in a larval and adult freshwater lamprey (*Lampetra planeri* Bl.). *Comp. Biochem. Physiol.* **60B**, 87–91.

Moore, J. W. and Beamish, F. W. H. (1973), Food of larval sea lamprey (*Petromyzon marinus*) and American brook lamprey (*Lampetra lamottei*). *J. Fish. Res. Bd. Can.* **30**, 7–15.

Moore, J. W. and Potter, I. C. (1976a), Aspects of feeding and lipid deposition and utilisation in the lampreys, *Lampetra fluviatilis* (L.) and *Lampetra planeri* (Bloch). *J. Anim. Ecol.*, **45**, 699–712.

Moore, J. W. and Potter, I. C. (1976b), A laboratory study in the feeding of larvae of brook lampreys, *Lampetra planeri* (Bloch). *J. Anim. Ecol.* **45**, 81–90.

Moore, G. W., Goodman, M., Callahan, C., Holmquist, R. and Moise, H. (1976), Stochastic versus augmented maximum parsimony method for estimating superimposed mutations in the divergent evolution of protein sequences. Methods tested on Cytochrome *c* amino acid sequences. *J. Mol. Biol.*, **105**, 15–37.

Moriarty, R. J., Logan, A. G. & Rankin, J. C. (1978). Measurement of single nephron filtration rate in the kidney of the river lamprey, *Lampetra fluviatilis* L., *J. exp. Biol.*, **77**, 57–69.

Morita, Y. and Dodt, E. (1973), Slow photic responses of the isolated pineal organ of lamprey. *Nova Acta Leopoldina*, **38**, 331–339.

Morris, R. (1956), The osmoregulatory ability of the lampern (*Lampetra fluviatilis* L.) in sea water during the course of its spawning migration. *J. exp. Biol.*, **23**, 235–248.

Morris, R. (1957), Some aspects of the structure and cytology of the gills of *Lampetra fluviatilis* L. *Q. J. microsc. Sci.*, **98**, 473–485.

Morris, R. (1958), The mechanisms of marine osmoregulation in the lamprey

(*Lampetra fluviatilis* L.) and the causes of its breakdown during the spawning migration. *J. exp. Biol.* **35**, 649–665.

Morris, R. (1960), General problems of osmoregulation with special reference to Cyclostomes. *Symp. Zool. Soc. London*, **1**, 1–16.

Morris, R. (1965), Studies on salt and water balance in *Myxine glutinosa* (L.) *J. exp. Biol.*, **42**, 359–371.

Morris, R. (1972), Osmoregulation. In: *The Biology of Lampreys.* (Hardisty, M. W. and Potter, I. C., eds.) Vol. 2, pp. 193–239. Academic Press, London.

Morris, R. and Islam, D. S. (1969a), Histochemical studies on the follicles of Langerhans of the ammocoete larva of *Lampetra planeri* Bloch. *Gen. comp. Endocrinol.* **12**, 72–80.

Morris, R. and Islam, D. S. (1969b), The effects of hormones and hormone inhibitors on blood sugar regulation and the follicles of Langerhans in ammocoete larvae. *Gen. comp. Endocrinol.*, **12**, 81–90.

Morris, R. and Pickering, A. D. (1975), Ultrastructure of the presumed ion-transporting cells in the gills of ammocoete lampreys, *Lampetra fluviatilis* (L.) and *Lampetra planeri* (Bloch). *Cell Tiss. Res.* **163**, 327–341.

Moss, M. L. (1968), The origin of vertebrate calcified tissues. In: *Current Problems in Lower Vertebrate Phylogeny.* (Orvig, T., ed.). pp. 359–371, Interscience, New York.

Moy-Thomas, J. and Miles, R. S. (1971), *Palaeozoic Fishes.* Chapman and Hall, London.

Müller, A. (1856), On the development of the lampreys. *Ann. Mag. nat. Hist.* **18**, 298–301.

Munz, F. W. and McFarland, W. N. (1964), Regulatory function of a primitive vertebrate kidney. *Comp. Biochem. Physiol.* **13**, 381–400.

Munz, F. W. and Morris, R. W. (1965), Metabolic rate of the hagfish *Eptatretus stouti* (Lockington) 1878. *Comp. Biochem. Physiol.*, **16**, 1–5.

Murtaugh, P. A., Halver, J. E., Lewis, M. S. and Gladner, J. A. (1974), Cross-linking reactions of lamprey fibrinogen and fibrin. *Biochim. biophys. Acta*, **359**, 415–420.

Nakao, T. (1976), An electron microscopic study of the neuromuscular junction in the myotomes of the larval lamprey, *Lampetra japonica*. *J. comp. Neur*, **165**, 1–16.

Nansen, F. (1888), A protandric hermaphrodite (*Myxine glutinosa* L.) amongst the vertebrates. *Bergens Mus. Aarsber.*, **7**, 3–34.

Nesterov, V. P. (1972), Functional specificity of the distribution of Na and K in the muscles of cyclostomes and marine bony fish. *Dokl. Akad. Nauk. USSR.*, **206**, 1022–1024.

Newth, D. R. and Ross, D. M. (1955), on the reaction to light of *Myxine glutinosa* L. *J. exp. Biol.* **32**, 4–21.

Nicolaysen, K. (1966) On the functional properties of the fast and slow cranial muscles of the Atlantic hagfish. *Acta physiol. scand.* **68** (Suppl. 277), 142.

Nicoll, R. A. (1977), Excitatory action of TRH on spinal motoneurones. *Nature*, **265**, 242–243.

Nicoll, C. S. and Bern, H. A. (1968), Further analysis of the occurrence of pigeon crop sac-stimulating activity (prolactin) in the vertebrate adeno-hypophysis. *Gen. comp. Endocrinol.*, **11**, 5–20.

Nieuwenhuys, R. (1967a), Comparative anatomy of the cerebellum. *Prog. Brain Res.*, **25**, 1–93.

Nieuwenhuys, R. (1967b), Comparative study of olfactory centres and tracts. *Prog. Brain Res.* **23**, 1–64.

Nieuwenhuys, R. (1972), Topographical analysis of the brain stem of the lamprey, *Lampetra fluviatilis*. *J. comp. Neur.*, **145**, 165–178.

Nieuwenhuys, R. (1977). The brain of the lamprey in a comparative perspective. *Ann. N. Y. Acad. Sci.* **299**, 97–145.

Nikol'skii, G. V. (1956), [Some data on the period of marine life of the Pacific lamprey, *Lampetra japonica*]. *Zool. Zh.*, **35**, 585–591. [In Russian].

Nilsson, A. (1973), Secretin-like and cholecystokinin-like activity in *Myxine glutinosa* L. In: *Acta Reg. Soc. Sci. Litt. Gothenburg. Zool.* (Fänge, R., ed.). Vol. 8, pp. 30–32.

Nilsson, A. and Fänge, R. (1970), Digestive proteases in the cyclostome, *Myxine glutinosa* (L.). *Comp. Biochem. Physiol.*, **32**, 237–250.

Nogusa, S. (1960), A comparative study of the chromosomes in fishes with particular considerations on taxonomy and evolution. *Mem. Hyogo Univ. Agr.*, **3**, 1–62.

Nosaki, M., Fernholm, B. and Kobayashi, H. (1975), Ependymal absorption of peroxidase into the third ventricle of the hagfish, *Eptatretus burgeri* (Girard). *Acta zool. (Stockh.)*, **56**, 265–269.

Noorden, S. van., Greenberg, J. and Pearse, A. G. E. (1972), Cytochemical and immunofluorescence investigations on polypeptide hormone localis-ation in the pancreas and gut of the larval lamprey. *Gen. comp. Endocrinol.*, **19**, 192–199.

Noorden, S. van. and Pearse, A. G. E. (1974), Immunoreactive polypeptide hormones in the pancreas and gut of the lamprey. *Gen. comp. Endocrinol.*, **23**, 311–324.

Noorden, S. van. and Pearse, A. G. E. (1976), The localisation of immuno-reactivity to insulin, glucagon and gastrin in the gut of Amphioxus (*Branchiostoma lanceolatus*). In: *Evolution of the Pancreatic Islets*. (Grillo, T. A. I., Leibson, L. C. and Epple, A., eds.) Pergamon Press, Oxford.

Northcutt, R. G. and Przbylski, R. J. (1973), Retinal projections in the lamprey, *Petromyzon marinus* L. *Anat. Rec.*, **175**, 400.

Notochin, Yu V. (1977), [Filtration, resorption and secretion in the evolution of kidney function]. *Zh. Evol. Biokhim. Fiziol.*, **13**, 607–613. [In Russian with English summary].

Nursall, J. R. (1956), The lateral musculature and the swimming of fish. *Proc. Zool. Soc. (Lond.)*, **126**, 127–143.

Nygten, A. and Jahnke, M. (1972), Chromosomes of *Myxine glutinosa*. In: *Myxine glutinosa* (Fänge, R., ed) pp. 80–81, Acta Reg. Soc. Sci. Litt. Gothenburg. Zool. Vol. 8.

O'Boyle, R. N. and Beamish, F. W. H. (1977), Growth and intermediary metabolism of larval and metamorphosing stages of the landlocked sea lamprey, *Petromyzon marinus* L. *Env. Biol. Fish*, **42**, 219–235.

Ohno, S. (1970), *Evolution by Gene Amplification*. Springer Verlag, Heidelberg.

Ohno, S. (1973), Ancient linkage groups and frozen accidents. *Nature*, **244**, 259–262.

Ohno, S., Klein, J., Poole, J., Harris, C., Destree, A. and Morrison, M. (1967), Genetic control of lactate dehydrogenase formation in the hagfish, *Eptatretus stouti. Science*, **156**, 96–98.

Ohno, S. and Morrison, M. (1966), Multiple gene loci for the monomeric haemoglobin of the hagfish (*Eptatretus stouti*). *Science*, **154**, 1034–1035.

Ohno, S., Wolf, U. and Atkin, N. B. (1968), Evolution from fish to mammal by gene duplication. *Hereditas*, **59**, 169–187.

Okkelberg, P. (1921), The early history of the germ cells in the brook lamprey, *Entosphenus wilderi* (Gage) up to and including the period of sex differentiation. *J. Morph.*, **35**, 1–151.

Olsson, J. Y. (1973), Activity of delta aminolevulinic acid dehydrase (ALA-D) in organs of *Myxine glutinosa*. In: *Myxine glutinosa*. (Fänge, R., ed.) *Acta Reg. Soc. Sci. Litt. Gothenburg. Zool.* Vol. 8, pp. 84–85.

Olsson, R. (1959), The neurosecretory hypothalamus system and the adeno-hypophysis of *Myxine. Z. Zellforsch.*, **51**, 97–107.

Olsson, R. (1968), Evolutionary significance of the 'prolactin' cells in teleostomean fishes. In: *Current Problems in Lower Vertebrate Phylogeny.* (Orvig, T., ed.). Nobel Symp. Vol. 4. pp. 455–472, Interscience, New York.

Oordt, P. G. W. van. (1968), The analysis and identification of hormone producing cells of the adenohypophysis. In: *Perspectives in Endocrinology.* (Barrington, E. J. W. and Jørgensen, C. B., eds.) pp. 405–467, Academic Press, New York.

Ooi, E. C. and Youson, J. H. (1977), Morphogenesis and growth of the definitive opisthonephros during metamorphosis of anadromous sea lamprey, *Petromyzon marinus* L. *J. Embryol. exp. Morph.*, **42**, 219–235.

Oota, Y. (1974), Electron microscopic studies on the adenohypophysis of the hagfish, *Eptatretus burgeri* (Girard). *Rep. Fac. Sci. Shizuoka Univ.*, **9**, 67–78.

Orvig, T. (1968), The dermal skeleton. In: *Current Problems of Lower Vertebrate Phylogeny*. (Orvig, T., ed.). Nobel Symp. Vol. 4, pp. 373–397, Interscience, New York.

Östberg, Y. (1976), The entero-insular endocrine organ in a Cyclostome, *Myxine glutinosa*. *Umea Univ. Med. Diss. N. S.*, **15**, 1–41.

Östberg, Y., Boquist, L. and Falkmer, S. (1972), Tumours or hamartomas in the endocrine pancreas of the hagfish, *Myxine glutinosa*. In: *Myxine glutinosa* (Fänge, R. ed.) *Acta Reg. Soc. Sci. Litt. Gothenburg. Zool.* Vol. 8, pp. 73–75.

Östberg, Y. and Boquist, L. (1976). Ultrastructural and fluorescence microscopical characterisation of the intestinal endocrine cells in a cyclostome, *Myxine glutinosa. Acta zool. (Stockh.)*, **57**, 41–51.

Östberg, Y., Van Noorden, S., and Pearse, A. G. E. (1975). Cytochemical, immunofluorescence and ultrastructural investigations on polypeptide hormone localisation in the islet parenchyma and bile duct mucosa of a cyclostome, *Myxine glutinosa. Gen. comp. Endocrinol.*, **25**, 274–291.

Östberg, Y., Van Noorden, S., Pearse, A. G. E. and Thomas, N. W. (1976a), Cytochemical, immunofluorescence and ultrastructural investigations on polypeptide hormone containing cells in the intestinal mucosa of a cyclostome, *Myxine glutinosa. Gen. comp. Endocrinol.*, **28**, 213–227.

Östberg, Y., Boquist, L., Van Noorden, S. and Pearse, A. G. E. (1976b), On the origin of islet parenchymal cells in a cyclostome, *Myxine glutinosa. Gen. comp. Endocrinol.*, **28**, 228–246.

Östberg, Y., Fänge, R., Mattison, A. and Thomas, N. W. (1976c), Light and electron microscopical characterisation of heterophilic granulocytes in the intestinal wall and islet parenchyma of the hagfish, *Myxine glutinosa* (Cyclostomata). *Acta zool. (Stockh.)*, **57**, 89–102.

Packard, G. C., Packard, M. J. and Gorbman, A. (1976), Serum thyroxine concentrations in the Pacific hagfish and lamprey and in the Leopard frog. *Gen. comp. Endocrinol.*, **28**, 365–367.

Paiement, J. M. and McMillan, D. B. (1976), The extra-cardiac chromaffin cells of larval lampreys. *Gen. comp. Endocrinol.*, **27**, 495–508.

Pajor, W. J. (1977). Comparative histoenzymological studies on the activity of some oxidative enzymes in thyroids of post-metamorphic lampreys of *Lampetra fluviatilis* L. *Folia Biol.* **25**, 409–414.

Paleus, S. and Braunitzer, G. (1972), Amino acid sequence in Cytochrome c from *Myxine glutinosa* L. In: *Structure and Function of Oxidation-Reduction Enzymes*. (Äkeson, A. and Ehrenberg, A., eds.) pp. 31–34, Pergamon Press, Oxford.

Paleus, S. and Liljequist, G. (1972), The haemoglobins of *Myxine glutinosa*. II. Amino acid analyses, end group determinations and further investigations. *Comp. Biochem. Physiol.*, **42B**, 611–617.

Paleus, S., Vesterberg, O. and Liljeqvuist, G. (1971), The haemoglobins of

Myxine glutinosa L. I. Preparation and crystallization. *Comp. Biochem. Physiol.*, **39B**, 551–557.

Parker, P. S. and Lennon, R. F. (1956), Biology of the sea lamprey in its parasitic phase. *Res. Rep. U.S. Fish Wildl. Serv.*, **44**, 1–32.

Parrington, F. R. (1958), On the nature of the Anaspida. In: *Studies on Fossil Vertebrates.* (Westoll, T. S., ed.) pp. 108–128, Athlone Press, London.

Patzner, R. A. (1974), Die früher Stadien der Oogenese bei *Myxine glutinosa* L. (Cyclostomata). Licht -und elektronmikroskopische Untersuchungen. *Norw. J. Zool.*, **22**, 81–93.

Patzner, R. (1978), Cyclical changes in the ovary of the hagfish, *Eptatretus burgeri* (Cyclostomata). *Acta Zool. (Stockh.)*, **59**, 57–62.

Patzner, R. A. and Ichikawa, T. (1977). Effects of hypophysectomy on the testis of the hagfish *Eptatretus burgeri* Girard (Cyclostomata). *Zool. Anz.* **5/6**, 371–380.

Pearse, A. G. E. (1968), Common cytochemical and ultrastructural characteristics of cells producing polypeptide hormones (the APUD series) and their relevance to thyroid and ultimobranchial C cells and calcitonin. *Proc. R. Soc. Lond. Ser. B.*, **170**, 71–80.

Pearson, R. and Pearson, L. (1976), *The Vertebrate Brain.* Academic Press, London.

Percy, R., Leatherland, J. F. and Beamish, F. W. H. (1975), Structure and ultrastructure of the pituitary gland in the sea lamprey, *Petromyzon marinus* at different stages in its life cycle. *Cell Tiss. Res.*, **157**, 141–164.

Percy, R. C. and Potter, I.C. (1976), Blood cell formation in the river lamprey, *Lampetra fluviatilis. J. Zool. (Lond.)*, **178**, 319–340.

Percy, R. C. and Potter, I. C. (1977), Changes in haemopoeitic sites during the metamorphosis of the lampreys, *Lampetra fluviatilis* and *Lampetra planeri. J. Zool. (Lond.)*, **183**, 111–123.

Percy, R. C. and Potter, I. C. (1978), The intestinal blood circulation in the river lamprey, *Lampetra fluviatilis. J. Zool. (Lond.)*.

Permitin, L. E. (1977), Species composition and zoogeographical analysis of the bottom fish fauna of the Scotia Sea. *Vop. Ikhtiol.*, **17**, 710–726.

Peters, A. and Mackay, B. (1961), The structure and innervation of the myotomes of the lamprey. *J. Anat.*, **95**, 575–585.

Petersen, J. D., Steiner, D. F., Emdin, S. O. and Falkmer, S. (1975), The amino acid sequences of the insulin from a primitive vertebrate, the Atlantic hagfish (*Myxine glutinosa*). *J. biol. Chem.*, **250**, 5183–5191.

Pfeiffer, W. and Fletcher, T. F. (1964), Club cells and granular cells in the skin of lamprey. *J. Fish. Res. Bd. Can.*, **21**, 1083–1087.

Phelps, P. (1975), Evidence that the endocrine pancreatic cells are not derived from the neural crest. *Anat. Rec.*, **181**, 449.

Phillips, J. W. and Hurd, F. J. R. (1977), Gluconeogenesis in vertebrate livers. *Comp. Biochem. Physiol.*, **57B**, 127–131.

Piavis, G. W., Dubin, N. H. and Nardell, B. (1975), The 'estradiol fraction' in the peripheral blood of the lamprey, *Petromyzon marinus*. *Anat. Rec.*, **181**, 449–450.

Pickering, A. D. (1972), Effect of hypophysectomy on the activity of the endostyle and thyroid gland in the larval and adult river lamprey, *Lampetra fluviatilis* L. *Gen. comp. Endocrinol.*, **18**, 335–343.

Pickering, A. D. (1976a), Iodide uptake by the isolated thyroid gland of the river lamprey, *Lampetra fluviatilis* L. *Gen. comp. Endocrinol.*, **28**, 358–364.

Pickering, A. D. (1976b), Stimulation of intestinal degeneration by oestradiol and testosterone implantation in the migrating river lamprey, *Lampetra fluviatilis* L. *Gen. comp. Endocrinol.*, **30**, 340–346.

Pickering, A. D. and Dockray, G. J. (1972), The effects of gonadodectomy on osmoregulation in the migrating river lamprey, *Lampetra fluviatilis* L. Comp. Biochem. Physiol., **41A**, 139–147.

Pickering, A. D. and Morris, R. (1970), Osmoregulation of *Lampetra fluviatilis* L. and *Petromyzon marinus* (Cyclostomata) in hyperosmotic solutions. *J. exp. Biol.*, **53**, 231–243.

Pickering, A. D. and Morris, R. (1973), Localisation of ion transport in the intestine of the migrating river lamprey, *Lampetra fluviatilis* L. *J. exp. Biol.*, **58**, 165–176.

Pickering, A. D. and Morris, R. (1976), Fine structure of the interplatelet area in the gills of the macrophthalmia stage of the river lamprey, *Lampetra fluviatilis* (L.). *Cell Tiss. Res.*, **168**, 433–443.

Pickering, A. D. and Morris, R. (1977), Sexual dimorphism in the gills of the spawning river lamprey, *Lampetra fluviatilis* L. *Cell Tiss. Res.*, **180**, 1–10.

Pictet, R., Rall, L. B. and Rutter, W. J. (1976), The neural crest and the origin of the insulin-producing and other gastrointestinal hormone-producing cells. *Science*, **191**, 191–192.

Pietschmann, V. (1933–4), Acrania (Cephalochordata)-Cyclostoma-Ichthya. In: *Handbuch der Zoologie*. (Kukenthal, W. and Krumbach, T., eds.) Vol. 6, 127–547, W. de Gruyter, Berlin.

Pikkarainen, J. (1968), The molecular structure of vertebrate skin collagens. *Acta physiol. scand.*, Suppl. **309**, 1–72.

Pikkarainen, J. and Kulonen, E. (1969), Comparative chemistry of collagen. *Nature*, **223**, 839–841.

Platt, J. B. (1894), Ontogenetic differentiation of the ectoderm in *Necturus*. Second preliminary note. *Zool. Anz.*, **9**, 51–56.

Plisetskaya, E. M. and Leibson, L. G. (1973), Influence of hormones on the glycogen synthetase activity of the liver and muscles of lampreys and scorpion fishes. *Dokl. Akad. Nauk. USSR.*, **210**, 1230–1232.

Plisetskaya, E. M. and Leibush, B. N. (1972), Insulin-like activity and immunoreactive insulin in the blood of the lamprey *Lampetra fluviatilis*.

Zh. Evol. Biokhim. Fiziol., **8**, 499–505.

Plisetskaya, E. M., Leibush, B. N. and Bondareva, V. (1976), The secretion of insulin and its role in cyclostomes and fishes. In: '*The Evolution of the Pancreatic Islets*'. (Grillo, T. A., Leibson, L. and Epple, A., eds.) pp. 251–269, Pergamon Press, Oxford.

Plisetskaya, E. M. and Prozorovskaya, M. P. (1971), Catecholamines in the blood and cardiac muscle of the lamprey, *Lampetra fluviatilis* during insulin hypoglycaemia. *Zh. Evol. Biokhim. Fiziol.*, **7**, 101–103.

Plisetskaya, E. M. and Zheludkova, Z. P. (1973), The effect of adrenalin on the amylase activity of the liver and muscles of the lamprey, *Lampetra fluviatilis*. *Zh. Evol. Biokhim. Fiziol.*, **9**, 611–613.

Polenov, A. I., Belenk'ii, M. A. and Konstantinova, M. S. (1974), The hypothalamo-hypophysial system of the lamprey, *Lampetra fluviatilis* L. *Cell Tiss. Res.*, **150**, 505–519.

Pollara, B., Finstad, J. and Good, R. A. (1966), The phylogenetic development of immunoglobulins. In: *The Phylogeny of immunity*. (Smith, R. T., Miescher, R. A. and Good, R. A., eds.) pp. 88–97, Univ. Florida Press, Gainsville.

Pollara, B., Litman, G., Finstad, J., Howell, J. and Good, R. A. (1970), The evolution of the immune response. III. Antibody to human 'O' cells and properties of immunoglobulins in lamprey. *J. Immunol.*, **105**, 738–745.

Potter, I. C. (1970), The life cycles and ecology of Australian lampreys of the genus *Mordacia*. *J. Zool. (Lond.)*, **161**, 487–511.

Potter, I. C. and Beamish, F. W. H. (1975), Lethal temperatures in ammocoetes of four species of lampreys. *Acta zool. (Stockh.)*, **56**, 88–91.

Potter, I. C. and Beamish, F. W. H. (1978), Changes in haematocrit and haemoglobin concentrations during the life cycle of the anadromous sea lamprey, *Petromyzon marinus* L. *Comp. Biochem. Physiol.*, **60A**, 431–434.

Potter, I. C., Hill, B. J. and Gentleman, S. (1970), Survival and behaviour of ammocoetes at low oxygen tensions. *J. exp. Biol.*, **53**, 59–73.

Potter, I. C. and Huggins, R. J. (1970), Observations on the morphology, behaviour and salinity tolerances of downstream migrating river lampreys (*Lampetra fluviatilis*). *J. Zool. (Lond.)*, **169**, 365–379.

Potter, I. C., Lanzing, W. J. R. and Strahan, R. (1968), Morphometric and meristic studies on populations of Australian lampreys of the genus *Mordacia*. *J. Linn. Soc. (Zool.)*, **47**, 533–546.

Potter, I. C. and Nicol, P. I. (1968), Electrophoretic studies on the haemoglobins of Australian lampreys. *Aust. J. exp. Biol. med. Sci.*, **46**, 639–641.

Potter, I. C., Percy, R. C. and Youson, J. H. (1978), A proposal for the adaptive significance of the development of the lamprey fat column. *Acta zool. (Stockh.)*, **59**, 63–67.

Potter, I. C., Prince, P. A. and Croxall, J. P. (in press). Data on the adult marine and migratory phases in the life cycle of the Southern Hemisphere lamprey, *Geotria australis* Gray. *Env. Biol. Fish.*

Potter, I. C. and Robinson, E. S. (1971), The chromosomes. In: *The Biology of Lampreys.* (Hardisty, M. W. and Potter, I. C., eds.) Vol. 1, pp. 279–293, Academic Press, London.

Potter, I. C. and Robinson, E. S. (1973), The chromosomes of the Cyclostomes. In: *Cytotaxonomy and Vertebrate Evolution.* (Chiarelli, A. B. and Capanna, E., eds.) pp. 179–203, Academic Press, London.

Potter, I. C., Robinson, E. S. and Walton, S. M. (1968), The mitotic chromosomes of *Mordacia mordax* (Agnatha: Petromyzonidae). *Experientia*, **24**, 966–967.

Potter, I. C. and Rogers, M. J. (1972), Oxygen consumption in burrowed and unburrowed ammocoetes of *Lampetra planeri* (Bloch). *Comp. Biochem. Physiol.*, **41A**, 427–432.

Potter, I. C. and Rothwell, B. (1970), The mitotic chromosomes of the lamprey, *Petromyzon marinus* L. *Experientia*, **26**, 429–430.

Potter, I. C., Wright, G. M. and Youson, J. H. (1978), Metamorphosis in the anadromous sea lamprey, *Petromyzon marinus* L. *Can. J. Zool.*, **56**, 561–570.

Poupa, O. and Oštádal, B. (1969), Experimental cardiomegalies and cardiomegalies in free living animals. *Ann. N. Y. Acad. Sci.*, **156**, 445–468.

Priede, I. G. (1977), Natural selection for energetic efficiency and the relation between activity level and mortality. *Nature*, **267**, 610–611.

Prosser, C. L. (ed.) (1973), *Comparative Animal Respiration*. Vol. 1, pp. 165–206, Saunders, Philadelphia.

Rahemtulla, F., Höglund, N-G. and Løvtrup, S. (1976), Acid mucopolysaccharides in the skin of some lower vertebrates (hagfish, lamprey and Chimaera). *Comp. Biochem. Physiol.*, **53B**, 295–298.

Rall, D. P. and Burger, J. W. (1967), Some aspects of hepatic and renal excretion in *Myxine*. *Am. J. Physiol.*, **212**, 354–356.

Rall, D. P., Schwab, P. and Zubrod, C. G. (1961), Alteration of plasma proteins at metamorphosis in the lamprey (*Petromyzon marinus dorsatus*). *Science*, **131**, 279–280.

Randall, D. J. (1970), Gas exchange in fish. In: *Fish Physiology* Vol. 4. 253–291. (Hoar, W. S. and Randall, D. J., eds.) Vol. 4, pp. 253–291, Academic Press, London.

Randall, D. J. (1972), Respiration. In: *The Biology of Lampreys* (Hardisty, M. W. and Potter, I. C., eds.) Vol. 2, pp. 287–306, Academic Press, London.

Rao, G. M. M. (1968), Oxygen consumption of rainbow trout (*Salmo gairdneri*) in relation to activity and salinity. *Can. J. Zool.*, **46**, 781–785.

Read, L. J. (1968), A study of ammonia and urea production and excretion in the freshwater adapted form of the Pacific lamprey, *Entosphenus tridentatus. Comp. Biochem. Physiol.*, **26**, 405–466.

Read, L. J. (1975), Absence of ureogenic pathways in the liver of the hagfish, *Bdellostoma cirrhatum. Comp. Biochem. Physiol.*, **51B**, 139–141.

Refetoff, S., Robin, N. I. and Fang, V. S. (1970), Parameters of thyroid function in serum of 16 selected vertebrate species: a study of PBI, serum T_4, free T_4 and the pattern of T_4 and T_3 binding to serum proteins. *Endocrinology*, **86**, 793–805.

Repetski, J. E. (1978), A fish from the Upper Cambrian of North America. *Science*, **200**, 529–531.

Retzius, G. (1890), Ueber Zellenteilung bei *Myxine glutinosa. Biol. Foren. Stockh. Forhandl.*, **11**, 80–90.

Reynolds, T. L. (1931), Hydrostatics of the suctorial mouth of the lamprey. *Univ. Calif. Publs. Zool.*, **37**, 15–34.

Riegel, J. A. (1978), Factors affecting glomerular function in the Pacific hagfish, *Eptatretus stouti* (Lockington). *J. exp. Biol.*, **73**, 261–277.

Riggs, A. (1970), Properties of fish haemoglobins. In: *Fish physiology.* (Hoar, W. S. and Randall, D. J., eds.) Vol. 2, pp. 209–252, Academic Press, London.

Riggs, A. (1972), The haemoglobins. In: *The Biology of Lampreys.* (Hardisty, M. W. and Potter, I. C., eds.) Vol. 2, pp. 261–286, Academic Press, London.

Ringham, G. L. (1975), Localisation and electrical characteristics of a giant synapse in the spinal cord of the lamprey. *J. Physiol.*, **251**, 398–407.

Ritchie, A. (1964), New light on the morphology of the Norwegian Anaspida. *Skr. norske VidenskAkad Oslo, Mat.-naturv. Kl.* 1–22.

Ritchie, A. (1968), New evidence on *Jamoytius kerwoodi* White, an important Ostracodern from the Silurian of Lanarkshire, Scotland. *Palaeontology*, **11**, 21–39.

Roberts, A. (1978), Pineal eye and behaviour in *Xenopus* tadpoles. *Nature*, **273**, 774–775.

Robertson, J. D. (1954), The chemical composition of the blood of some aquatic Chordates, including members of the Tunicata, Cyclostomata and Osteichthyes. *J. exp. Biol.*, **31**, 420–442.

Robertson, J. D. (1957), The habitat of the early vertebrates. *Biol. Rev.*, **32**, 156–187.

Robertson, J. D. (1959), The origin of the vertebrates – marine or freshwater. *Adv. Sci.*, **61**, 516–520.

Robertson, J. D. (1974), Osmotic and ionic regulation in Cyclostomes. In *Chemical Zoology* (Florkin, M. and Scheer, B. T., eds.) Vol. 8, pp. 149–193, Academic Press, London.

Robertson, J. D. (1976), Chemical composition of the body fluids and muscle of the hagfish, *Myxine glutinosa* and the rabbit fish, *Chimaera monstrosa.* *J. Zool. (Lond.)*, **178**, 261–277.

Robinson, E. S. and Potter, I. C. (1969), Meiotic chromosomes of *Mordacia praecox* and a discussion of chromosome numbers in lampreys. *Copeia*, **1969**, 824–828.

Robinson, E. S., Potter, I. C. and Webb, C. J. (1974), Homogeneity of holarctic lamprey karyotypes. *Caryologia*, **27**, 443–454.

Robinson, E. S., Potter, I. C. and Atkin, N. B. (1975), The nuclear DNA content pf lampreys. *Experientia*, **31**, 912–9; 3.

Romer, A. A. (1968), *Notes and Comments on Vertebrate Palaeontology.* Univ. Chicago Press.

Ross, D. M. (1963), The sense organs of *Myxine glutinosa.* In: *The Biology of Myxine.* (Brodal, A. and Fänge, R., eds.) pp. 150–160, Universitets- forlaget, Oslo.

Rovainen, C. M. (1967), Physiological and anatomical studies on larger neurons of the central nervous system of the sea lamprey (*Petromyzon marinus*). I. Muller and Mauthner cells. *J. Neurophysiol.* **30**, 1000–1023.

Rovainen, C. M. (1974a), Synaptic interactions of identified nerve cells in the spinal cord of the sea lamprey. *J. comp. Neur.*, **154**, 189–206.

Rovainen, C. M. (1974b), Synaptic interactions of reticulospinal neurones and nerve cells in the spinal cord of the sea lamprey. *J. comp. Neur*, **154**, 207–224.

Rovainen, C. M. (1974c), Respiratory motoneurones in lampreys. *J. comp. Physiol.*, **94**, 57–68.

Rovainen, C. M. (1976), Vestibulo-ocular reflexes in the adult sea lamprey. *J. comp. Physiol.*, **112**, 159–164.

Rovainen, C. M., Johnson, P. A., Roach, E. A. and Mankovsky, J. A. (1973), Projections of individual axons in lamprey spinal cord determined by tracing through serial sections. *J. comp. Neur.*, **149**, 193–202.

Rovainen, C. M. and Schieber, M. H. (1975), Ventilation of larval lampreys. *J. comp. Physiol.*, **104**, 185–203.

Rozhkova, E. K. (1972), Characteristics of cholinoreception in somatic muscle of the lamprey, *Lampetra fluviatilis. Comp. gén. Pharmac.*, **3**, 410–422.

Rudy, P. P. and Wagner, R. C. (1970), Water permeability in the Pacific hagfish, *Polistotrema stouti* and the staghorn sculpin, *Leptococcus armatus. Comp. Biochem. Physiol.*, **34**, 399–403.

Rühle, H. J. and Sterba, G. (1966), Zur Histologie der Hypophyse des Flüssneunauges (*Lampetra fluviatilis* L.) *Z. Zellforsch. mikrosk. Anat.*, **70**, 136–168.

Rumen, N. M. and Love, W. E. (1963), The six haemoglobins of the sea lamprey (*Petromyzon marinus*). *Archs. Biochem. Biophys.*, **103**, 24–35.

Rurak, D. W. and Perks, A. M. (1974), The pharmacological characterisation of arginine vasotocin in the pituitary of the Pacific hagfish (*Polistotrema stouti*). Gen. comp. Endocrinol., **22**, 480–488.

Rurak, D. W. and Perks, A. M. (1976), The neurohypophysial principle of the Western brook lamprey, *Lampetra richardsonii*. Studies in the adult. *Gen. comp. Endocrinol.*, **29**, 301–312.

Rurak, D. W. and Perks, A. M. (1977), The neurohypophysial principle of the Western brook lamprey, *Lampetra richardsonii*. Studies in the ammocoete larva. *Gen. comp. Endocrinol.* **31**, 91–100.

Salvatore, G., Sena, L., Viscidi, E. and Salvatore, M. (1965), The thyroid iodoproteins 12 S, 19 S and 27 S in various animal species and their physiological significance. In: *Current topics in thyroid research*. Cassano, C. and Andreoli, M., (eds.), pp. 193–206, Academic Press, New York.

Sanders, H. L. (1968), Marine benthic diversity: a comparative study. *Am. Nat.*, **102**, 243–282.

Sasaki, M. and Hitotsumachi, S. (1967), Notes on the chromosomes of a freshwater lamprey, *Entosphenus reissneri* (Cyclostomata). *Chromosome Inf. Serv.* **8**, 22–24.

Savina, M. V. and Plisetskaya, E. M. (1976), Synthesis of glycogen from glycerin in isolated tissues of the river lamprey, *Zh. Evol. Biokhim. Fiziol*, **12**, 282–284.

Savina, M. V. and Wojtczak, A. B. (1977), Enzymes of gluconeogenesis and the synthesis of glycogen from glycerol in various organs of lamprey (*Lampetra fluviatilis*). *Comp. Biochem. Physiol.*, **57B**, 185–196.

Sawyer, W. H. (1957), Unpublished data In: Black 1957 *Physiology of Fishes* (Brown, M. E. ed.). Vol. 1, pp. 103–205, Academic Press, London.

Sawyer, W. H. (1973), Discussion: endocrines and osmoregulation among fishes. *Am. Zool.*, **13**, 819–821.

Sawyer, W. H. and Roth, W. D. (1954), The storage of biliverdin by the liver of the migrating sea lamprey, *Petromyzon marinus*. *Anat. Rec.*, **120**, 741–742.

Schirner, H. (1963), The pancreas. In: *The Biology of Myxine*. (Brodal, A. and Fänge, R., eds.) pp. 481–487, Universitetsforlaget, Oslo.

Schmidt-Nielson, K. (1972), Locomotion; energy cost of swimming, flying and running. *Science*, **177**, 222–227.

Schober, W. (1966), Vergleichende Betrachtungen am Telencephalon niederer Wirbeltiere. In: *Phylogenesis and Organogenesis of the Forebrain*. (Hassler, R. and Stephan, H., eds.) Georg. Thieme Verlag, Stuttgart.

Schreiner, K. (1955), Studies on the gonad of *Myxine glutinosa* L. *Univ. i. Bergen Arbok Nat. rek.*, **8**, 1–40.

Schreiner, A. and Schreiner, K. E. (1904), Die Reiferteilungen bei den Wirbeltieren. Ein Beitrag zur Frage nach des Chromatinreduktion. *Anat. Anz.*, **24**, 561–578.

Schreiner, A. and Schreiner, K. E. (1904b), Ueber die Entwicklung der mannlichen Geschlechtszellen von *Myxine glutinosa* (L.). Vermehrungsperiod, Reifungsperiod und Reifungsteilungen. *Arch. Biol.*, **21**, 183–355.

Schreiner, A. and Schreiner, K. E. (1908), Zur Spermienbildung der Myxinoiden. Ueber die Entwicklung der mannlichen Geschlechtszellen von *Myxine glutinosa* (L.) *Arch. Zool.*, **1**, 152.

Schroll, F. (1959), Zur Ernährungsbiologie der Steirischen Ammoceten *Lampetra planeri* (Bloch) und *Eudontomyzon danfordi* (Regan). *Int. Rev. Hydrobiol.*, **44**, 395–429.

Schwab, M. E. (1973), Some new aspects about the prosencephalon of *Lampetra fluviatilis* (L.). *Acta Anat.*, **86**, 353–375.

Seiler, K., Seiler, R. and Sterba, G. (1970), Histochemische Untersuchungen am Interrenal System des Bachneunauges (*Lampetra planeri* Bloch). *Acta biol. med. germ.*, **24**, 553–554.

Selzer, M. E. (1978), Mechanisms of functional recovery and regeneration after spinal cord transection. *J. Physiol. (Lond.)*, **277**, 395–408.

Seno, N., Akiyama, F. and Anno, K. (1972), A novel dermatan polysulphate from hagfish skin containing trisulphated disaccharide residues. *Biochim. biophys. Acta*, **264**, 229–233.

Shapovalov, A. I. (1972), Evolution of neuronal systems of suprasegmental motor control. *Neurofiziologiya*, **4**, 453–470.

Shelton, G. (1970), The regulation of breathing. In: *Fish physiology*. (Hoar, W. S. and Randall, D. J., eds.) Vol. 4, pp. 293–359, Academic Press, London.

Shidoji, Y. and Muto, Y. (1977), Vitamin A transport in plasma of the non-mammalian vertebrates; isolation and partial characterisation of piscine retinol-binding protein. *J. Lipid Res.*, **18**, 679–691.

Sjöberg, K. (1977), Locomotor activity of the river lamprey, *Lampetra fluviatilis* L during the spawning season. *Hydrobiol.*, **55**, 265–270.

Smith, B. R. (1971), Sea lampreys in the Great Lakes of North America. In: *The Biology of Lampreys*. (Hardisty, M. W. and Potter, I. C. eds.) Vol. 1, pp. 207–247, Academic Press, London.

Smith, K. L. (1978), Metabolism of the abyssopelagic rattail, *Coryphaenoides armatus* measured *in situ*. *Nature*, **274**, 362–363.

Smith, K. L. and Hessler, R. R. (1974), Respiration of benthopelagic fishes: *in situ* measurements at 1230 metres. *Science*, **184**, 72–73.

Somlyo, A. V. and Somlyo, A. P. (1968), Vasotocin-magnesium interaction in vascular smooth muscle of the hagfish (*Eptatretus stouti*). *Comp. Biochem. Physiol.*, **24**, 267–270.

Spjeldnaes, N. (1967), The palaeoecology of the Ordovician vertebrates of the Harding formation. *Coll. Int. Centre Nat. Rech. Sc.*, **163**, 11–20.

Spencer, R. P., Schleig, R. L. and Binder, H. J. (1966), Lipids of the alimentary canal of the hagfish, *Eptatretus stouti. Comp. Biochem. Physiol.*, **19,** 139–144.

Spitzer, R. H., Downing, S. W., Koch, E. A. and Kaplan, M. A. (1976), Haemagglutinins in the mucus of the Pacific hagfish, *Eptatretus stouti. Comp. Biochem. Physiol.*, **54B,** 409–411.

Stabrovsky, E. M. (1967), The distribution of adrenalin and nor-adrenalin in the organs of the Baltic lamprey (*Lampetra fluviatilis*) at rest and under various functional stresses. *Zh. Evol. Biokhim. Fisiol.*, **3,** 216–221.

Steiner, D. E., Peterson, J. D., Tager, H., Emdin, S., Östberg, Y. and Falkmer, S. (1973), Comparative aspects of proinsulin and insulin structure and biosynthesis. *Am. Zool.*, **13,** 591–604.

Stensio, E. A. (1927), The Downtonian and Devonian vertebrates of Spitsbergen. 1. Family Cephalaspidae. *Skr. Svalbard Nordishavet*, **12,** 1–391.

Stensio, E. A. (1958), Les Cyclostomes fossiles ou ostracodermes. In: *Traité de Zoologie.* (Grassé, P. P., ed.) Vol. 13, pp. 171–425, Masson, Paris.

Stensio, E. A. (1968), The cyclostomes with special reference to the diphyletic origin of the Petromyzontida and the Myxinoidea. In: *Current Problems in Lower Vertebrate Phylogeny* (Orvig, T., ed.) pp. 13–71, Almquist and Wiksell, Stockholm.

Sterba, G. (1953), Die Physiologie und Histogenese der Schilddrüse und des Thymus beim Bachneunauge (*Lampetra planeri* Bloch = *Petromyzon planeri* Bloch) als Grundlage phylogenetischer Studien uber die Evolution der innersekretorischen Kiemendarmderivate. *Wiss. Z. Friedrich-Schiller Univ. Jena.*, **2,** 239–298.

Sterba, G. (1955), Das Adrenal und Interrenal System in Lebenslauf von *Petromyzon planeri* Bloch. *Zool. Anz.*, **155,** 151–168.

Sterba, G. (1962), Die Neunaugen (Petromyzonidae). In: *Handbuch der Binnenfischerei Mitteleuropas* (Demoll. R. and Maier, H. N., eds.) Vol. 3, 263–352, Schweizerbart, Stuttgart.

Sterba, G. (1972), Neuro- and gliasecretion. In: *The Biology of Lampreys* (Hardisty, M. W. and Potter, I. C., eds.). Vol. 2, pp. 69–89, Academic Press, London.

Sterba, G., Hoheisel, G., Ruhle, H-J. and Engelmann, W. E. (1973), Extrahypothalamische peptiderge Neurosekretion. 1. Neurosekretion im Mittelhirn der Neunaugen. *Z. Zellforsch.*, **142,** 329–345.

Sterling, J. A., Meranze, D. R., Windsten, S. and Krieger, M. K. (1967), Observations of lamprey liver during its life cycle. *J. A. Einstein med. Centre*, **15,** 107–116.

Steven, D. M. (1955), Experiments on the light sense of the hag, *Myxine glutinosa* L. *J. exp. Biol.*, **32,** 22–38.

Stokes, M. (1939), Thyroid treatment and the cyclostome endostyle. *Proc. Soc. exp. Biol. Med.*, **42,** 810–811.

Stollte, H. and Eisenbach, G. M. (1973), Single nephron filtration rate in the hagfish, *Myxine glutinosa. Bull. Mt. Desert Isl. Biol. Lab.,* **13,** 120–121.

Strahan, R. (1958), Speculations on the evolution of the agnathan head. *Proc. cent. and bicent. Congr. Biol. Singapore,* 83–94.

Strahan, R. (1959a), Slime production in *Myxine glutinosa. Copeia,* **1959,** 165–166.

Strahan, R. (1959b), Pituitary hormones in *Myxine* and *Lampetra. Trans. First Asia Ocean. Reg. Congr. Endocrinol.,* **24.**

Strahan, R. (1963a), The behaviour of myxinoids, *Acta zool. (Stockh.),* **24,** 1–30.

Strahan, R. (1963b), The behaviour of *Myxine* and other myxinoids. In: *The Biology of Myxine.* (Brodal, A. and Fänge, R., eds.) pp. 22–32, Universitetsforlaget, Oslo.

Strahan, R. (1966), In: *Textbook of Zoology,* (Parker and Haswell, eds.), Vol. 2, Macmillan, London.

Strahan, R. (1975), *Eptatretus longipinnis,* n. sp., a new hagfish (family Eptatretidae) from South Australia, with a key to the 5–7 gilled Eptatretidae. *Aust. Zool.,* **19,** 137–148.

Strahan, R. and Honma, Y. (1960), Notes on *Paramyxine atami* Dean (Fam. Myxinidae) and its fishery in Sado Strait, Sea of Japan. *Hong Kong Univ. Fish. J.,* **3,** 27–35.

Strahan, R. and Maclean, J. L. (1969), A pancreas-like organ in the larva of the lamprey, *Mordacia mordax. Aust. J. Sci.,* **32,** 54–55.

Susuki, S. and Gorbman, A. (1973), Properties of an iodoprotein from thyroid tissue of the Pacific hagfish, *Eptatretus stouti. Gen. Comp. Endocrinol.,* **22,** 312–314.

Susuki, S., Gorbman, A., Rolland, M., Montfort, M. F. and Lississky, A. (1975), Thyroglobulins of cyclostomes and elasmobranchs. *Gen. comp. Endocrinol.* **26,** 56–69.

Susuki, S. and Kondo, Y. (1973), Thyroidal morphogenesis and biosynthesis of thyroglobulins before and after metamorphosis in the lamprey, *Lampetra reissneri. Gen. comp. Endocrinol.,* **21,** 451–460.

Szabó, Zs., Molnár, B. and Mihail, N. (1965), Histologische Kennzeichnen der Hypophyse von *Eudontomyzon danfordi Reg. Anat. Anz.,* **70,** 136–168.

Tambs-Lyche, H. (1969), Notes on the distribution and ecology of *Myxine glutinosa* L. *FiskDir. Skr. Ser. HavUnders.,* **15,** 279–284.

Tarlo, L. B. H. and Whiting, H. P. (1965), A new interpretation of the internal anatomy of the Heterostraci (Agnatha), *Nature,* **206,** 148–150.

Taylor, K. M. (1967), The chromosomes of some lower chordates. *Chromosoma,* **21,** 181–188.

Teravainen, H. (1971), Anatomical and physiological studies on muscles of lamprey. *J. Neurophysiol.,* **34,** 954–973.

Teravainen, H. and Rovainen, C. M. (1971a), Fast and slow motoneurones to body muscle of the sea lamprey, *J. Neurophysiol.*, **34**, 990–998.

Teravainen, H. and Rovainen, C. M. (1971b), Electrical activity of myotomal muscle fibres, motoneurones and sensory dorsal cells during spiral reflexes in lampreys. *J. Neurophysiol.*, **34**, 999–1009.

Theisen, B. (1973), The olfactory system in the hagfish, *Myxine glutinosa*. I. Fine structure of the apical part of the olfactory epithelium. *Acta zool. (Stockh.).*, **154**, 271–284.

Theisen, B. (1976), The olfactory system in the Pacific hagfishes, *Eptatretus stouti, Eptatretus deani and Myxine circifrons. Acta zool. (Stockh.)*, **57**, 167–173.

Thoenes, G. H. (1972), The hagfish at the phylogenetic juncture towards immunological response. *Inst. Nat. Sante Rech. Med.*, **1**, 69–74.

Thoenes, G. H. and Hildemann, W. H. (1970), Immunological responses of Pacific hagfish. II. Serum antibody production to soluble antigens. In: *Developmental Aspects of Antibody Formation and Structure.* (Sterzl, F. and Riha, A., eds.) Vol. 2, p. 711.

Thomas, N. W., Östberg, Y. and Falkmer, S. (1973), A second granular cell in the endocrine pancreas of the hagfish, *Myxine glutinosa*? *Acta zool. (Stockh.)*, **54**, 201–207.

Thornhill, R. A. (1972), The development of the labyrinth of the lamprey (*Lampetra fluviatilis* Linn.). *Proc. R. Soc. Lond. Ser. B.*, **181**, 175–198.

Tonder, O., Larsen, B., Aarskog, D. and Haneberg, B. (1978), Natural and immune antibodies to rabbit erythrocyte antigens. *Scand. J. Immunol.*, **7**, 245–250.

Tong, W., Kerkoff, P. and Chaikoff, I. L. V. (1961), ^{131}I utilisation by thyroid tissue of the hagfish. *Biochim. Biophys. Acta*, **52**, 299–304.

Trochard, M. C., Vaudry, H., Leboulenger, F., Delarne, C. and Vaillant, R. (1977), Is TRH an MSH releasing hormone in the frog? *Acta Endocrinol.* Suppl. **212**, 85.

Tsuneki, K. (1974), Distribution of monoamine oxidase and acetylcholinesterase in the hypothalamo-hypophysial system of the lamprey, *Lampetra japonica. Cell Tiss. Res.*, **154**, 17–27.

Tsuneki, K. (1976), Effects of estradiol and testosterone in the hagfish, *Eptatretus burgeri. Acta zool. (Stockh.)*, **57**, 137–146.

Tsuneki, K., Adachi, T., Ishii, S. and Oota, Y. (1976), Morphometric classification of neurosecretory granules in the neurohypophysis of the hagfish, *Eptatretus burgeri. Cell Tiss. Res.*, **166**, 145–157.

Tsuneki, K. and Fernholm, B. (1975), Effect of thyrotropin-releasing hormone on the thyroid of a teleost, *Chasmichthys dolichognathus* and a hagfish, *Eptatretus burgeri. Acta zool. (Stockh.)*, **56**, 61–65.

Tsuneki, K. and Gorbman, A. (1975a), Ultrastructure of the anterior neurohypophysis and the pars distalis of the lamprey, *Lampetra tride-*

ntata. Gen. comp. Endocrinol., **25**, 487–508.

Tsuneki, K. and Gorbman, A. (1975b), Ultrastructure of pars nervosa and pars intermedia of the lamprey, *Lampetra tridentata. Cell Tiss. Res.*, **157**, 165–184.

Tsuneki, K. and Gorbman, A. (1977a), Ultrastructure of the testicular interstitial tissue of the hagfish, *Eptatretus stouti. Acta zool. (Stockh).*, **58**, 17–25.

Tsuneki, K. and Gorbman, A. (1977b), Ultrastructure of the ovary of the hagfish, *Eptatretus stouti. Acta zool. (Stockh.)*, **58**, 27–40.

Tsuneki, K., Kobayashi, H., Yanagisawa, M. and Bando, T. (1975), Histochemical distribution of monoamines in the hypothalamo-hypophysial region of the lamprey, *Lampetra japonica. Cell Tiss. Res.*, **161**, 25–32.

Tsuneki, K., Urano, A. and Kobayashi, H. (1974), Monoamine oxidase and acetylcholinesterase in the neurohypophysis of the hagfish, *Eptatretus burgeri. Gen. comp. Endocrinol.*, **24**, 240–256.

Uthe, J. F. and Tsuyuki, H. (1967), Comparative zone electropherograms and muscle myogens and blood proteins of adult and ammocoete lamprey. *J. Fish. Res. Bd. Can.*, **24**, 1269–1273.

Uzzelli, T., Nolan, C., Fitch, W. M. and Margoliash, E. (1968), Data submitted to *Atlas of Protein Sequence and Structure*. Nat. Biomed. Res. Found. Washington. 1972.

Veselkin, N. P. (1963), (Localisation of electrical responses to visual stimulation in the lamprey brain). *Sechenov physiol. J. USSR*, **49**, 181–185. (In Russian).

Veselkin, N. P. (1966), Electrical reactions in midbrain, medulla and spinal cord of the lamprey to visual stimulation. *Fedn. Proc. Fedn. Am. Socs. exp. Biol.*, **25**, 957–960.

Vladykov, V. D. and Kott, E. (1976a), Tentative list of holarctic lampreys (Petromyzonidae) and their broad distribution. *Contr. Wilfred Laurier Univ. Mus. Zool.* Waterloo, Ontario. **1**, 1–6.

Vladykov, V. D. and Kott, E. (1976b), A non-parasitic species of lamprey of the genus *Entosphenus* Gill, 1862 (Petromyzonidae) from south central California. *Bull. S. Calif. Acad. Sci.*, **75**, 60–67.

Vladykov, V. D., Kott, E. and Pharand-Coad, S. (1975), A new non-parasitic species of lamprey, genus *Lethenteron* (Petromyzonidae) from eastern tributaries of the gulf of Mexico, U. S. A. *Nat. Mus. Can. Publ. zool.* No. **12**, 1–36.

Vuk-Pavlovic, S. (1975), Evolution of the haem-haem interaction in vertebrate haemoglobin – a hypothesis. *J. molec. Evol.*, **6**, 209–214.

Walvig, F. (1963), The gonads and the formation of the sexual cells. In: *The Biology of Myxine.* (Brodal, A. and Fänge, R., eds.) pp. 530–580, Universitetsforlaget, Oslo.

Wangsjö, G. (1952), The Downtonian and Devonian vertebrates of Spitsbergen. In: *Morphology and Systematic Studies of the Spitsbergen cephalaspids*. Skr. norsk. Polarinst. **97**, 1–611.

Watson, D. M. S. (1954), A consideration of ostracoderms. *Phil. Trans. R. Soc. Lond. Ser. B.*, **238**, 1–25.

Watts, R. L. and Watts, D. C. (1974). Nitrogen metabolism in fishes. In: *Chemical Zoology*. (Florkin, M. and Scheer, B. T., eds.) Vol. 8, pp. 369–446, Academic Press, London.

Webb, J. E. (1973), The role of the notochord in forward and reverse swimming and burrowing in the amphioxus, *Branchiostoma lanceolatus*. *J. Zool. (Lond.)*, **170**, 325–338.

Webster, R. D. and Pollara, B. (1969), Isolation and partial characterisation of transferrin in the sea lamprey, *Petromyzon marinus*. *Comp. Biochem. Physiol.*, **30**, 509–527.

Weinstein, B. (1968), On the relationship between glucagon and secretin. *Experientia*, **24**, 406–408.

Weinstein, B. (1972), A generalised homology correlation for various hormones and proteins. *Experientia*, **28**, 1517–1522.

Weisbart, M. (1975), *In vitro* incubations of presumptive adrenocortical cells from the opisthonephros of the adult sea lamprey, *Petromyzon marinus*. *Gen. comp. Endocrinol.*, **26**, 368–373.

Weisbart, M. and Idler, D. R. (1970), Re-examination of the presence of corticosteroids in two cyclostomes, the Atlantic hagfish (*Myxine glutinosa* L.) and the sea lamprey (*Petromyzon marinus* L.). *J. Endocrinol.*, **46**, 29–43.

Weisbart, M. and Youson, J. H. (1975), Steroid formation in the larval and parasitic adult sea lamprey, *Petromyzon marinus* L. *Gen. comp. Endocrinol.*, **27**, 517–526.

Weisbart, M. and Youson, J. H. (1977), *In vivo* formation of steroids from [1, 2, 6, 7–^{3}H] progesterone by the sea lamprey, *Petromyzon marinus* L. *J. Steroid Biochem.*, **8**, 1249–1252.

Weisbart, M., Youson, J. H. and Wiebe, J. P. (1978), Biochemical, histochemical and ultrastructural analyses of presumed steroid-producing tissues in the sexually mature sea lamprey, *Petromyzon marinus* L. *Gen. comp. Endocrinol.*, **34**, 26–37.

White, E. I. (1946), *Jamoytius kerwoodi*, a new chordate from the Silurian of Lanarkshire. *Geol. Mag.*, **83**, 89–97.

White, E. I. (1958), Original environment of the craniates. In: *Studies on Fossil Vertebrates*. (Westoll, T. S., ed.) pp. 212–234, Athlone Press, London.

Whiting, H. P. (1957), Mauthner neurones in young larval lampreys. *Q. J. microsc. Sci.*, **98**, 163–178.

Whiting, H. P. (1972), Cranial anatomy of the Ostracoderms in relation to the organisation of larval lampreys. In: *Studies in Vertebrate Evolution*.

(Joysey, K. A. and Kemp, T. S., eds.), pp. 1–19, Oliver and Boyd, London.

Whiting, H. P. (1977), Cranial nerves in lampreys and cephalaspids. In: *Problems in Vertebrate Evolution*. Linn. Soc. Symp. Ser. **4**, 1–23.

Whiting, H. P. and Tarlo, L. B. H. (1965), The brain of the Heterostraci (agnatha). *Nature*, **207**, 829–831.

Wickelgren, W. O. (1977). Physiological and anatomical characteristics of reticulospinal neurones in lamprey. *J. Physiol.* **270**, 89–114.

Wickelgren, W. O. (1977), Post-tetanic potentiation, habituation and facilitation of synaptic potentials in reticulospinal neurones of lamprey. *J. Physiol.* **270**, 115–131.

Wickstead, J. H. (1969), Some further comments on *Jamoytius kerwoodi* White. *Zool. J. Linn. Soc.* **48**, 421–422.

Wikgren, B. J. (1953), Osmotic regulation in some aquatic animals with special reference to the influence of temperature. *Acta zool. fenn.*, **71**, 1–102.

Wilson, J. A. (1972), *Principles of Animal Physiology*. Collier-Macmillan, London.

Winbladh, L. (1976), Follicles in the endocrine pancreas of some myxinoid species. *Acta zool. (Stockh.)*, **57**, 7–11.

Winbladh, L. and Horstedt, (1975), Follicles in the endocrine pancreas of *Myxine glutinosa* studied by scanning electron microscopy. *Acta zool. (Stockh.)*, **56**, 213–216.

Wingstrand, K. G. (1966), Comparative anatomy and evolution of the hypophysis. In: *The Pituitary Gland*. (Harris, G. W. and Donovan, B. T., eds.) Vol. 1, pp. 58–126. Butterworths, London.

Wolff, T. (1977), Diversity and faunal composition of the deep sea benthos. *Nature*, **267**, 780–785.

Wright, G. M., Filosa, M. F. and Youson, J. H. (1978a), Light and electron microscopic and immunocytochemical localisation of thyroglobulin in the thyroid gland of the anadromous sea lamprey, *Petromyzon marinus* L. during its upstream migration. *Cell Tiss. Res.*, **187**, 473–478.

Wright, G. M., Filosa, M. F. and Youson, J. H. (1978b), Immunocytochemical localisation of thyroglobulin in the endostyle of the anadromous sea lamprey, *Petromyzon marinus* L. *Am. J. Anat.*, **152**, 263–268.

Wright, G. M. and Youson, J. H. (1976), Transformation of the endostyle of the anadromous sea lamprey, *Petromyzon marinus* L. during metamorphosis. 1. Light microscopy and autoradiography with ^{125}I. *Gen. comp. Endocrinol.*, **30**, 243–257.

Wright, G. M. and Youson, J. H. (1977), Serum thyroxine concentrations in larval and metamorphosing anadromous sea lampreys. *Petromyzon marinus* L. *J. exp. Zool.* 202, 27–32.

Young, J. Z. (1935), The photoreceptors of lampreys. III. Control of colour change by the pineal and pituitary. *J. exp. Biol.*, **12**, 258–270.

Young, J. Z. (1971), Preface to *The Biology of Lampreys*. (Hardisty, M. W. and Potter, I. C. eds.) Academic Press, London.

Youson, J. H. (1972), Structure and distribution of interrenal cells (presumptive interrenal cells) in the opisthonephric kidney of larval and adult sea lamprey, *Petromyzon marinus* L. *Gen. comp. Endocrinol.*, **19**, 56–68.

Youson, J. H. (1973a), A comparison of presumptive interrenal tissue in the opisthonephric kidney and dorsal vessel region of the sea lamprey, *Petromyzon marinus* L. *Can. J. Zool.*, **51**, 769–799.

Youson, J. H. (1973b), Effects of mammalian corticotrophin on the ultrastructure of presumptive interrenal cells in the opisthonephros of the lamprey, *Petromyzon marinus* L. *Am. J. Anat.*, **138**, 235–252.

Youson, J. H. (1975a), Radiography of presumptive interrenal cells in the sea lamprey after ^{3}H-cholesterol injection. *Acta zool. (Stockh.)*, **56**, 219–223.

Youson, J. H. (1975b), Absorption and transport of exogenous protein in the archinephric duct of the opisthonephric kidney of the sea lamprey, *Petromyzon marinus* L. *Comp. Biochem. Physiol.*, **50A**, 2–5.

Youson, J. H. (1975c), Absorption and transport of ferritin and exogenous horseradish peroxidase in the opisthonephric kidney of the sea lamprey. 1. The renal corpuscle. *Can. J. Zool.*, **53**, 571–581.

Youson, J. H. and Connelly, K. L. (in press), Development of longitudinal mucosal folds in the intestine of the anadromous sea lamprey, *Petromyzon marinus* L. during metamorphosis. *J. Fish. Res. Bd. Can.*

Youson, J. H. and Freeman, P. A. (1976), Morphology of the gills of larval and parasitic adult sea lampreys, *Petromyzon marinus*. *J. Morph.*, **149**, 73–104.

Youson, J. H. and McMillan, D. B. (1970a), The opisthonephric kidney of the sea lamprey of the Great Lakes, *Petromyzon marinus* L. I. The renal corpuscle. *Am. J. Anat.*, **127**, 207–232.

Youson, J. H. and McMillan, D. B. (1970b), The opisthonephric kidney of the sea lamprey of the Great Lakes, *Petromyzon marinus* L. II. Neck and proximal segments of the tubular nephron. *Am. J. Anat.*, **127**, 233–258.

Youson, J. H. and McMillan, D. B. (1971a), The opisthonephric kidney of the sea lamprey of the Great Lakes, *Petromyzon marinus* L. III. Intermediate, distal and collecting segments of the adult. *Am. J. Anat.*, **130**, 281–304.

Youson, J. H. and McMillan, D. B. (1971b), The opisthonephric kidney of the sea lamprey of the Great Lakes, *Petromyzon marinus* L. IV. The archinephric duct. *Am. J. Anat.*, **131**, 289–314.

Youson, J. H., Wright, G. M. and Ooi, E. C. (1977), The timing of changes in several internal organs during metamorphosis of anadromous larval sea lamprey, *Petromyzon marinus* L. *Can. J. Zool.*, **55**, 469–473.

Zanandrea, G. (1954), Corrispondenza tra forme parassite e non parassite nei generi *Ichthyomyzon* è *Lampetra* (problemi di speciazione). *Boll. Zool.*, **21**, 461–466.

Zanandrea, G. (1956), Neotenia in *Lampetra planeri* zanandreai (Vladykov) é l'endocrinologia sperimentale dei Ciclostomi. *Boll. Zool.*, **22**, 412–427.

Zanandrea, G. (1958), Altri casi di lamprede neoteniche é il loro apparato naso-faringeo. *Atti. Inst. Ven. Sci. Lett. Art.*, **116**, 179–191.

Zanandrea, G. and Capanna, E. (1964), Contributo della cariologia del genere *Lampetra. Boll. Zool.*, **31**, 669–677.

Zelnik, P. R., Hornsey, D. J. and Hardisty, M. W. (1977), Insulin and glucagon-like immunoreactivity in the river lamprey, *Lampetra fluviatilis. Gen. comp. Endocrinol.*, **33**, 53–60.

Zottoli, S. J. (1977), Correlation of the startle reflex and Mauthner cell auditory responses in untrained goldfish. *J. exp. Biol.*, **66**, 243–254.

Zottoli, S. J. (1978), Comparative morphology of the Mauthner cell in fish and amphibians. In: *Neurobiology of the Mauthner cell.* (Faber, D. S. and Korn, H. eds.) pp. 13–45, Raven Press, New York.

Author index

399

Subject index

Page numbers in italics refer to text figures